D0138771

Population and Evolutionary Genetics: A Primer

The Benjamin/Cummings Series in the Life Sciences

F. J. Ayala
Population and Evolutionary Genetics: A Primer (1982)

F. J. Ayala and J. A. Kiger, Jr.
Modern Genetics (1980)

F. J. Ayala and J. W. Valentine
Evolving: The Theory and Process of Organic Evolution (1979)

M. G. Barbour, J. H. Burk, and W. D. Pitts
Terrestrial Plant Ecology (1980)

L. L. Cavalli-Sforza
Elements of Human Genetics, second edition (1977)

R. E. Dickerson and I. Geis
Hemoglobin (1982)

L. E. Hood, I. L. Weissman, W. B. Wood, and J. H. Wilson
Immunology, second edition (1983)

L. E. Hood, J. H. Wilson, and W. B. Wood
Molecular Biology of Eucaryotic Cells (1975)

A. L. Lehninger
Bioenergetics: The Molecular Basis of Biological Energy Transformations, second edition (1971)

S. E. Luria, S. J. Gould, and S. Singer
A View of Life (1981)

A. P. Spence
Basic Human Anatomy (1982)

A. P. Spence and E. B. Mason
Human Anatomy and Physiology (1979)

G. J. Tortora, B. R. Funke, and C. Case
Microbiology: An Introduction (1982)

J. D. Watson
Molecular Biology of the Gene, third edition (1976)

I. L. Weissman, L. E. Hood, and W. B. Wood
Essential Concepts in Immunology (1978)

N. K. Wessells
Tissue Interactions and Development (1977)

W. B. Wood, J. H. Wilson, R. M. Benbow, and L. E. Hood
Biochemistry: A Problems Approach, second edition (1981)

Population and Evolutionary Genetics: A Primer

Francisco J. Ayala

University of California, Davis

The Benjamin/Cummings Publishing Company, Inc.
Menlo Park, California • Reading, Massachusetts
London • Amsterdam • Don Mills, Ontario • Sydney

Sponsoring Editor: Jim Behnke
Production Editor: Bonnie Garmus
Cover Designer: John Edeen

Library of Congress Cataloging in Publication Data

Ayala, Francisco José, 1934–
Population and evolutionary genetics.

Bibliography: p.
Includes index.
1. Population genetics. 2. Evolution.
I. Title.
QH455.A94 575.1′5 81-21623

ISBN 0-8053-0315-4 AACR2

12 13 14-MW-96 95

The Benjamin/Cummings Publishing Company, Inc.
2727 Sand Hill Road
Menlo Park, California 94025

To José and Carlos

The smiles, the tears of boyhood's years,
The words of love then spoken

Thomas Moore, *Oft in the Stilly Night*

Contents

Preface

The living world is incredibly diverse: there are more than two million species, from lowly bacteria to orchids, sequoias, iguanas, and humans. The stupendous diversity conceals remarkable unity, for all organisms use similar nucleic acids as carriers of hereditary information, perform vital functions by means of enzymes made up of the same twenty amino acids, use the same genetic code, and have identically organized membranes. The diversity as well as the unity of life are the result of the evolutionary process. There is unity because all organisms are genetically continuous, derived from the same ancestors. The differences reflect adaptation to different conditions of life, different places, and different history. The aphorism uttered by the eminent geneticist Theodosius Dobzhansky is correct: "Nothing in biology makes sense except in the light of evolution."

Hereditary continuity and change underlie the evolutionary process. *Population and Evolutionary Genetics: A Primer* introduces the concepts and theories that account for the evolutionary process at the genetic level. These should be of interest to all biologists; not only to students of the process of evolution itself, but also to ecologists, physiologists, anthropologists, and paleontologists.

The *Primer* is an elementary treatment of population and evolutionary genetics. The required concepts of general genetics are reviewed in Chapter 1. The mathematics have been kept to a minimum and can be understood with knowledge of high school algebra. Yet, the *Primer* covers not only the classical principles of population genetics, but proceeds into the exciting new field of molecular evolution. Indeed, during the last decade the most significant advances in the understanding of evolution have come about through the study of DNA and proteins—the information-laden macromolecules of life.

Chapters 2 and 7 deal with molecular evolution; they can be understood without previous knowledge of biochemistry.

The *Primer* may serve as text for elementary courses in population genetics, but it can be useful as supplementary reading in other courses, such as evolution, genetics, ecology, paleontology, population biology, anthropology, and behavioral biology. The style is clear and all fundamental principles are illustrated with examples. The text is, thus, comprehensible even when the material has not been covered in lecture.

Special Features

- The book is generously illustrated to enhance the clarity of the text.
- Examples and empirical data are used to illustrate and support all major concepts.
- About a dozen special topics of various kinds are set apart as "boxes," which can be skipped without loss of comprehension of the text.
- The principles of quantitative inheritance are included in Chapter 6.
- The problems at the end of the chapters are an integral part of the book; some contain new information considered subsidiary or too detailed to be included in the text proper. The answers to the problems are given at the end of the book.
- Students who have not taken a course in statistics will find in the appendix the concepts and methods needed to understand the text and solve the problems.
- A glossary is provided as an aid in reviewing both new and familiar concepts.
- The bibliography lists basic references that document the material covered in the text; the credit lines for figures and tables are additional sources of information.

Acknowledgments

This book is indebted in a most fundamental way to all scientists whose brilliant efforts have produced the body of theory and empirical information that make up the exciting fields of population and evolutionary genetics. Special thanks are due to those who gave permission to reproduce illustrations and data, and to those who reviewed the text (see list below). The reviewers' comments have greatly contributed to enhance clarity and reduce errors.

I thank my collaborators who assisted in the preparation of the text, with typing, proofreading, and indexing: Lorraine Barr, David Friedman, Candy Miller, and Elizabeth Toftner. Bonnie Garmus has been an effective and always cheerful editor. Jim Behnke, biology editor of Benjamin/Cummings, is not only a distinguished exemplar of professionalism in the publishing industry, but also a cherished friend. The contributions of Fred Raab, as manuscript editor, and Georg Klatt, as illustrator, to *Modern Genetics* persist in the present book; I appreciatively acknowledge their dedication and skill.

Readers familiar with *Modern Genetics* by F. J. Ayala and J. A. Kiger, will recognize that most of the text and illustrations of this *Primer* are reproduced from that book. Chapters 2 to 5 and 7 of the *Primer* correspond almost exactly to Chapters 18 to 22 of *Modern Genetics*, and Chapter 6 incorporates about two-thirds of Chapter 15. Chapter 1 has been newly prepared, but it uses materials from diverse chapters of the earlier book.

I owe special thanks to John Kiger, co-author of *Modern Genetics*, for allowing me to use the lion's share of my contribution from our collaborative enterprise in the *Primer*. I value our friendship and have the highest admiration for his uncompromisingly rigorous approach to the teaching of genetics.

Francisco J. Ayala
Davis, California

List of Reviewers

James P. Collins, Arizona State University
David C. Culver, Northwestern University
Yun-Tzu-Kiang, University of New Hampshire
Roger Milkman, University of Iowa
Jeffry B. Mitton, University of Colorado
Henry E. Schaffer, North Carolina State University

Population and Evolutionary Genetics: A Primer

1

Introduction: Basic Concepts of Genetics

Mendel's Theory of Heredity

The fundamental principles of heredity were discovered by Gregor Mendel (1822–1884), who was a monk in the Augustinian monastery in Brünn, Austria (now Brno, Czechoslovakia). Starting around 1856, he used garden peas (*Pisum sativum*) to investigate how individual traits are inherited.

In one experiment, Mendel studied the inheritance of seed shape by crossing plants yielding round seeds with plants yielding wrinkled seeds. The results were clear-cut: all progeny plants of the first filial generation (F_1) produced round seeds, regardless of whether the round-seed plant had been the female or male parent. Mendel saw that wrinkling was suppressed by the dominance of roundness. He found that all seven characters he had selected for study behaved in this way: in each case only one of the two contrasting traits appeared in the F_1 progeny. Mendel called such traits (round seeds, yellow peas, etc.) *dominant,* and their alternatives (wrinkled seeds, green peas, etc.) he called *recessive.*

Dominance of one trait over another is a common but not universal phenomenon. In some cases there is *incomplete dominance:* the F_1 progeny is intermediate between the two parents, although it may resemble one more than the other parent. There are also cases in which the traits of both parents are exhibited in their F_1 progeny; this is called *codominance.* In humans, for example, the characteristics of blood group A and blood group B are expressed equally in individuals who have inherited the two, one from each parent.

Mendel planted seeds from the F_1 progeny and allowed each plant that grew from these seeds to self-fertilize. Both round and wrinkled seeds appeared side-by-side in the same pods in the second generation (F_2) of the cross between round-seed and wrinkled-seed plants. He counted the seeds: 5474 were round and 1850 were wrinkled. (Figure 1.1). The ratio was very close to 3:1 (in fact, 2.96:1). A similar ratio appeared in each of the seven crosses that he made. In every case the dominant trait was about three times as common in the F_2 as the recessive trait.

He then planted the F_2 seeds and allowed each plant that grew from them to self-fertilize. In every case, the F_2 plants showing the recessive trait bred true: their seeds produced F_3 plants identical to their parents. The F_2 plants showing the dominant trait were of two kinds, however: one-third bred true while the other two-thirds produced F_3 progenies in which the dominant and recessive traits appeared in the ratio 3:1.

Mendel advanced the following hypothesis to explain the results of his experiments. Contrasting traits, such as the roundness or wrinkling of peas, are determined by "factors" (now called *genes*) that are transmitted from parents to offspring through the gametes. Each factor may exist in alternative forms (now called *alleles*) responsible for the alternative forms that the character may take. For each character, every pea plant has two genes, one inherited from the male parent and the other inherited from the female. Thus, each pea plant has two genes for seed shape. These genes may exist in the form that determines round seeds (allele for roundness), or in the form that determines wrinkled seeds (allele for wrinkling).

It is appropriate at this point to introduce two other terms of the genetics vocabulary. A *homozygote* (adjective: homozygous) is an individual in which the two genes of a pair (one gene inherited from each parent) are identical, i.e., an individual with two identical alleles. A *heterozygote* (adjective: heterozygous) is an individual in which the two genes of a given pair are different, i.e., an individual with two different alleles. Thus, the true-breeding round-seed plants are homozygous for roundness, and the F_1 hybrids from the cross round $\times$ wrinkled are heterozygous for roundness and wrinkling.

From the phenomenon of dominance, Mendel inferred that in heterozygous individuals one allele is dominant and the other is recessive. From the reappearance of the two parental traits in the progeny of hybrids (heterozygotes), he concluded that *the two factors (genes) for each trait do not fuse or blend in any way, but remain distinct throughout the life of the individual and segregate in the formation of gametes,* so that half the gametes carry one gene and the other half carry the other gene. This conclusion is known as Mendel's *law of segregation.*

Pairs of genes are often symbolized by letters: the dominant allele by a capital italic letter, the recessive allele by the same letter in lowercase italic. For example, the allele for roundness may be represented as *R* and

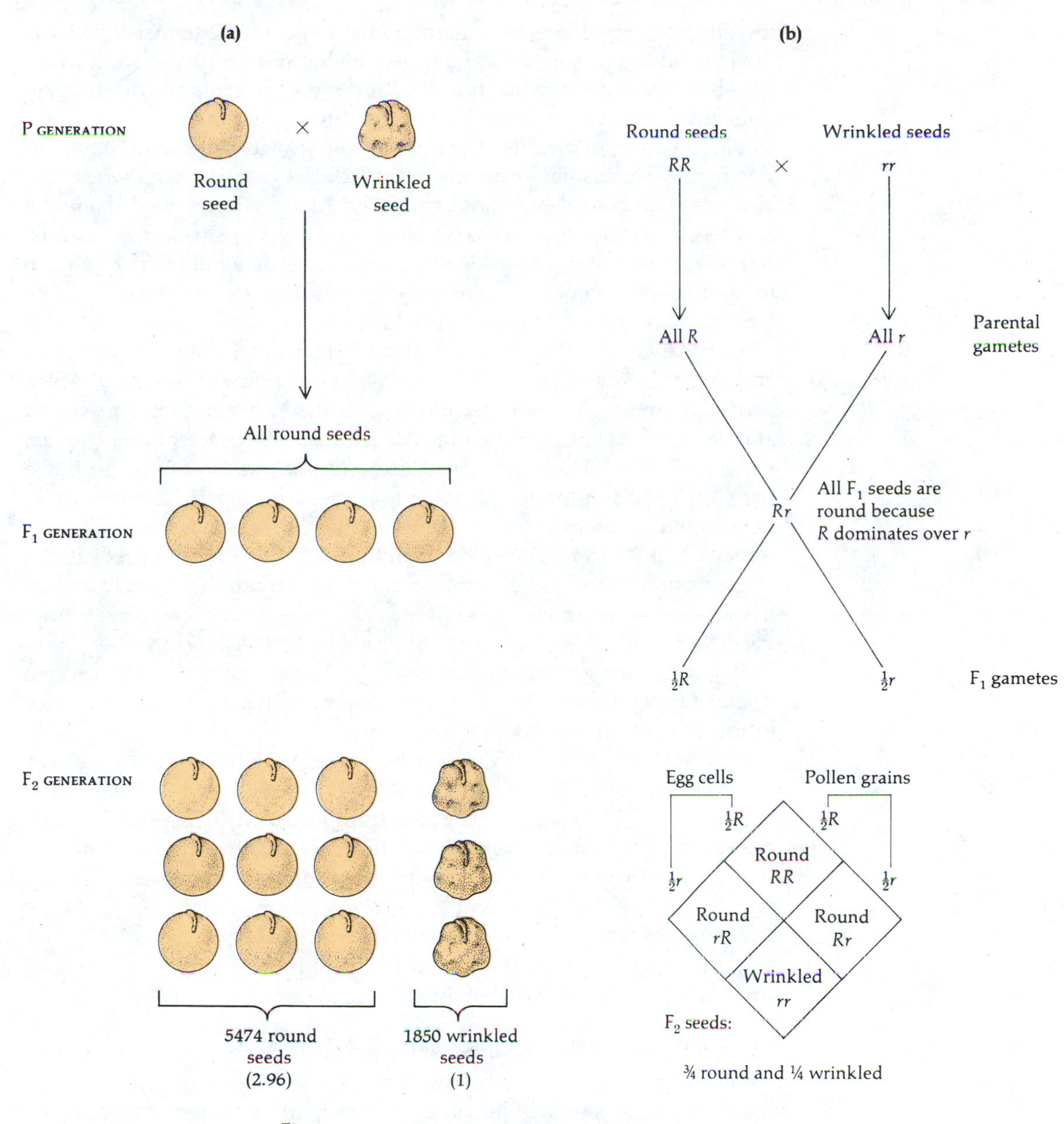

Figure 1.1
Progeny of a cross between a round-seed and a wrinkled-seed pea plant. **(a)** When the round-seed plants of the F_1 generation are self-fertilized or crossed to each other, the seeds of the F_2 generation are, approximately, three-quarters round and one-quarter wrinkled. **(b)** The explanation proposed by Mendel. R and r represent the alternative factors (alleles) determining roundness and wrinkling, respectively. The probability with which each kind of plant is expected is obtained by multiplying the probabilities of the types of gametes that must unite in order to form that kind of plant. Thus, one-quarter of the F_2 plants are RR because the probability of R is ½ among F_1 male gametes, and also ½ among F_1 female gametes: $(½)(½) = ¼$.

the allele for wrinkling as *r*. The genetic makeup of Mendel's plants is then as follows (Figure 1.1): the true-breeding round plants are *RR*; the true-breeding wrinkled plants are *rr*; and the F_1 hybrids are *Rr* and produce two kinds of gametes, *R* and *r*, in equal amounts.

Mendel's experiments described so far concern the inheritance of alternative expressions of a single character. What happens when two characters are considered simultaneously? Mendel formulated the *law of independent assortment*, which says that *genes for different characters are inherited independently of one another*. (This law, however, would later be shown to apply only to genes on different chromosomes.) Mendel derived this law from the results of crosses between plants that were different with respect to two separate characters. In one experiment, plants having round and yellow seeds were crossed with plants having wrinkled and green seeds (Figure 1.2). As expected, all peas in the F_1 generation were round and yellow, but the interesting results came in the F_2 generation. Mendel had already considered two possibilities: (1) that traits derived from one parent are transmitted together, and (2) that they are transmitted independently of each other. He formulated the expectations from these alternatives. If (1) were true, there would be only two kinds of seeds in the F_2 generation: round-yellow and wrinkled-green, and they should appear in the ratio 3:1 according to the law of segregation. If (2) were correct, however, there should be four kinds of seeds: round-yellow (two dominant traits), round-green (dominant-recessive), wrinkled-yellow (recessive-dominant), and wrinkled-green (two recessive traits), which should appear in the proportions 9:3:3:1.

Mendel found that the peas in the F_2 generation were of four kinds: 315 round-yellow, 108 round-green, 101 wrinkled-yellow, and 32 wrinkled-green. This was reasonably close to the proportions 9:3:3:1 expected from the second alternative. Mendel concluded that the genes determining different characters are transmitted independently.

Multiple Alleles

Mendel's experiments involved only two alleles of each gene. However, many and possibly all genes have *multiple alleles*, i.e., they exist in more than two allelic forms, although any one diploid organism can carry no more than two alleles.

Many examples of multiple alleles are known. One series of multiple alleles underlies the ABO blood groups, discovered by Karl Landsteiner (1868–1943) in 1900. Blood groups must be considered when matching donors with recipients for blood transfusions, to prevent the donor's red blood cells from becoming agglutinated in the recipient's blood stream. There are four common blood groups within the ABO system: O, A, B, and AB. These are determined by three alleles: I^A, I^B, and i. Alleles I^A

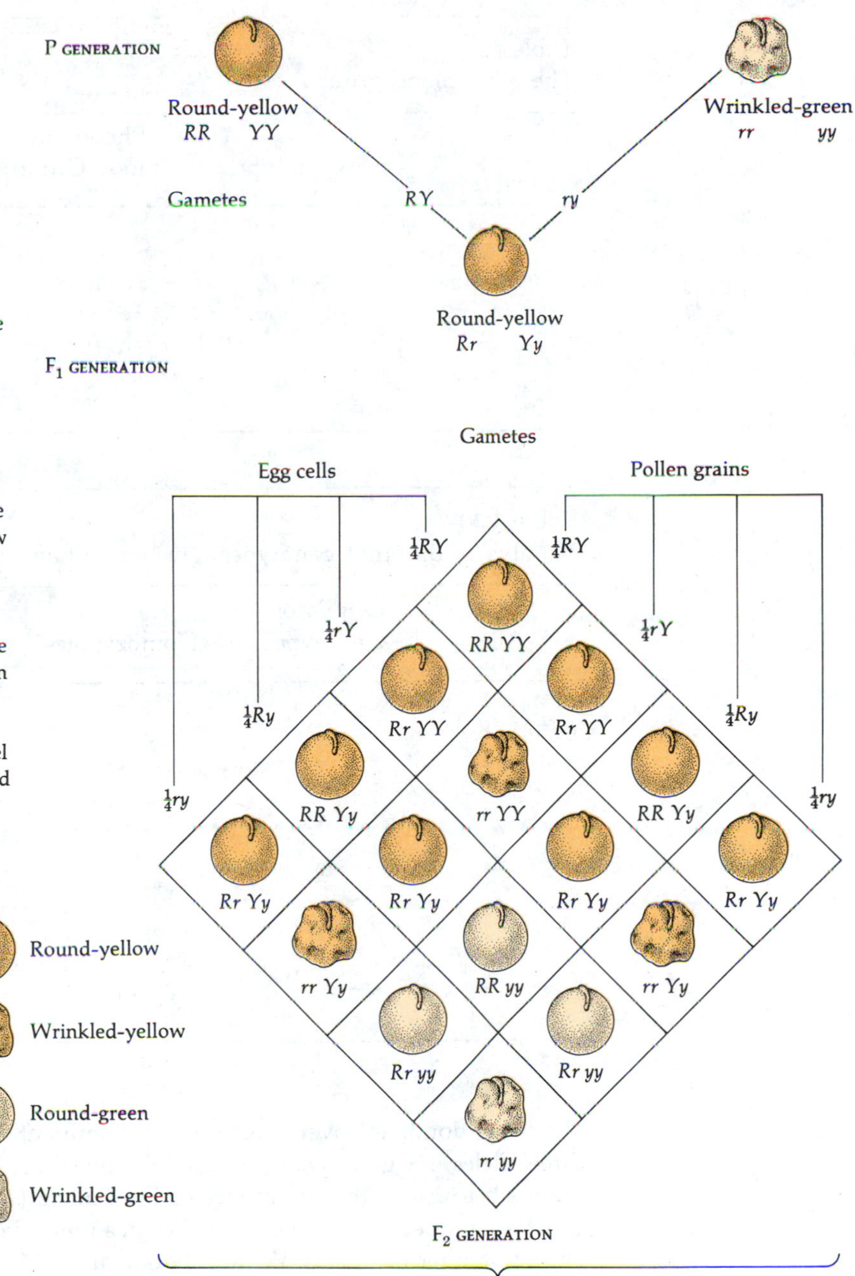

Figure 1.2
Independent assortment. Round-yellow (*RR YY*) plants crossed with wrinkled-green (*rr yy*) plants produce F_1 hybrids that are all round-yellow (*Rr Yy*). The F_1 hybrids produce four kinds of gametes, each with a frequency of ¼. Random association between female and male gametes of the four kinds produces F_2 seeds belonging to nine different genetic classes. When the F_2 seeds are grouped by their appearance, there are nine squares with round-yellow seeds, three squares with wrinkled-yellow seeds, three squares with round-green seeds, and one square with wrinkled-green seeds; these four kinds of seeds are expected in the proportions 9:3:3:1. Mendel obtained 315, 108, 101, and 32 seeds of the respective kinds, in good agreement with the expectations.

$\frac{1}{16}(RR\ YY) + \frac{2}{16}(Rr\ YY) + \frac{2}{16}(RR\ Yy) + \frac{4}{16}(Rr\ Yy) = \frac{9}{16}$ round-yellow seeds

$\frac{1}{16}(RR\ yy) + \frac{2}{16}(Rr\ yy) = \frac{3}{16}$ round-green seeds

$\frac{1}{16}(rr\ YY) + \frac{2}{16}(rr\ Yy) = \frac{3}{16}$ wrinkled-yellow seeds

$\frac{1}{16}(rr\ yy) = \frac{1}{16}$ wrinkled-green seeds

Table 1.1
The ABO blood groups.

Genotype	Phenotype (Blood Group)
I^AI^A, I^Ai	A
I^BI^B, I^Bi	B
I^AI^B	AB
ii	O

Table 1.2
Number of different genotypes, given a certain number of alleles.

Alleles	Kinds of Genotypes	Kinds of Homozygotes	Kinds of Heterozygotes
1	1	1	0
2	3	2	1
3	6	3	3
4	10	4	6
5	15	5	10
n	$\frac{n(n+1)}{2}$	n	$\frac{n(n-1)}{2}$

and I^B are dominant over allele i, but codominant with each other. With three alleles, six genotypes are possible, but these reduce to four blood groups because of the recessivity of the i allele (Table 1.1).

The number of possible genotypes in a multiple-allele series depends on the number of alleles. With only one allele, A, only one genotype is possible, AA. With two alleles, A_1 and A_2, three genotypes are possible: the two homozygotes A_1A_1 and A_2A_2 and one heterozygote, A_1A_2. With three alleles, A_1, A_2, and A_3, six genotypes are possible: the three homozygotes A_1A_1, A_2A_2, and A_3A_3 and three heterozygotes, A_1A_2, A_1A_3, and A_2A_3. In general, given n alleles, there are $n(n + 1)/2$ genotypes, of which n are homozygotes and $n(n - 1)/2$ are heterozygotes (Table 1.2).

Genotype and Phenotype

In 1909 Wilhelm Johannsen (1857–1927) introduced the important distinction between phenotype and genotype. The *phenotype* of an organism is its appearance—what we can observe: its morphology, physiology, and behavior. The *genotype* is the genetic constitution it has inherited. During the lifetime of an individual, the phenotype may change; the genotype, however, remains constant.

The distinction between phenotype and genotype must be kept in mind, since the relation between the two is not fixed. This is because the phenotype results from complex networks of interactions between different genes, and between the genes and the environment. In general, individuals do not have identical phenotypes, although the phenotypes may be similar when only one or a few traits are considered. Moreover, organisms having similar phenotypes with respect to a given trait do not necessarily have identical genotypes. For example, yellow peas can be either homozygous for the yellow allele, or heterozygous for the yellow and green alleles.

An illustration of environmental effects on the phenotype is shown in Figure 1.3. Three plants of the cinquefoil, *Potentilla glandulosa,* were collected in California—one on the coast at about 100 feet above sea level, the second at about 4600 feet, and the third in the Alpine zone of the Sierra Nevada at about 10,000 feet. Each plant was cut into three parts, and the parts were planted in three experimental gardens at different altitudes, using the same gardens for all three plants. The division of one plant insured that all three parts planted at different altitudes had the same genotype. Comparison of the plants in any row shows how a given genotype gives rise to different phenotypes in different environments. Besides the obvious divergence in appearance, there are differences in fertility, growth rates, etc. Plants from different altitudes are known to be genetically different. Hence, comparison of the plants in any column shows that in a given environment different genotypes result in different phenotypes. An important observation derived from this experiment is that there is no single genotype that is "best" in all environments. For example, the plant from near sea level that prospers there fails to develop at 10,000 feet. Likewise, the plant collected at 10,000 feet prospers at that altitude, but withers at sea level.

The interaction between the genotype and the environment is further illustrated in Figure 1.4. Two strains of rats were selected; one for brightness at finding their way through a maze and the other for dullness. Selection was done in the bright strain by using the brightest rats of each generation to breed the following generation, and in the dull strain by breeding the dullest rats every generation. After many generations of selection, bright rats made only about 120 errors running through the maze, whereas dull rats averaged 165 errors. However, the differences

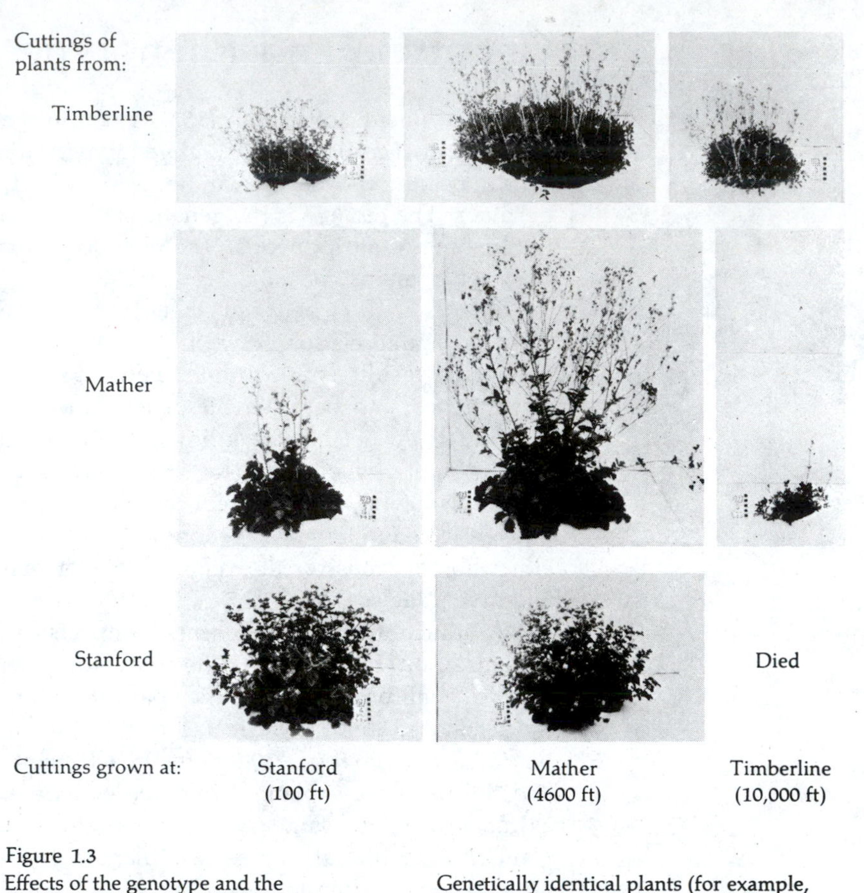

Figure 1.3
Effects of the genotype and the environment on the phenotype. Cuttings from *Potentilla glandulosa* plants collected at different altitudes were planted together in three different experimental gardens. Plants in the same row are genetically identical because they have been grown from cuttings of a single plant; plants in the same column are genetically different but have been grown in the same environment. Genetically identical plants (for example, those in the bottom row) may prosper or die, depending on the environmental conditions. Genetically different plants (for example, those in the first column) may have quite different phenotypes, even when grown in the same environment. (Photographs courtesy of Dr. William M. Hiesey, Carnegie Institution of Washington, Palo Alto, Cal.)

between the strains disappeared when rats of both strains were raised in an unfavorable environment of severe deprivation, and they nearly disappeared when the rats were raised with abundant food and other unusually favorable conditions. As with the cinquefoil plants, we see (1) that a given genotype gives rise to different phenotypes in different environments, and (2) that the *differences* in phenotype between two genotypes change from one environment to another—the genotype that is best in one environment may not be best in another.

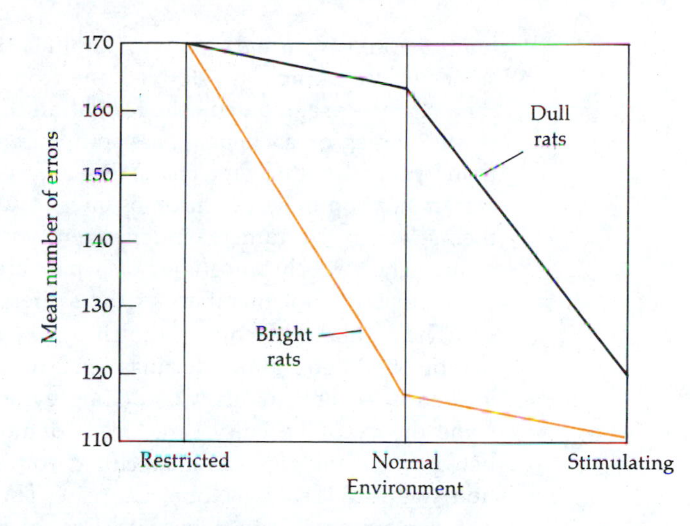

Figure 1.4
Results of an experiment with two strains of rats, one selected for brightness, the other for dullness. When raised in the same environment as that in which the selection was practiced ("normal"), bright rats made about 45 fewer errors than dull rats in the maze used for the tests. However, when they were raised in an impoverished ("restricted") environment, bright and dull rats made the same number of errors. When raised in an abundant ("stimulating") environment, the two strains performed nearly equally well. [After R. M. Cooper and J. P. Zubek, *Can. J. Psychol.* 12:159 (1958).]

Because of the variable interactions it may have with different environments, the genotype of an organism does not unambiguously specify its phenotype. Rather, the genotype determines the range of phenotypes that *may* develop. This is called the *range of reaction,* or norm of reaction, of the genotype. Which phenotype will be realized depends on the environment in which development takes place. For this reason, the entire range of reaction of a genotype is never known, since this would require that individuals with that genotype be exposed to all possible kinds of environments, which are virtually infinite.

Genes and Chromosomes

In 1902 two investigators—Walter S. Sutton in the United States and Theodor Boveri in Germany—independently suggested that genes are contained in chromosomes. This idea came to be known as the *chromosome theory of heredity.* Their argument was based on the parallel behavior, at meiosis and fertilization, between chromosomes on the one hand and genes on the other. The existence of two alleles for a given character, one inherited from each parent, parallels the existence of two chromosomes of each kind, also derived one from each parent. The two alleles for a character segregate in the formation of the gametes because the two chromosomes of each pair pass into different gametes during meiosis. Some genes for different characters assort independently because they are in nonhomologous chromosomes, and these chromosomes assort themselves in the gametes independently of the parent from which they came. (The two chromosomes of a pair are called *homologous.* Chromosomes that are not members of the same pair are called *nonhomologous.*)

The parallel behavior of the chromosomes and the genes in the formation of gametes and in fertilization strongly suggested that genes were located in chromosomes. Compelling evidence supporting the chromosome theory of heredity came from demonstrations of the association between specific genes and specific chromosomes. An early demonstration was provided by Nobel Laureate Thomas Hunt Morgan in 1910. The experiments were done with *Drosophila melanogaster,* the small, yellowish brown fruit fly that hovers around fallen fruit in summer and fall.

Morgan had a strain of *D. melanogaster* flies that had white, rather than normal red, eyes. The strain was true-breeding: the offspring of white-eyed flies also had white eyes. However, when white-eyed flies were crossed to red-eyed flies, the progenies were not as expected according to Mendelian heredity.

When a red-eyed female and a white-eyed male are crossed, all the F_1 flies are red-eyed, exactly as expected if red-eye is dominant over white-eye (Figure 1.5). When the F_1 red-eyed flies are bred together, they produce F_2 flies that are three quarters red-eyed and one-quarter white-eyed, also as expected if red-eye is dominant and white-eye recessive. However, *all the F_2 females are red-eyed, whereas half the males are red-eyed and half are white-eyed.* This is *not* expected according to the principles of Mendelian heredity. Other unexpected results appear when the F_2 flies are bred. The males are all true-breeding: the red-eyed males carry only red-eye genes, and the white-eyed males carry only white-eye genes. The F_2 females are of two kinds: half produce only red-eyed offspring, while the other half produce male offspring, half of which are red-eyed and half white-eyed.

The results are different when white-eyed females are crossed with red-eyed males. Not all the F_1 offspring are red-eyed, as would be expected according to Mendelian principles if red-eye is dominant over white-eye. Rather, half the offspring are red-eyed and half are white-eyed. Moreover, all the red-eyed flies are females, and the white-eyed flies are males. When these are bred together, the F_2 offspring consist of half (rather than one-quarter) white-eyed flies and half red-eyed flies, this time in equal numbers in both sexes.

Morgan saw that the results could be explained if (1) the eye-color gene was in the sex (X) chromosome, and (2) the male sex chromosome (Y) contained *no* gene for eye color. Chromosomes occur in pairs, but

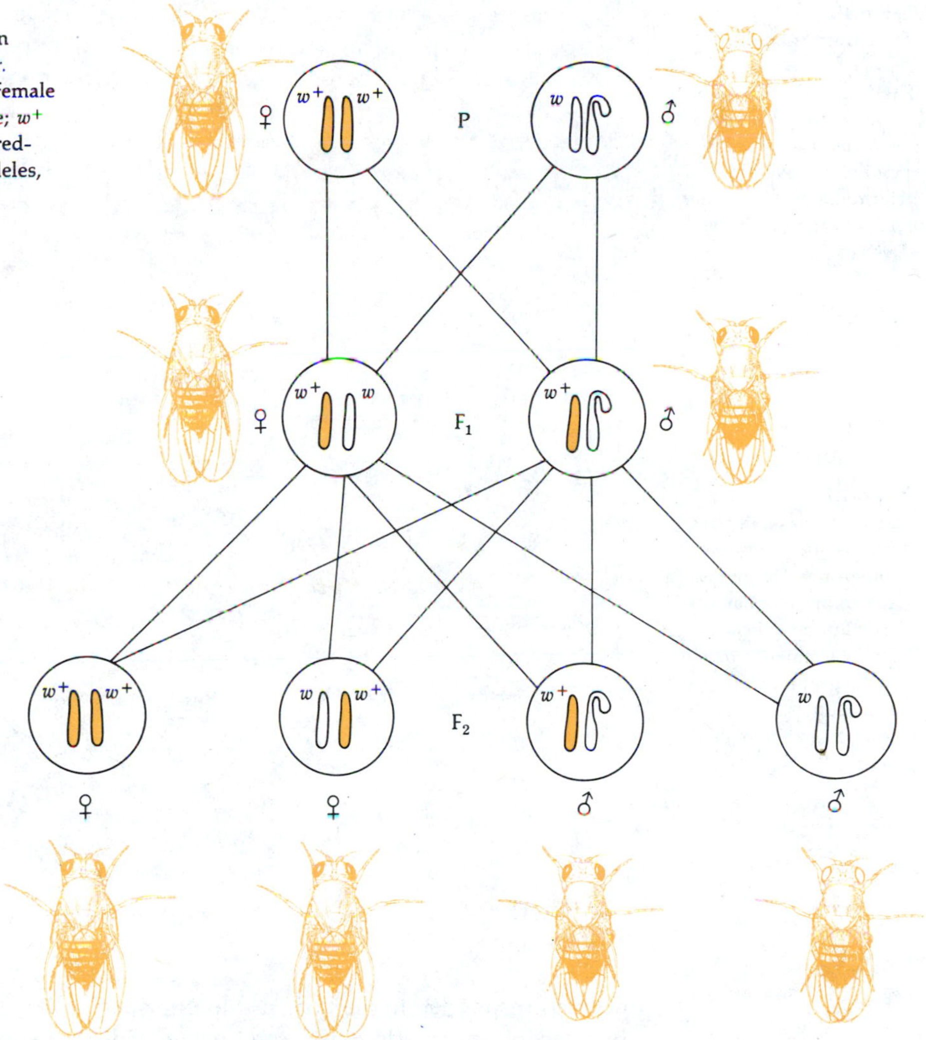

Figure 1.5
Sex-linked heredity in *Drosophila melanogaster*. Cross of a red-eyed female by a white-eyed male; w^+ and w represent the red-eye and white-eye alleles, respectively.

males and females differ with respect to the one pair of chromosomes that are associated with sex determination. *Drosophila* females carry two identical X chromosomes, but males carry two different chromosomes, one X and one Y (Figure 1.6). Females inherit one X chromosome each from their father and mother, and transmit their X chromosomes to their sons as well as to their daughters. Males, on the other hand, receive their only X chromosome from their mother and transmit it only to their daughters (Figure 1.7).

Morgan concluded that the eye-color gene is *sex-linked*, i.e., it is carried in the X chromosome. His demonstration that the behavior of one specific

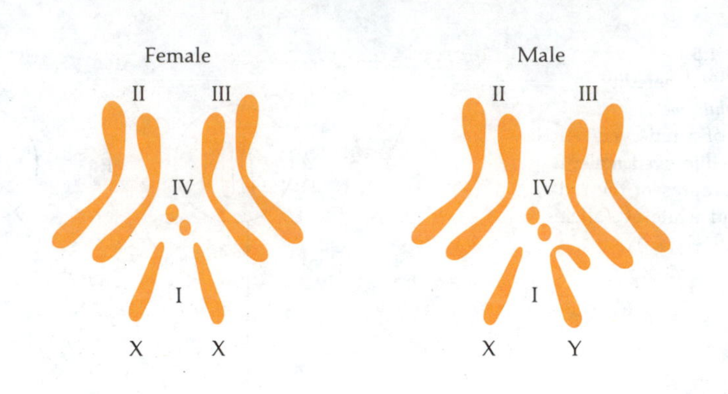

Figure 1.6
The chromosomes of *Drosophila melanogaster*. The X and the Y chromosomes are called *sex chromosomes*, because they are involved in sex determination. The other chromosomes are called *autosomes*.

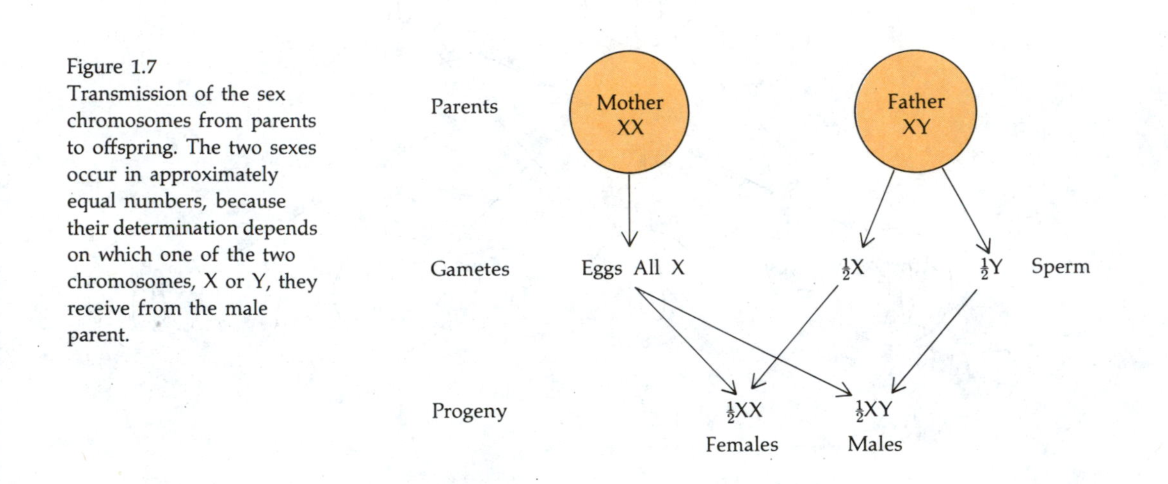

Figure 1.7
Transmission of the sex chromosomes from parents to offspring. The two sexes occur in approximately equal numbers, because their determination depends on which one of the two chromosomes, X or Y, they receive from the male parent.

gene corresponded to the behavior of one specific chromosome visible in microscopic preparations, was convincing evidence in favor of the chromosome theory of heredity, a theory which was later confirmed by many other experiments.

Sex-linked Heredity in Humans and Other Organisms

The pattern of sex-linked inheritance described for *Drosophila* is also observed in all animals and plants in which the males carry two different sex chromosomes. The males are said to be *hemizygous* with respect to the genes in the X chromosome; indeed, they are neither homozygous nor heterozygous with respect to such genes. One example of human sex-linked inheritance is hemophilia, a serious disease characterized by the inability of blood to clot. In normal persons, bleeding after a moderate injury is limited by the clotting of blood. In hemophiliacs, even a small injury can lead to death from bleeding. There are at least three kinds of hemophilia; two of them are caused by recessive sex-linked genes, and a very rare kind is caused by an autosomal recessive gene. Each affects a different factor normally required for blood to clot. A famous case of sex-linked hemophilia that has affected some royal houses of Europe can be traced to Queen Victoria. Since hemophilia is not known in her family background, it is likely that the hemophilia allele appeared by mutation in one of the gametes from which she was conceived.

The pattern of inheritance of sex-linked traits described for *Drosophila* and humans is reversed in birds, moths, butterflies, and some fish, in which females are the sex carrying two different sex chromosomes. In these organisms the females are hemizygous with respect to sex-linked traits and transmit such traits only to their sons, whereas the males transmit sex-linked traits to both their sons and daughters.

Linkage and Crossing Over

Mendel observed independent assortment for the seven characters he studied in peas because the seven genes that he studied each happened to be in a different nonhomologous chromosome. Genes that are in the same chromosome are said to be *linked.* However, it is not the case that linked genes are always kept together in the formation of the gametes. This is because homologous chromosomes exchange parts during meiosis when the two homologous chromosomes of each pair first align with each other (Figure 1.8). The exchange of genes between homologous chromosomes is called *crossing over.* As a consequence of crossing over, linked genes may be transmitted to the progeny in combinations different from those in which they are present in the parents.

In *Drosophila melanogaster,* normal body color (yellowish brown) is determined by an allele that is dominant over the allele for black body; normal wings are determined by an allele that is dominant over the allele producing very short ("vestigial") wings. These genes for body color and for wing length are linked; they are both in the second chromosome. We can represent the allele for normal body color as b^+, which is dominant over the allele for black body color, b; and the allele for normal long wings as vg^+, which is dominant over the allele for vestigial wings, vg. Thomas H. Morgan crossed females having black body color and long wings ($b\ b\ vg^+vg^+$) with males having normal body color and vestigial wings ($b^+b^+\ vg\ vg$). The F_1 flies were heterozygous for both genes ($b^+b\ vg^+vg$) but had normal body color and wings, because normal body color is dom-

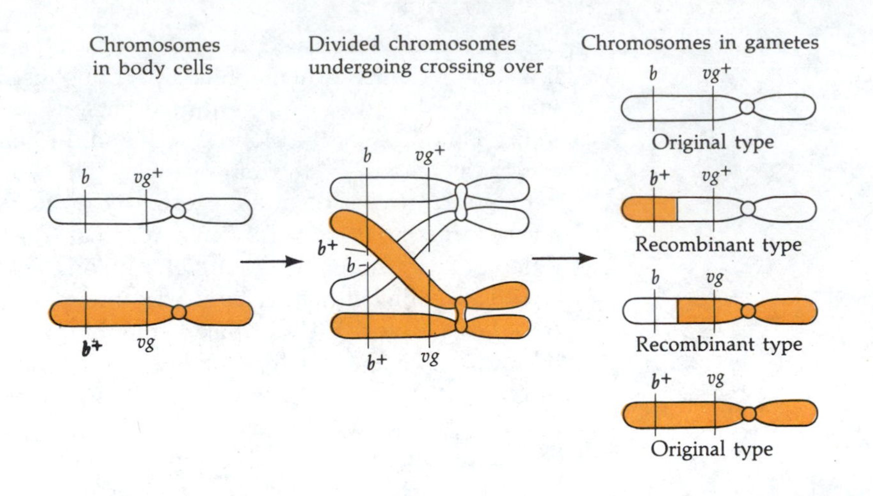

Figure 1.8
Crossing over. Left, two homologous chromosomes are represented, one carrying the alleles b and vg^+, the other carrying the alleles b^+ and vg. During meiosis the duplicated homologous chromosomes are aligned with each other and may exchange parts. Four kinds of chromosomes, and therefore four kinds of gametes, result whenever crossing over takes place; two of them have the original combinations ($b\ vg^+$ and b^+vg) while the other two have complementary combinations ($b\ vg$ and b^+vg^+).

inant over black and long wing is dominant over vestigial (Figure 1.9). Heterozygous females ($b^+b\ vg^+vg$) of the F_1 generation were then crossed to males homozygous for the two recessive alleles ($b\ b\ vg\ vg$). In this cross, the males only contribute gametes carrying recessive alleles; thus all alleles, whether recessive or dominant, contributed by the heterozygous females show in the following generation. Four types of progeny were obtained in the following proportions:

1. Brown vestigial 41.5%
2. Black long 41.5%
3. Brown long 8.5%
4. Black vestigial 8.5%

If the genes for body color and wing length were completely linked, only two kinds of flies would have appeared: brown-vestigial and black-long. If the two genes assorted independently, four kinds of flies would be expected but in equal proportions. Four kinds of flies were indeed observed, but those representing the parental allele combinations were considerably more common than the flies representing alternative combinations. The gene for body color and the gene for wing length are linked but not completely linked. The nonparental classes of gametes are called *recombinant* gametes.

The phenomenon of crossing over has made possible an important achievement of genetics, namely the construction of genetic maps (Figure 1.10). The principle used in genetic mapping is simple. If genes are

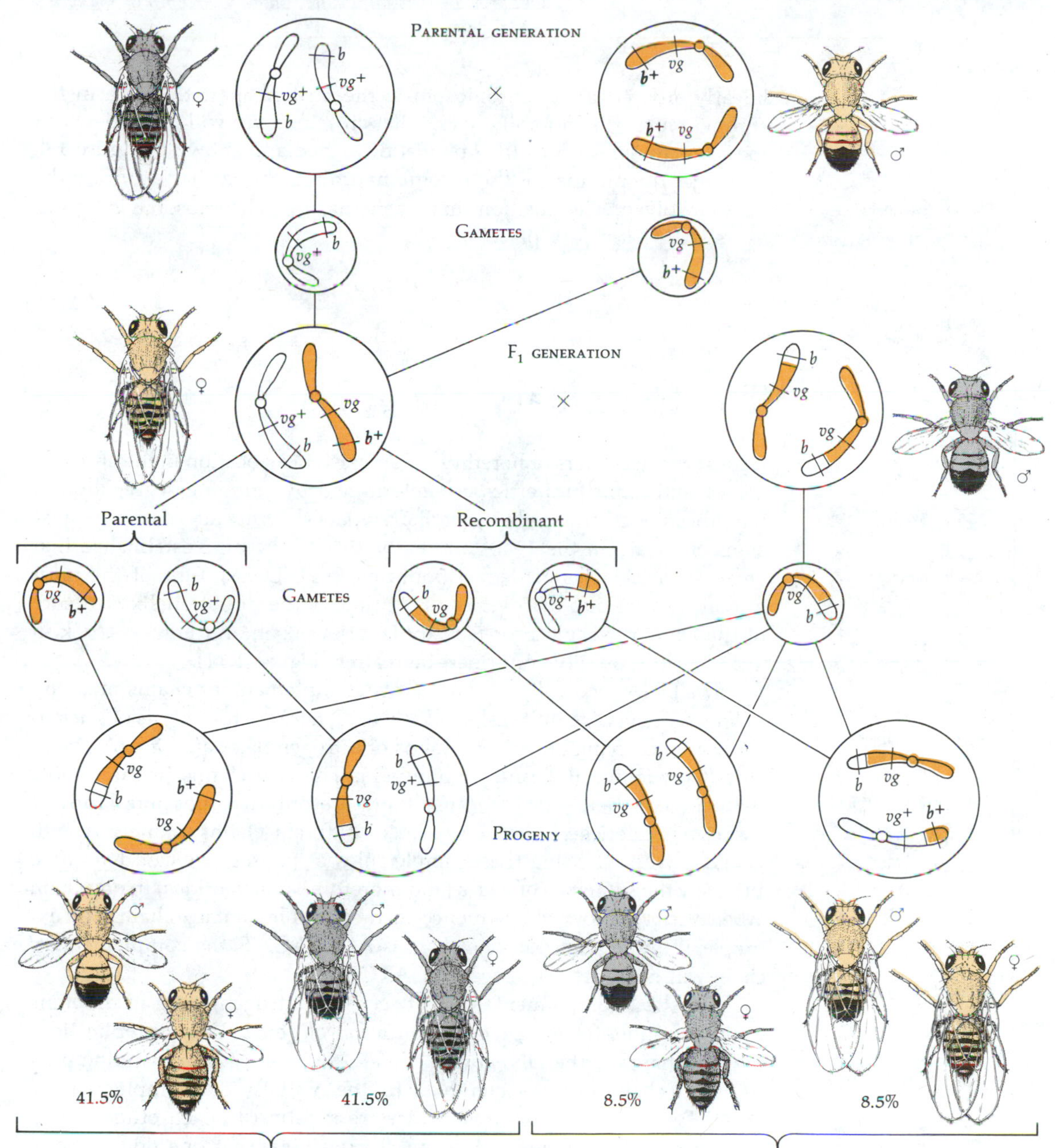

Figure 1.9
Recombination in *Drosophila melanogaster*. The parental flies carry different alleles for two traits, body color and wing size; the genes controlling these traits are on the same chromosome. The F_1 females (b^+b vg^+vg) produce four kinds of gametes, two with the parental combinations (b^+vg and b vg^+) and two with recombined genes (b vg and b^+vg^+). This is made manifest by a cross between F_1 females and males homozygous for the two recessive alleles (b b vg vg). In the following generation, the frequency of flies having a trait from one parent and a trait from the other parent reflects the frequency of recombination between the gene for body color and the gene for wing size.

linearly ordered along chromosomes, the farther apart they are in the chromosomes, the more likely it is that crossing over will occur between them. In crosses such as the one just described and shown in Figure 1.9, the larger the number of the recombinant flies, the farther apart are the genes involved. The position that a gene has in a chromosome is known as its *locus* (Latin for "place").

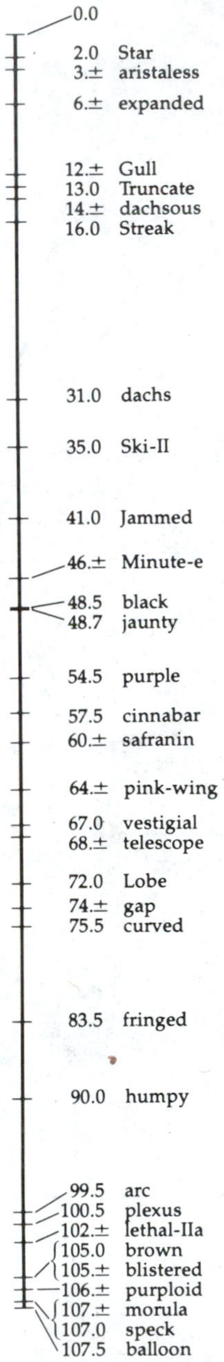

0.0
2.0 Star
3.± aristaless
6.± expanded
12.± Gull
13.0 Truncate
14.± dachsous
16.0 Streak
31.0 dachs
35.0 Ski-II
41.0 Jammed
46.± Minute-e
48.5 black
48.7 jaunty
54.5 purple
57.5 cinnabar
60.± safranin
64.± pink-wing
67.0 vestigial
68.± telescope
72.0 Lobe
74.± gap
75.5 curved
83.5 fringed
90.0 humpy
99.5 arc
100.5 plexus
102.± lethal-IIa
105.0 brown
105.± blistered
106.± purploid
107.± morula
107.0 speck
107.5 balloon

Figure 1.10
A partial map of the second chromosome of *Drosophila melanogaster.* This is the map published by T. H. Morgan in 1926.

The Hereditary Materials

Genes are the carriers of heredity. Their existence, location in the chromosomes, and other properties are determined by studying segregation in the progenies of crosses between individuals showing alternative expressions of a trait. In the 1940s and early 1950s it became established that genes are molecules of deoxyribonucleic acid (DNA). One of the most important scientific discoveries of all time is the double-helix structure of the DNA molecule proposed by James Watson and Francis Crick in 1953 and thoroughly confirmed thereafter (Figure 1.11).

The DNA molecule consists of two complementary chains made up of long sequences of units called *nucleotides.* There are four kinds of nucleotides, each carrying one of four kinds of nitrogen base: adenine (A), cytosine (C), guanine (G), and thymine (T). The two chains in the double helix are complementary because there are only two possible kinds of associations between nitrogen bases in different chains, namely A with T and C with G. Using these simple rules of pairing, we can determine precisely the sequence of bases (and therefore of nucleotides) in one chain whenever we know the sequence in the complementary chain. For example, if the sequence in one chain is ACCTAGAT, the complementary chain will have the sequence TGGATCTA.

The strict complementarity between the nitrogen bases in different chains accounts for the precise replication of genes. During replication, the two chains of the DNA double helix separate by an unwinding process. Each chain serves as a template for the synthesis of a complementary chain. Two double chains result that are identical to each other and to the parental double chain. Assume, for example, that one double helix starts with the sequence

ACCTAGAT
| | | | | | | |
TGGATCTA

. In the process of replication the two chains separate, and each determines the sequence in a complementary chain. If we represent the nitrogen bases in the newly synthesized chains with boldface letters, the two resulting double helices will be

ACCTAGAT
| | | | | | | |
TGGATCTA

and

ACCTAGAT
| | | | | | | |
TGGATCTA

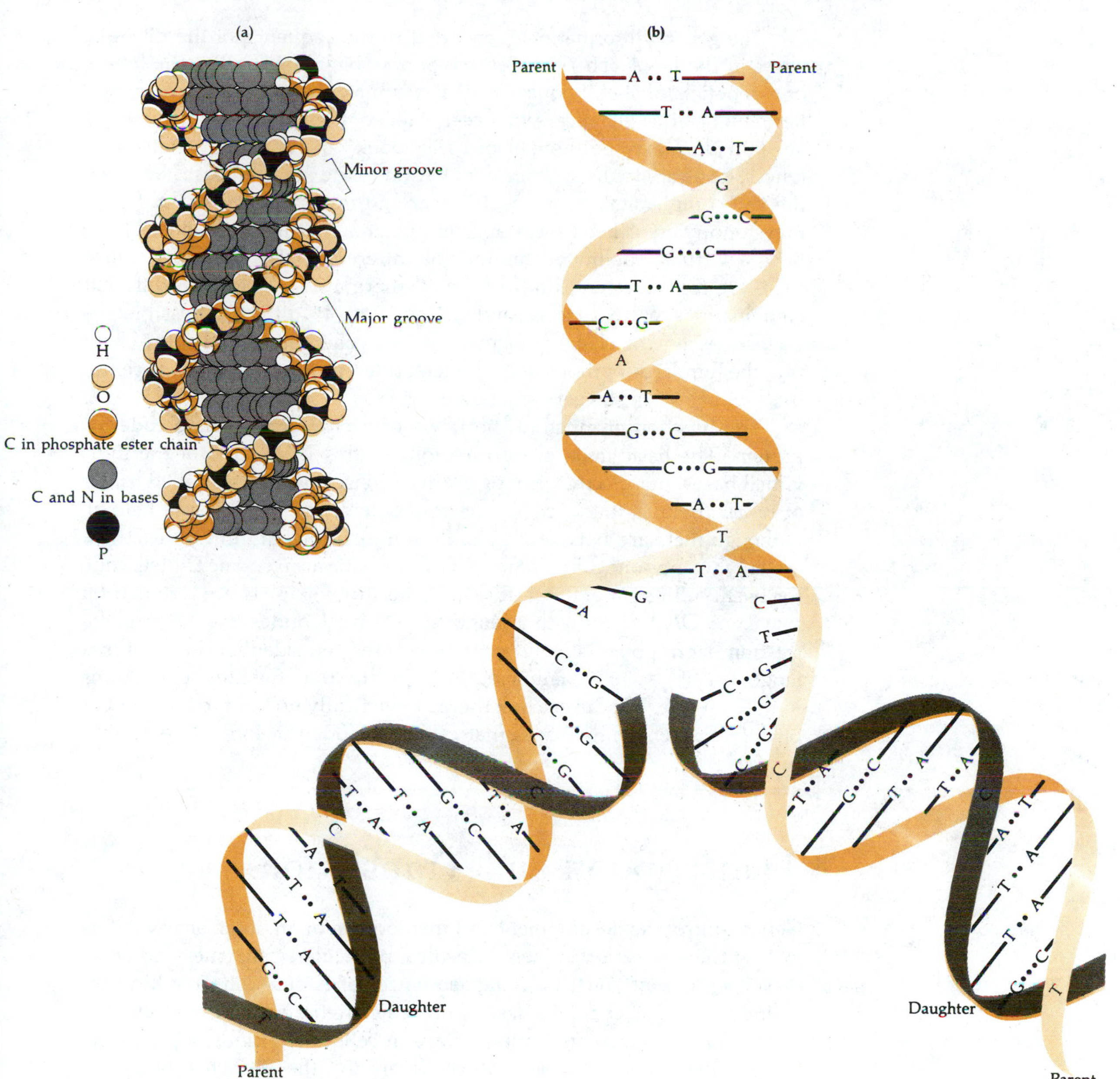

Figure 1.11
The Watson-Crick model of the DNA double helix. **(a)** A space-filling model of the molecule. **(b)** Mode of replication.

The genetic information is encoded in the sequence of the nitrogen bases in the DNA. The nitrogen bases may be considered as the letters of a genetic alphabet. Sequences of bases make up "words"; genes may be thought of as genetic "sentences." Hence, the genetic endowment of an individual may be thought of as a "book" made up of genetic sentences. Contrary to the strict determination of the nitrogen bases between the two complementary chains, there are no restrictions as to what base may follow any other base along one chain. This makes it possible to have a virtually unlimited number of different DNA molecules. Since there are four different kinds of bases, there are $4 \times 4 = 16$ different combinations of two bases, and, in general, 4^n different possible sequences each with a length of n nucleotides. When the length of the chain is in the hundreds or thousands of nucleotides, as it is in the case of genes, the number 4^n is staggeringly large.

The function of much of the DNA of each organism is to code for protein. The basic units of information for this DNA are not the individual bases, but discrete groups of three consecutive bases called *triplets* or *codons* (because they "code" for amino acids, as we shall see below). Although there are sixty-four possible combinations of three bases, there are only twenty-one different units of information: sixty-one triplets code for twenty different amino acids; the other three triplets are termination signals. A DNA chain with a length of 900 nucleotides has 300 triplets; the number of potentially different "messages" encoded by chains of that length is $21^{300} = 10^{397}$, a number much greater than the number of atoms in the universe. We can see that there is practically no limit to the number of different genetic messages that can be encoded in long DNA chains.

Transcription and Translation

Genes control the development and metabolism of an organism by determining the synthesis of other molecules, particularly enzymes and other proteins. Proteins consist of long sequences of twenty different kinds of amino acids. One common class of proteins are the enzymes, which control the chemical activities taking place in cells. Two kinds of genes can be identified: *structural genes,* which determine the sequence of amino acids in a protein, and *regulatory genes,* which control the activity of other genes by, for example, turning on and off structural genes. In addition, there are genes that, in a strict sense, are neither structural nor regulatory, such as the genes coding for ribosomal RNA and transfer RNA.

The information contained in the nucleotide sequence of a structural gene determines the synthesis of a particular protein through the processes *transcription* and *translation* (Figure 1.12). Transcription is a process by which the information contained in the base sequence of a DNA molecule is transferred to a complementary RNA molecule (called *messenger RNA*). Translation is the process by which the information contained in

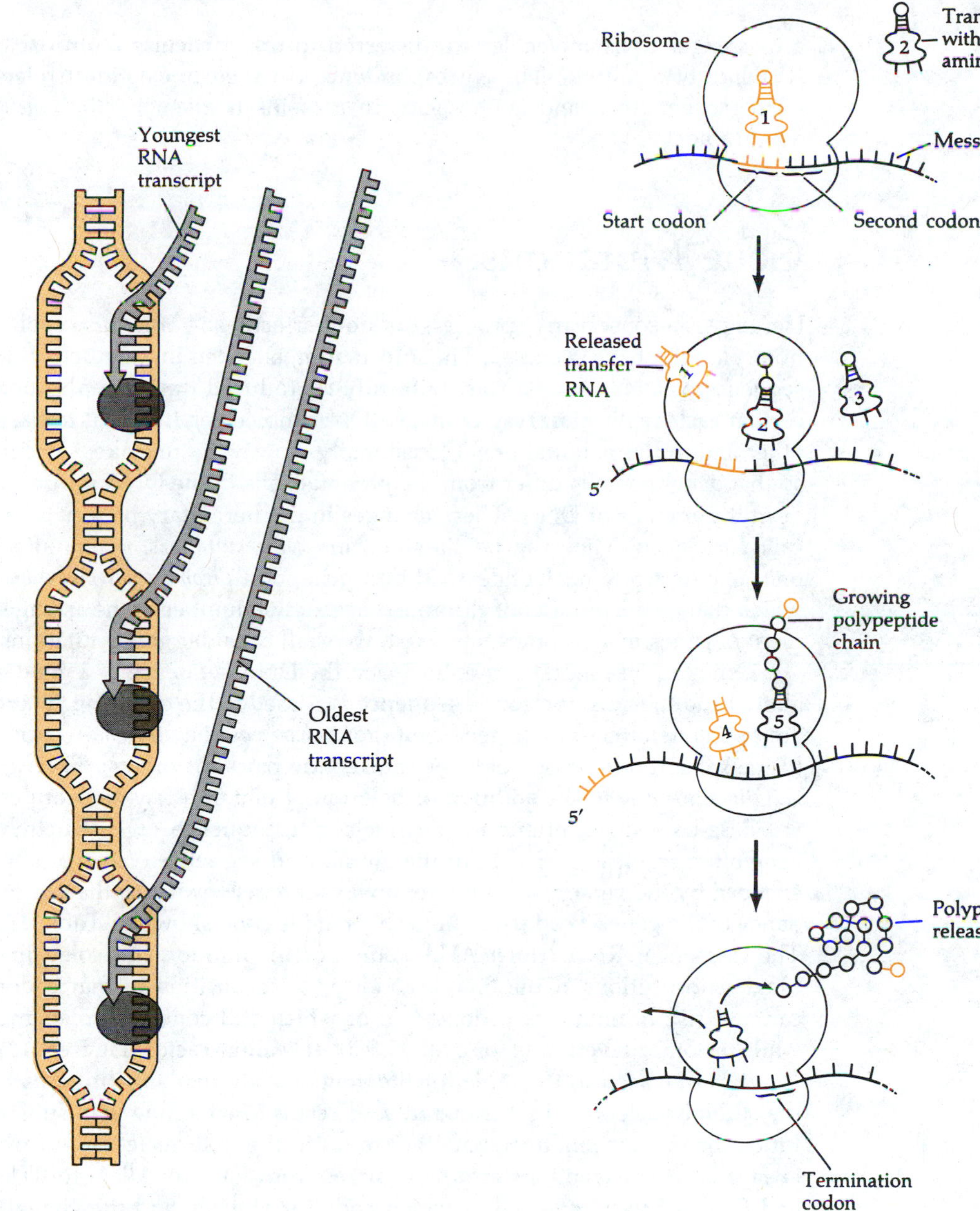

Figure 1.12
Transcription and translation. **(a)** The order of bases in the DNA dictates the order of nucleotides in the messenger RNA transcribed from a DNA strand. Several messenger RNA molecules may be formed simultaneously, one following another down the DNA strand. **(b)** Translation occurs in the ribosomes of the cell, to which the messenger RNA becomes attached. Different kinds of transfer RNA molecules, each carrying a different kind of amino acid, recognize the codons in the messenger RNA. As the ribosome moves along the messenger RNA, the amino acids become attached to each other; when the termination codon is reached, the protein (polypeptide) is released.

a messenger RNA molecule is transferred to the particular amino acid sequence of a protein. The correspondence between nucleotide triplets in messenger RNA and amino acids in proteins is given by the *genetic code* (Table 1.3).

Gene Mutations

Heredity is a conservative process, but not perfectly so—otherwise evolution could not have occurred. The information encoded in the nucleotide sequence of DNA is, as a rule, faithfully reproduced during replication so that each replication results in two DNA molecules identical to each other and to the parental one. Occasionally, however, "mistakes" occur, so that daughter cells differ from the parental cells in the DNA sequence or in the amount of DNA. These changes in the hereditary materials are called *mutations.* They can be classified into *gene mutations,* which affect only one or a few nucleotides within a gene, and *chromosomal mutations,* which change the number of chromosomes, or the number or the arrangement of genes in a chromosome. First, we shall consider gene mutations.

Gene or point mutations occur when the DNA sequence of a gene is altered and the new nucleotide sequence is passed to the offspring. There are two general classes of gene mutations: *base-pair substitutions* are due to the substitution of one or a few nucleotide pairs for others; *frameshift mutations* are due to the addition or deletion of one or a few nucleotides.

Base-pair substitutions in the nucleotide sequence of a structural gene often result in a change in the amino acid sequence of the protein encoded by the gene, but this is not always the case owing to the redundancy of the genetic code. Consider the genetic code shown in Table 1.3. The messenger RNA triplet AUA codes for the amino acid isoleucine. Single substitutions at the first, second, or third position in that codon can give rise to nine new codons, two of which still code for isoleucine, while the other seven code for a total of six new amino acids (Figure 1.13).

It can be seen in the table for the genetic code that substitutions in the second nucleotide of a triplet always result in an amino acid substitution (or in a terminating signal); changes in the first nucleotide nearly always do (the exceptions are the changes from UUA or UUG to CUA and CUG, or vice versa, all of which code for leucine, and the changes from AGA or AGG to CGA and CGG, or vice versa, all of which code for arginine). However, substitutions in the third nucleotide of a triplet often do not lead to amino acid substitutions in the encoded protein because most of the redundancy of the genetic code affects the third position of triplets. Triplets that code for the same amino acid are called "synonymous" codons.

Some base-pair substitutions may change a triplet coding for an amino acid into a terminating triplet, or vice versa (for example, a muta-

Table 1.3
The genetic code: correspondence between the 64 possible codons in messenger RNA and the amino acids (or termination signals).*

First Position	Second Position: U		Second Position: C		Second Position: A		Second Position: G		Third Position
U	UUU	Phe	UCU	Ser	UAU	Tyr	UGU	Cys	U
U	UUC	Phe	UCC	Ser	UAC	Tyr	UGC	Cys	C
U	UUA	Leu	UCA	Ser	UAA	*Stop*	UGA	*Stop*	A
U	UUG	Leu	UCG	Ser	UAG	*Stop*	UGG	Trp	G
C	CUU	Leu	CCU	Pro	CAU	His	CGU	Arg	U
C	CUC	Leu	CCC	Pro	CAC	His	CGC	Arg	C
C	CUA	Leu	CCA	Pro	CAA	Gln	CGA	Arg	A
C	CUG	Leu	CCG	Pro	CAG	Gln	CGG	Arg	G
A	AUU	Ile	ACU	Thr	AAU	Asn	AGU	Ser	U
A	AUC	Ile	ACC	Thr	AAC	Asn	AGC	Ser	C
A	AUA	Ile	ACA	Thr	AAA	Lys	AGA	Arg	A
A	AUG	Met	ACG	Thr	AAG	Lys	AGG	Arg	G
G	GUU	Val	GCU	Ala	GAU	Asp	GGU	Gly	U
G	GUC	Val	GCC	Ala	GAC	Asp	GGC	Gly	C
G	GUA	Val	GCA	Ala	GAA	Glu	GGA	Gly	A
G	GUG	Val	GCG	Ala	GAG	Glu	GGG	Gly	G

*The first position of a codon is given in the column on the left. The nitrogen base thymine does not exist in RNA, where uracil (U) takes its place; the other three nitrogen bases in messenger RNA are the same as in DNA: adenine (A), cytosine (C), and guanine (G). The 20 amino acids making up proteins are as follows: alanine (Ala), arginin (Arg), asparagine (Asn), aspartic acid (Asp), cysteine (Cys), glycine (Gly), glutamic acid (Glu) , glutamine (Gln), histidine (His), isoleucine (Ile), leucine (Leu), lysine (Lys), methionine (Met), phenylalanine (Phe), proline (Pro), serine (Ser), threonine (Thr), tyrosine (Tyr), tryptophane (Trp), and valine (Val).

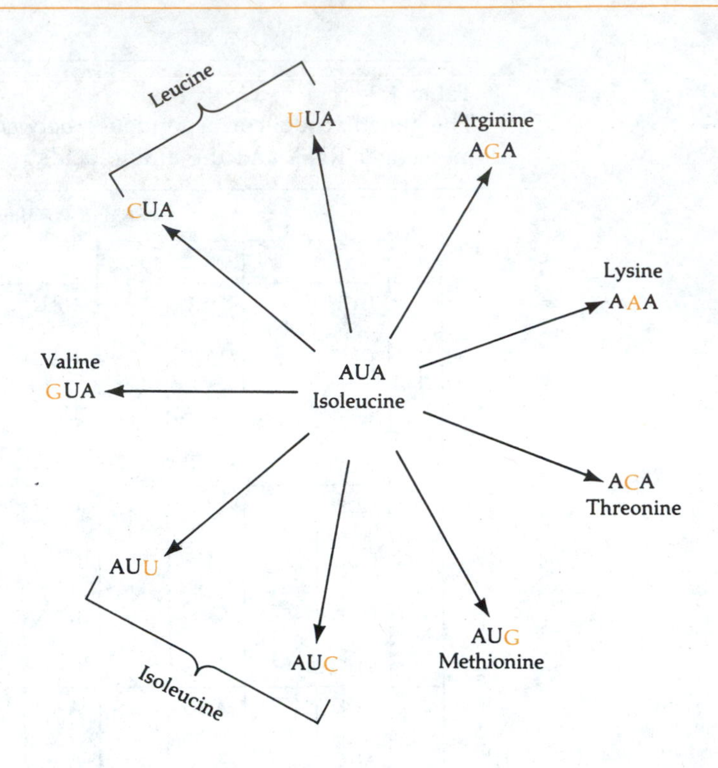

Figure 1.13
Point mutations. Substitutions at the first, second, or third position in the messenger-RNA codon for the amino acid isoleucine can give rise to nine new codons that code for six different amino acids. The effects of the mutation depend on what change takes place: Arginine, threonine, and lysine have chemical properties that differ sharply from those of isoleucine.

tion that changes the messenger RNA triplet UAU, coding for tyrosine, into the triplet UAA, which is a terminating signal). This type of base substitution will result in proteins with changed lengths, since the nucleotide sequence is not translated beyond a terminating signal.

Frameshift mutations often result in a very altered sequence of amino acids in the translated protein. The addition or deletion of one or more (other than exact multiples of three) nucleotide pairs shifts the "reading frame" of the nucleotide sequence from the point of an insertion or deletion to the end of the molecule (Figure 1.14). If one nucleotide pair is inserted at some point and another is deleted at another point, the original reading frame and the corresponding amino acid sequence are restored after the second mutational change.

Gene mutations may occur spontaneously due to the molecular dynamics involved in the replication or maintenance of the DNA. Mutations may also be induced by natural or artificial ultraviolet light, X rays, and other high-frequency radiations, as well as by exposure of the organisms to certain chemicals, such as mustard gas and many others, called *mutagens.*

Gene mutations may have effects ranging from negligible to lethal. A base-pair substitution that results in no amino acid changes in the encoded protein may have little or no effect on the ability of an organism to survive and reproduce. Mutations changing one or even several amino acids may also have small or no detectable effect on the organism if the

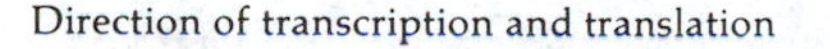
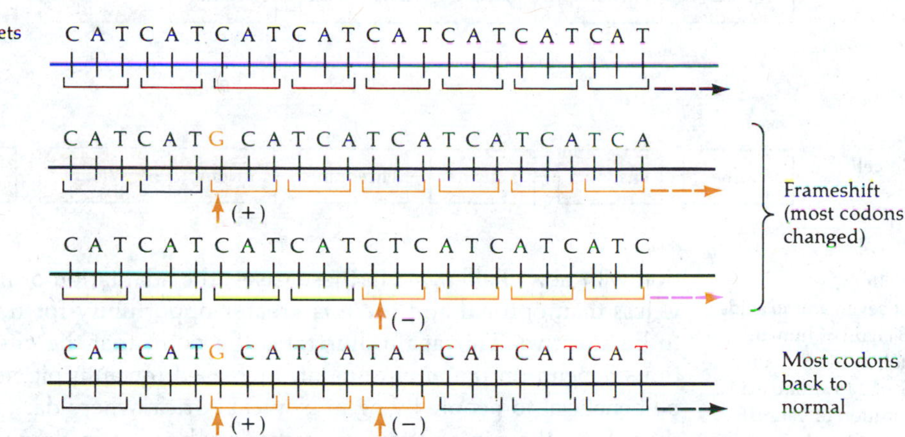

Figure 1.14
Frameshift mutations are due to the addition or deletion of nucleotides in the DNA sequence. If one nucleotide is added at one position and another one removed somewhere else, all the codons outside the segment between the two changes will be the same as in the original sequence.

essential biological function of the coded protein is not affected. However, the consequences of an amino acid substitution may be severe when the substitution affects the active site of an enzyme or modifies in some other way an essential function of a protein (Figure 1.15).

The deleterious effects of mutations often depend on particular environmental conditions. For example, in *Drosophila* there is a class of mutants known as "temperature-sensitive." At standard temperatures of 20°C to 25°C, flies homozygous for these mutants live and reproduce more or less normally. But at temperatures about 28°C these flies become paralyzed or die, although wild-type flies can still function normally. In humans, phenylketonuria (PKU) is a severe disease caused by homozygosis for a recessive allele. However, individuals homozygous for the PKU allele may be effectively normal if they maintain a diet free of phenylalanine, since their problems arise from an inability to metabolize this amino acid.

Newly arisen mutations are more likely to be deleterious than beneficial to their carriers because mutations are random events with respect to adaptation. In other words, they occur independent of whether they have beneficial or harmful consequences. However, the allelic variants already present in a population have been subject to natural selection. If they occur in substantial frequencies in a population, it is because they improve, or improved at one time, the adaptation of their carriers relative to alternative alleles that have been eliminated or kept at low frequencies by natural selection. A newly arisen mutant is likely to have arisen by mutation in the previous history of a population. If it does not exist in substantial frequencies, this is because it is not beneficial to the organisms.

Occasionally, however, a newly arisen mutation may increase adaptation. The probability of such an event is greater when organisms colonize a new territory, or when environmental changes confront a popula-

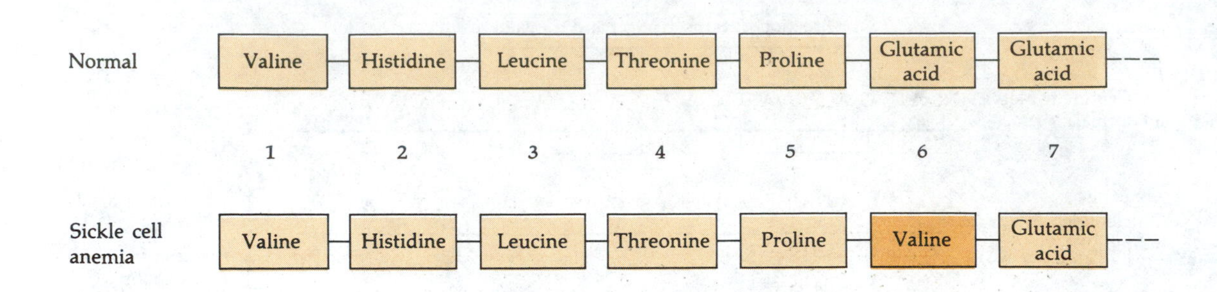

Figure 1.15
The first seven amino acids of the β chain of human hemoglobin; the β chain consists of 146 amino acids. A substitution of valine for glutamic acid at the sixth position is responsible for the severe disease known as sickle-cell anemia.

tion with new challenges. In these cases, the adaptation of the organisms is less than optimal and there is greater opportunity for new mutations to be adaptive. This again illustrates the point that the effects of mutations depend on the environment. Increased melanin pigmentation may be beneficial to people living in tropical Africa, where dark skin protects them from the sun's ultraviolet radiation, but not in Scandinavia where the intensity of sunlight is low and light skin facilitates the synthesis of vitamin D. Mutations to drug resistance in microorganisms are additional examples: a mutant making bacteria resistant to streptomycin may be beneficial to the bacteria in the presence of the drug but not in its absence.

Rates of Mutation

Mutation rates have been measured in a great variety of organisms, although mostly for mutants having conspicuous effects. Mutation rates are generally lower in bacteria and other microorganisms than in multicellular organisms. In humans and other multicellular organisms, mutants typically appear at about one per 100,000 (1×10^{-5}) to one per 1,000,000 (1×10^{-6}) gametes. There is, however, considerable variation from gene to gene, as well as from organism to organism, as shown in Table 1.4.

Although mutation rates are low, new mutants appear continuously in nature. This is because there are many individuals in any species and many gene loci in each individual. For example, a typical insect species may consist of about 100 million (10^8) individuals. If we assume that the average mutation rate per locus is one per 100,000 gametes (1×10^{-5}), the average number of mutations newly appeared in a given generation of an insect species would be $2 \times 10^8 \times 10^{-5} = 2000$ per locus. (The mutation rate is multiplied by the number of individuals, and then by two because each individual results from the union of two gametes.) Therefore, the process of mutation provides species with plenty of new genetic variation every generation.

The probability that a given individual will have a new mutation at a certain locus is low. This probability is simply the mutation rate multiplied by two. This is because each diploid individual arises from two gametes. However, the probability that a given individual will have some

Table 1.4
Mutation rates of specific genes in various organisms.

Organism and Trait	Mutations per Genome per Generation
Bacteriophage T2 (virus)	
Host range	3×10^{-9}
Lysis inhibition	1×10^{-8}
Escherichia coli (bacterium)	
Streptomycin resistance	4×10^{-10}
Streptomycin dependence	1×10^{-9}
Resistance to phage T1	3×10^{-9}
Lactose fermentation	2×10^{-7}
Salmonella typhimurium (bacterium)	
Tryptophan independence	5×10^{-8}
Chlamydomonas reinhardi (alga)	
Streptomycin resistance	1×10^{-6}
Neurospora crassa (fungus)	
Adenine independence	4×10^{-8}
Inositol independence	8×10^{-8}
Zea mays (corn)	
Shrunken seeds	1×10^{-6}
Purple seeds	1×10^{-5}
Drosophila melanogaster (fruit fly)	
Electrophoretic variants	4×10^{-6}
White eye	4×10^{-5}
Yellow body	1×10^{-4}
Mus musculus (mouse)	
Brown coat	8×10^{-6}
Piebald coat	3×10^{-5}
Homo sapiens (human)	
Huntington's chorea	1×10^{-6}
Aniridia (absence of iris)	5×10^{-6}
Retinoblastoma (tumor of retina)	1×10^{-5}
Hemophilia A	3×10^{-5}
Achondroplasia (dwarfness)	$4\text{–}8 \times 10^{-5}$
Neurofibromatosis (tumor of nerve tissue)	2×10^{-4}

new mutation *anywhere in the genome* is not low. Consider, for example, *Drosophila melanogaster*. If we assume that it has 10,000 (10^4) gene loci and that the average mutation rate per locus per gamete is 10^{-5}, the probability that a fly will carry a new mutation is $2 \times 10^4 \times 10^{-5} = 0.2$. Humans may have as many as 100,000 (10^5) gene loci. If we assume the same mutation rate as for *Drosophila*, the probability that each human being will have an allelic variant not present in its parents is $2 \times 10^5 \times 10^{-5} = 2$. In other words, on the average each human being carries about two new mutations.

The calculations just made are based on rates of mutations with visible effects. For the genome as a whole, mutation rates are estimated to be no less than 7×10^{-9} per nucleotide pair per year. In mammals, the number of nucleotide pairs per diploid genome is about 4×10^9. Hence, nucleotide substitutions in mammals occur at a rate no less than $4 \times 10^9 \times 7 \times 10^{-9} = 28$ per diploid genome per year. Because not all nucleotide substitutions have phenotypic effects, this rate is roughly consistent with the values obtained for visible mutations. In any case, it is clear that the potential of the mutation process to generate new hereditary variation is indeed enormous.

Chromosomal Mutations

Different cells of the same organism and different individuals of the same species have, as a rule, the same number of chromosomes, except that gametic cells have only half as many chromosomes as somatic cells. Homologous chromosomes are, also as a rule, uniform in the number and order of genes they carry. These rules have exceptions known as *chromosomal mutations*, abnormalities, or aberrations. Chromosomal mutations can be subdivided as follows (Figure 1.16):

A. **Changes in the Structure of Chromosomes.** These may be due to changes in the *number of genes* in chromosomes (deletions and duplications) or in the *location of genes* on the chromosomes (inversions and translocations).

1. *Deletion*, or deficiency. A chromosome segment is lost from a chromosome.
2. *Duplication*, or repeat. A chromosome segment is present more than once in a set of chromosomes.
3. *Inversion*. A chromosome segment is reversed. If the inverted segment includes the centromere, the inversion is called *pericentric* ("around" the centromere); if not, the inversion is *paracentric*.
4. *Translocation*. The location of a chromosome segment is changed. The most common forms of translocations are *reciprocal*, involving

Figure 1.16
A deletion has a chromosome segment missing. A duplication has a chromosome segment represented twice. Inversions and translocations are chromosomal mutations that change the locations of genes in the chromosomes. Centric fusions are the joining of two chromosomes at the centromere to become one single chromosome. Centric fissions, or dissociations, are the reciprocal of fusions: one chromosome splits into two chromosomes.

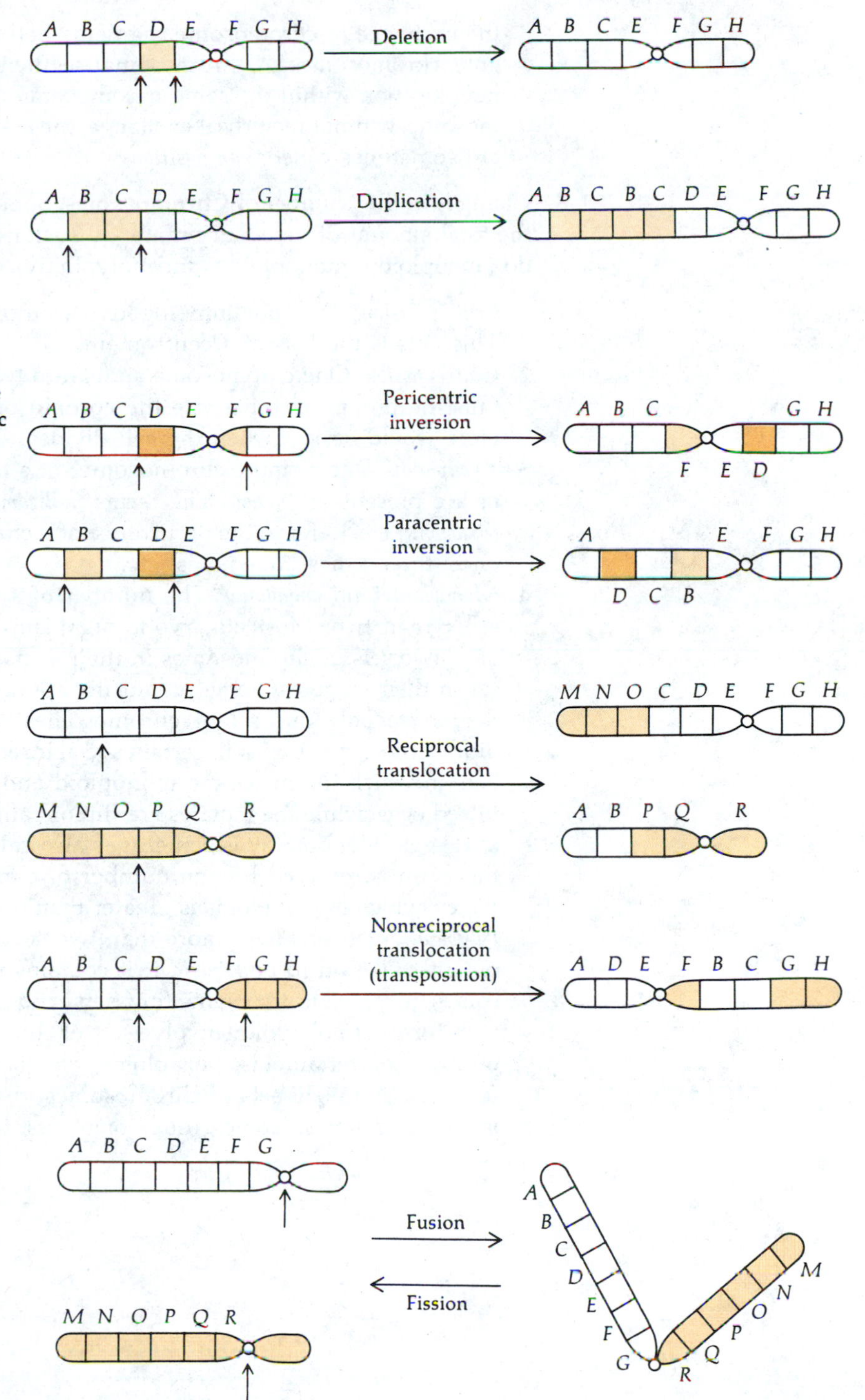

the exchange of chromosome segments between two nonhomologous chromosomes. A chromosomal segment may also move to a new location within the same chromosome or in a different chromosome, without reciprocal exchange; these kinds of translocations are sometimes called *transpositions.*

B. **Changes in the Number of Chromosomes.** Some changes do not alter the total amount of hereditary material (fusions and fissions); others do (aneuploidy, monoploidy, and polyploidy).

1. *Centric fusion.* Two nonhomologous chromosomes fuse into one. This entails the loss of a centromere.
2. *Centric fission.* One chromosome splits into two. A new centromere must be produced; otherwise the chromosome without a centromere would be lost when the cell divides.
3. *Aneuploidy.* One or more chromosomes of a normal set are lacking or are present in excess. The terms *nullisomic, monosomic, trisomic, tetrasomic,* etc., refer to the occurrence of a chromosome zero times, once, three times, four times, etc.
4. *Monoploidy* and *polyploidy.* The number of *sets* of chromosomes is other than two. Most eukaryotic organisms are *diploid,* i.e., they have two sets of chromosomes in their somatic cells, but only one set in their gametes. Some organisms are normally *monoploid,* i.e., they have only one set of chromosomes. Both monoploid and diploid individuals exist in certain social insects, such as the honeybee, in which the males are monoploid and develop from unfertilized eggs, while the females are diploid and develop from fertilized eggs. Monoploidy is sometimes also called *haploidy,* although this term is reserved for the number of chromosomes in the gametes, which in polyploids is greater than the monoploid number. *Polyploid* organisms have more than two sets of chromosomes; the organism is said to be *triploid* if it contains three sets of chromosomes, *tetraploid* if it contains four sets, and so on. The more common forms of polyploidy involve sets of chromosomes in multiples of two, i.e., tetraploids, hexaploids, and octoploids, which have four, six, and eight sets of chromosomes, respectively. Polyploidy is very common in some groups of plants, but is rare in animals.

2

Genetic Structure of Populations

Population Genetics

Genetics in general concerns the genetic constitution of organisms and the laws governing the transmission of this hereditary information from one generation to the next. *Population genetics* is that branch of genetics concerned with heredity in groups of individuals, i.e., in populations. Population geneticists study the genetic constitution of populations and how this genetic constitution changes from generation to generation.

Hereditary changes through the generations underlie the evolutionary process. Hence, population genetics may also be considered as *evolutionary genetics.* However, these two concepts can be distinguished. It is often understood that population genetics deals with populations of a given species, while evolutionary genetics deals with heredity in any populations, whether of the same or of different species. By these definitions, evolutionary genetics is a broader subject than population genetics—it includes population genetics as one of its parts.

Populations and Gene Pools

The most obvious unit of living matter is the individual organism. In unicellular organisms, each cell is an individual; multicellular organisms consist of many interdependent cells, many of which die and are replaced by other cells throughout the life of an individual. In evolution, the relevant unit is not an individual but a population. A *population* is a community of individuals linked by bonds of mating and parenthood; in other words, a population is a community of individuals of the same species. The bonds of parenthood that link members of the same population are always present, but mating is absent in organisms that reproduce asexually. A *Mendelian population* is a community of interbreeding, sexually reproducing individuals, i.e., Mendelian populations are those in which reproduction involves mating.

The reason why the individual is not the relevant unit in evolution is that the genotype of an individual remains unchanged throughout its life; moreover, the individual is ephemeral (even though some organisms, such as conifer trees, may live up to several thousand years). A population, on the other hand, has continuity from generation to generation; moreover, the genetic constitution of a population may change—evolve—over the generations. The continuity of a population through time is provided by the mechanism of biological heredity.

The most inclusive Mendelian population is the *species* (Chapter 7). As a rule, the genetic discontinuities between species are absolute; sexually reproducing organisms of different species are kept from interbreeding by reproductive isolating mechanisms. Species are independent evolutionary units: genetic changes taking place in a local population can be extended to all members of the species, but are not ordinarily transmitted to members of a different species.

The individuals of a species are not usually homogeneously distributed in space; rather, they exist in more or less well-defined clusters, or local populations. A *local population* is a group of individuals of the same species living together in the same territory. The concept of local population may seem clear, but its application in practice entails difficulties because the boundaries between local populations are often fuzzy (Figure 2.1). Moreover, the organisms are not homogeneously distributed within a cluster, even when the clusters are quite discrete, as is true of organisms living in lakes or on islands; the lakes or islands may be sharply distinct, but individuals are not evenly distributed within a lake or on an island. Animals often migrate from one local population to another, and the

Figure 2.1 (*opposite*)
Geographic distribution of *Lacerta agilis*. **(a)** This lizard exists over a broad area encompassing large parts of Europe and western Asia, but its distribution is far from homogeneous. **(b)** *Lacerta agilis* has greater density along streams and rivers than in the intermediate areas. **(c)** The lizards occur in small family groups consisting of a few individuals each. Demes consist of about 20 to 40 family groups. Local populations consist of several demes each. Within a deme, matings occur rather freely between members of different family groups; however, fewer than 4% of all matings occur between individuals from different demes within the same local population. Fewer than 0.01% of all matings involve individuals from different local populations. (Data courtesy of Prof. Alexis B. Yablokov, Institute of Developmental Biology, USSR Academy of Sciences, Moscow.)

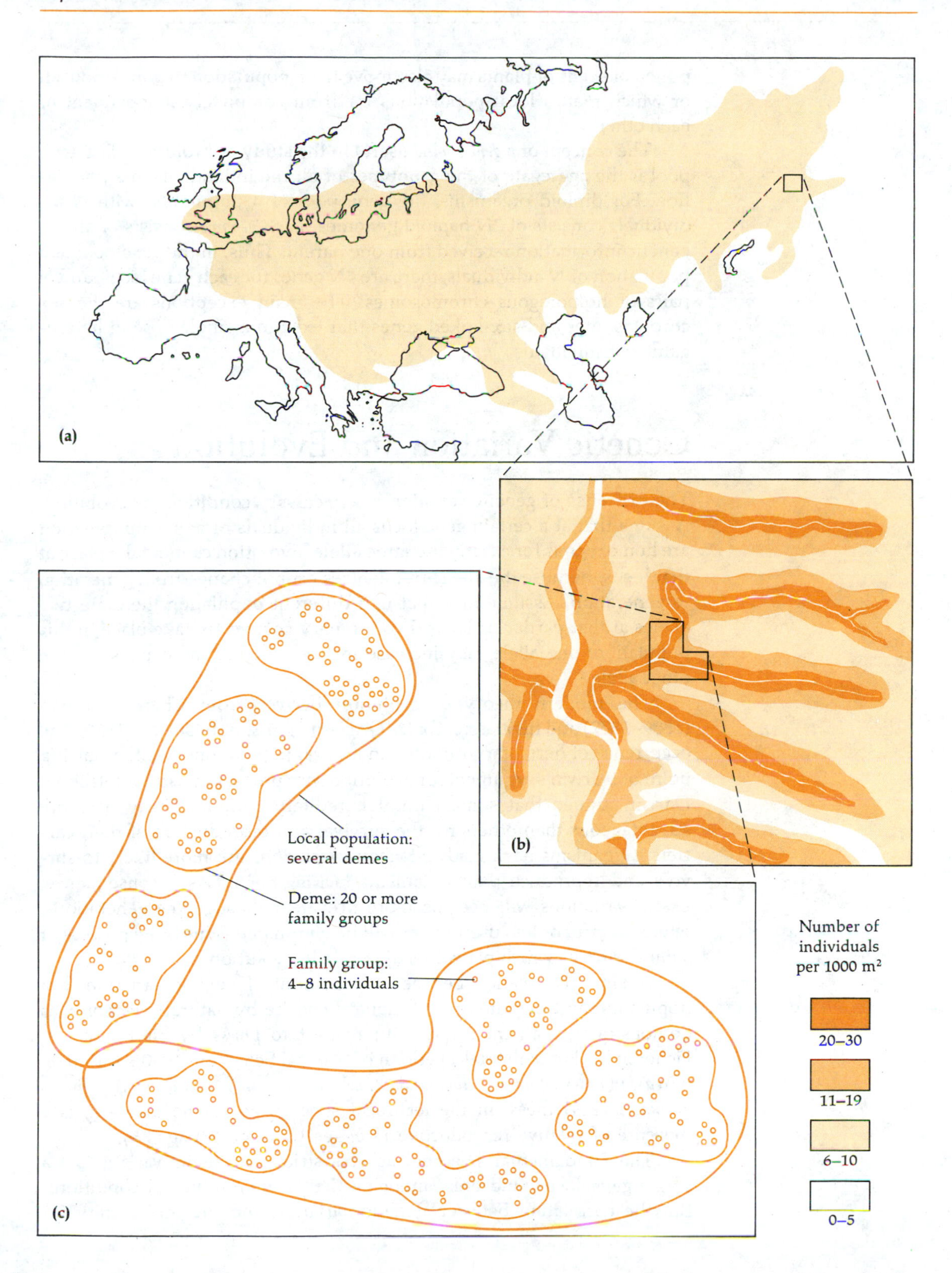
(a)
(b)
(c)
Local population:
several demes
Deme: 20 or more
family groups
Family group:
4–8 individuals
Number of
individuals
per 1000 m²
20–30
11–19
6–10
0–5

pollen or seeds of plants may also move from population to population, all of which makes local populations far from completely independent of each other.

The concept of a *gene pool* is useful in the study of evolution. The gene pool is the aggregate of the genotypes of all the individuals in a population. For diploid organisms, the gene pool of a population with N individuals consists of $2N$ haploid genomes. Each *genome* consists of all the genetic information received from one parent. Thus, in the gene pool of a population of N individuals, there are $2N$ genes for each gene locus, and N pairs of homologous chromosomes. The main exceptions are the sex chromosomes and sex-linked genes that exist in a single dose in heterogametic individuals.

Genetic Variation and Evolution

The existence of genetic variation is a necessary condition for evolution. Assume that at a certain gene locus all individuals of a given population are homozygous for exactly the same allele. Evolution cannot take place at that locus, because the allelic frequencies cannot change from generation to generation. Assume now that in a different population there are two alleles at that particular locus. Evolutionary change *can* take place in this population: one allele may increase in frequency at the expense of the other allele.

The modern theory of evolution derives from Charles Darwin (1809–1882) and his classic, *On the Origin of Species,* published in 1859. The occurrence of hereditary variation in natural populations was the starting point of Darwin's argument for evolution by a process of natural selection. Darwin argued that some natural hereditary variations may be more advantageous than others for the survival and reproduction of their carriers. Organisms having advantageous variations are more likely to survive and reproduce than organisms lacking them. As a consequence, useful variations will become more prevalent through the generations, while harmful or less useful ones will be eliminated. This is the process of *natural selection,* which plays a leading role in evolution.

A direct correlation between the amount of genetic variation in a population and the rate of evolutionary change by natural selection was demonstrated mathematically with respect to *fitness* by Sir Ronald A. Fisher in his Fundamental Theorem of Natural Selection (1930): *The rate of increase in fitness of a population at any time is equal to its genetic variance in fitness at that time.* (Fitness, in the technical sense used in the theorem, is a measure of relative reproductive rate; see Chapter 4, page 88).

The Fundamental Theorem applies strictly to allelic variation at a single gene locus, and only under particular environmental conditions. But the correlation between genetic variation and the opportunity for

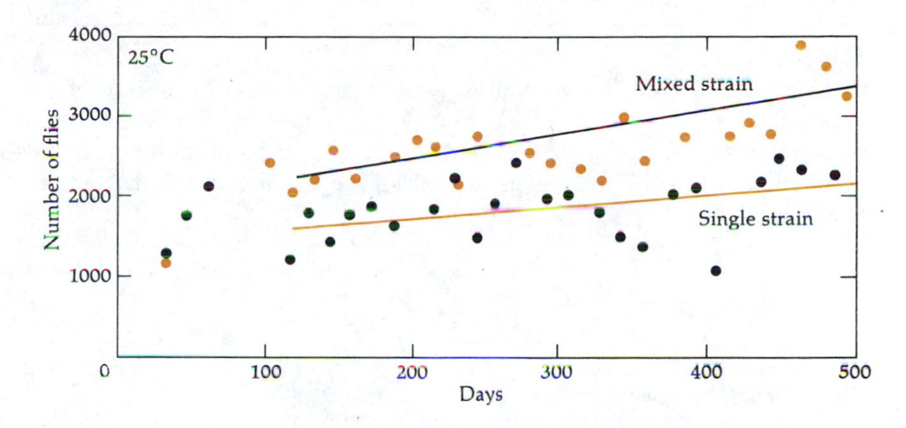

Figure 2.2
Correlation between amount of genetic variation and rate of evolution in laboratory populations of *Drosophila serrata* exposed to a new environment. The graph shows the change in number of flies during approximately 25 generations. The mixed-strain population initially had greater genetic variation than the single-strain population. Both populations increased in numbers throughout the experimental period, but the average rate of increase was substantially greater in the mixed-strain population than in the single-strain population. Increases in the number of flies over the generations reflect the increasing adaptation of the population to the experimental environment, which is promoted by evolution.

evolution is intuitively obvious. The greater the number of variable loci and the more alleles there are at each variable locus, the greater the possibility for change in the frequency of some alleles at the expense of others. This requires, of course, that there be selection favoring the change of some trait(s) and that the variation be relevant for the trait(s) being selected. Figure 2.2 and Table 2.1 give an experimental demonstration of the correlation between the amount of genetic variation and the rate of evolution when these conditions are met.

Genotypic and Genic Frequencies

We observe phenotypes directly, but not genotypes or genes. Variation in gene pools is expressed in terms of either genotype frequencies or gene frequencies. If we know the relationship between specific genotypes and the corresponding phenotypes, we are able to transform phenotypic frequencies into genotypic frequencies. Let us consider the M-N blood groups. There are three blood groups, M, N, and MN, which are determined by two alleles, L^M and L^N, at a single locus, according to the simple relationship shown in Table 2.2.

Table 2.1
Correlation between the amount of genetic variation and the rate of evolution in experimental populations of *Drosophila serrata* from Popondetta, New Guinea, and Sydney, Australia. The rate of evolution was measured by the rate at which the number of flies increased in the populations over about 25 generations. Figure 2.2 shows the data for the experiment at 25° C.

Population	Mean Number of Flies in Population	Mean Increase in Number of Flies per Generation
Experiment at 25°C		
Single strain (Popondetta)	1862 ± 79	31.5 ± 13.8
Mixed strain (Popondetta × Sydney)	2750 ± 112	58.5 ± 17.4
Experiment at 19°C		
Single strain (Popondetta)	1724 ± 58	25.2 ± 9.9
Mixed strain (Popondetta × Sydney)	2677 ± 102	61.2 ± 13.8

After F. J. Ayala, *Science 150*:903 (1965).

Examination of 730 Australian Aborigines produced the following results: 22 had blood group M, 216 had MN, and 492 had N. The frequencies of the blood groups and the corresponding genotypes are obtained by dividing the number of each kind observed by the total. For example, the frequency of blood group M is 22/730 = 0.030.

We can describe the variation at the M-N gene locus in this group of people by giving the frequencies of the three genotypes. If we assume that the 730 individuals are a *random sample* of Australian Aborigines, we may take the observed frequencies as characteristic of Australian Aborigines in general. A random sample is a representative, or unbiased, sample of a population.

It is convenient for some purposes to describe genetic variation at a locus using not the genotypic frequencies but the allelic frequencies. Allelic frequencies can be calculated either from the genotypic numbers observed or from the genotypic frequencies.

In order to calculate allelic frequencies directly from the genotype *numbers*, we simply count the number of times each allele is found and divide it by the total number of alleles in the sample. An $L^M L^M$ individual contains two L^M alleles; an $L^M L^N$ individual contains one L^M allele. Therefore, the number of L^M alleles in the sample described above is $(2 \times 22) + 216 = 260$. The total number of alleles in the sample is twice the number of individuals, because each individual has two alleles: $2 \times 730 = 1460$. The frequency of the L^M allele in the sample is therefore $260/1460 = 0.178$. Similarly, the frequency of the L^N allele is $[(2 \times 492) + 216]/1460 = 0.822$.

Table 2.2
M-N blood group and genotypic frequencies in a population of Australian Aborigines.

Blood Group	Genotype	Number	Frequency
M	$L^M L^M$	22	0.030
MN	$L^M L^N$	216	0.296
N	$L^N L^N$	492	0.674
Total:		730	1.000

Allelic frequencies can also be calculated from the genotypic *frequencies*, by observing as before that all alleles in homozygotes are of a given kind, whereas only half the alleles of a heterozygote are of a given kind. Thus the frequency of an allele is the frequency of individuals homozygous for that allele plus half the frequency of heterozygotes for that allele. Among the Australian Aborigines, the frequency of L^M is 0.030 + ½(0.296) = 0.178; similarly, the frequency of L^N is 0.674 + ½(0.296) = 0.822. Table 2.3 gives genotypic and allelic frequencies for the M-N gene locus in three human populations. It is apparent that human populations are quite different from one another with respect to this locus.

The calculation of gene frequencies when the number of alleles at a locus is greater than two is based on the same rules that apply for two alleles: homozygotes carry two copies of one allele, heterozygotes carry one each of two alleles. For example, in a certain natural population of *Drosophila willistoni*, six different genotypes were found at the *Lap-5* locus

Table 2.3
Genotypic and allelic frequencies for the M-N gene locus in three human populations.

	Number Having Blood Group:				Genotypic Frequency			Allelic Frequency	
Population	M	MN	N	Total	$L^M L^M$	$L^M L^N$	$L^N L^N$	L^M	L^N
Australian Aborigines	22	216	492	730	0.030	0.296	0.674	0.178	0.822
Navaho Indians	305	52	4	361	0.845	0.144	0.011	0.917	0.083
U.S. Caucasians	1787	3039	1303	6129	0.292	0.496	0.213	0.539	0.461

Table 2.4
Genotypic frequencies observed at the *Lap-5* locus in a population of *Drosophila willistoni*.

Genotype	Number	Frequency
98/98	2	0.004
100/100	172	0.344
103/103	54	0.108
98/100	38	0.076
98/103	20	0.040
100/103	214	0.428
Total:	500	1.000

in the numbers shown in Table 2.4. (The *Lap-5* gene codes for a leucine aminopeptidase enzyme; each allele is identified by a number that refers to the mobility of the corresponding polypeptide under electrophoresis—see Box 2.1, page 44).

Genotypic frequencies are obtained by dividing the number of times each genotype is observed by the total number of genotypes. Thus the frequency of the *98/98* genotype is 2/500 = 0.004. The frequency of a given allele can be obtained from the genotypic frequencies by adding the frequency of the homozygotes for that allele and half the frequency of each of the heterozygotes for that allele. Thus the frequency of the allele *98* is the frequency of the homozygote *98/98* plus half the frequencies of the heterozygotes *98/100* and *98/103*, or 0.004 + ½(0.076) + ½(0.040) = 0.062. Similarly, the frequencies of alleles *100* and *103* are calculated to be 0.596 and 0.342, respectively. The sum of these three frequencies is, of course, 1.000.

The allele frequencies can also be calculated by counting the number of times each allele appears and dividing it by the total number of alleles in the sample. Allele *98* appears twice in *98/98* homozygotes and once each in *98/100* and *98/103* heterozygotes, or (2 × 2) + 38 + 20 = 62 times; because the number of alleles in the sample is 2 × 500 = 1000, the frequency of allele *98* is 0.062. The number of times each allele appears in the sample and the allelic frequencies for the data in Table 2.4 are shown in Table 2.5.

One reason why it is often preferable to describe genetic variation at a locus using allelic frequencies rather than genotypic frequencies is

Table 2.5
Allelic frequencies observed at the *Lap-5* locus in a population of *D. willistoni*.

Allele	Number	Frequency
98	62	0.062
100	596	0.596
103	342	0.342
Total:	1000	1.000

because usually there are fewer alleles than genotypes. With two alleles, the number of possible genotypes is three; with three alleles, it is six; with four alleles, it is ten. In general, if the number of different alleles is k, the number of different possible genotypes is $k(k + 1)/2$.

Two Models of Population Structure

Two conflicting hypotheses were advanced during the nineteen forties and fifties concerning the genetic structure of populations. The *classical* model argues that there is very little genetic variation, the *balance* model, that there is a great deal (Figure 2.3).

According to the classical model, the gene pool of a population consists, at the great majority of loci, of a wild-type allele with a frequency very close to 1, plus a few deleterious alleles arisen by mutation but kept at very low frequencies by natural selection. A typical individual would be homozygous for the wild-type allele at nearly every locus, but at a few loci it would be heterozygous for the wild allele and a mutant. The "normal," ideal genotype would be an individual homozygous for the wild-type allele at every locus. Evolution would occur because occasionally a beneficial allele arises by mutation. The beneficial mutant would gradually increase in frequency by natural selection and become the new wild-type allele, with the former wild-type allele being eliminated or reduced to a very low frequency.

According to the balance model, there is often no single wild-type allele. Rather, at many—perhaps most—loci, the gene pool consists of an array of alleles with various frequencies. Hence, individuals are heterozygous at a large proportion of these loci. There is no single "normal" or ideal genotype; instead, populations consist of arrays of genotypes that

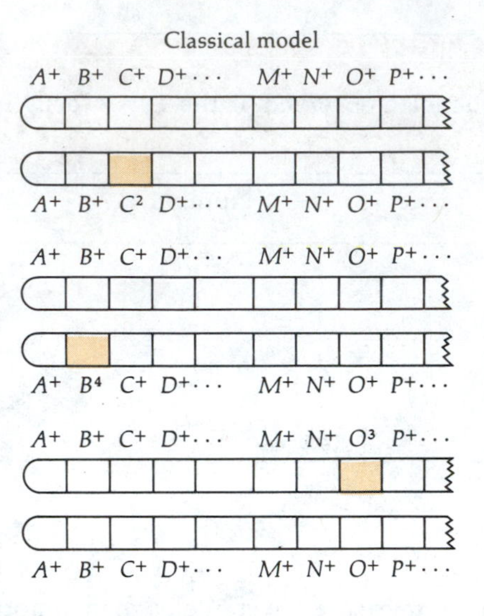

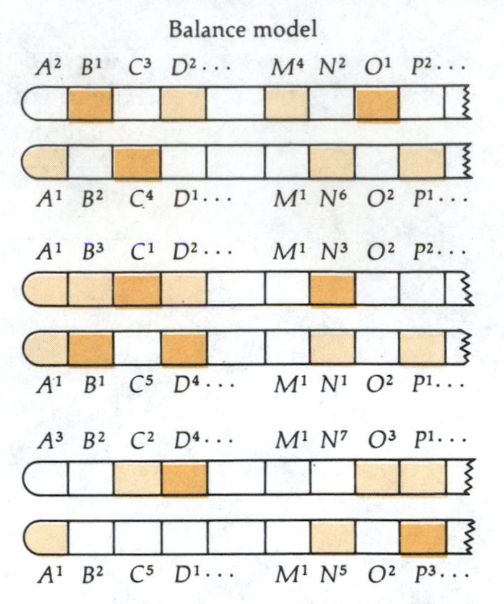

Figure 2.3
Two models of the genetic structure of populations. The hypothetical genotypes of three typical individuals are shown according to each model. Capital letters symbolize gene loci, and each number represents a different allele; the wild-type allele postulated by the classical model is represented by a + sign. According to the classical model, individuals are homozygous for the wild-type allele at nearly every locus, although they may be heterozygous for the wild allele and a mutant allele at an occasional locus (*C* in the first individual, *B* in the second, *O* in the third). According to the balance model, individuals are heterozygous at many gene loci.

differ from one another at many loci but are satisfactorily adapted to most environments encountered by the population.

The balance model sees evolution as a process of gradual change in the frequencies and kinds of alleles at many loci. Alleles do not act in isolation; rather, the fitness conferred by one allele depends on which other alleles exist in the genotype. The set of alleles present at any one locus is coadapted with the sets of alleles at other loci; hence, allelic changes at one locus are accompanied by allelic changes at other loci. However, like the classical model, the balance model accepts that many mutants are unconditionally harmful to their carriers; these deleterious alleles are eliminated or kept at low frequencies by natural selection, but play only a secondary, negative role in evolution.

Looking at Variation

It is now known that natural populations possess a great deal of genetic variation. Definitive evidence for this was not obtained until the late 1960s,

Figure 2.4
Variations in facial features, skin pigmentation, height, and other traits are apparent in human populations. (Owen Franken/Stock, Boston.)

however. It was known before that time that variation was a common phenomenon in nature. Whether the evidence suggested that allelic variation existed at many loci or only at a few was appraised differently by the balance school and the classical school. In any case, the evidence did not allow one to tell what proportion of the gene loci of an organism consisted of several alleles. The kind of evidence available before the 1960s was as follows.

Individual variation is a conspicuous phenomenon whenever organisms of the same species are carefully examined. Human populations, for example, exhibit variation in facial features, skin pigmentation, hair color and shape, body configuration, height and weight, blood groups, etc. (Figure 2.4). We notice human differences more readily than variation in other organisms, but morphological variation has been carefully recorded in many cases, e.g., with respect to color and pattern in snails, butterflies, grasshoppers, lady beetles, mice, and birds (Figure 2.5). Plants often differ in flower and seed color and in pattern, as well as in growth habit. A difficulty is that it is not immediately clear how much of this morphological variation is due to genetic variation and how much to environmental effects.

Geneticists have discovered that there is much more genetic variation than is apparent when organisms living in nature are observed. This has been accomplished by inbreeding, i.e., by mating close relatives, which increases the probability of homozygosis; recessive genes thus become expressed. Inbreeding has shown, for example, that virtually every *Drosophila* fly has allelic variants that in homozygous condition result in abnormal phenotypes, and that plants carry many alleles that when homozygous result in abnormal chlorophyll or none at all. Inbreeding has also shown that organisms carry alleles that in homozygous condition affect their fitness, i.e., that modify their fertility and survival probability (Table 2.6).

Figure 2.5
Morphological variation within a species is apparent in the color patterns of the wing covers (elytra) of the lady beetle *Harmonia axyridis*. The species occurs in Siberia, China, Korea, and Japan. The almost completely black phenotype (1) predominates in west-central Siberia, but farther eastward populations are more polymorphic, with increasing frequency of the black-spots-on-yellow-background phenotypes (2–8). The phenotypes 9–12 have yellow spots on a black background. Phenotypes 13–16 have red spots on a black background; these are found exclusively in the Far East.

Table 2.6
Frequencies of wild chromosomes of *Drosophila pseudoobscura* from the Sierra Nevada, California, that in homozygous condition have various effects on viability. A majority of the chromosomes have alleles that in homozygous condition reduce viability.

	Frequency (%) of Chromosome:		
Effect on Viability	Second	Third	Fourth
Lethal or semilethal	33.0	25.0	25.9
Subvitality (significantly lower viability than wild flies)	62.6	58.7	51.8
Normal (viability not significantly different from wild flies)	4.3	16.3	22.3
Supervitality (significantly higher viability than wild flies)	<0.1	<0.1	<0.1

After Th. Dobzhansky and B. Spassky, *Genetics* 48:1467 (1963).

A convincing source of evidence indicating that genetic variation is pervasive comes from artificial selection experiments. In artificial selection the individuals chosen to breed the next generation are those that exhibit the greatest expression of the desired characteristic (Chapter 6, page 169). For example, if we want to increase the yield of wheat, we choose in every generation the wheat plants with the greatest yield and use their seed to produce the next generation. If, over the generations, the selected population changes in the direction of the selection, it is clear that the original organisms had genetic variation with respect to the selected trait.

The changes obtained by artificial selection are often impressive. For example, the egg production in a flock of White Leghorn chickens increased from 125.6 eggs per hen per year in 1933 to 249.6 eggs per hen per year in 1965 (Figure 2.6). Artificial selection can also be practiced in opposite directions. Selection for high protein content in a variety of corn increased the protein content from 10.9 to 19.4%, while selection for *low* protein content reduced it from 10.9 to 4.9%. Artificial selection has been successful for innumerable commercially desirable traits in many domesticated species, including cattle, swine, sheep, poultry, corn, rice, and wheat, as well as in many experimental organisms, such as *Drosophila*, in which artificial selection has succeeded for more than 50 different traits. The fact that artificial selection succeeds virtually every time it is tried was taken by proponents of the balance model as showing that genetic variation exists in populations for virtually every characteristic of the organism.

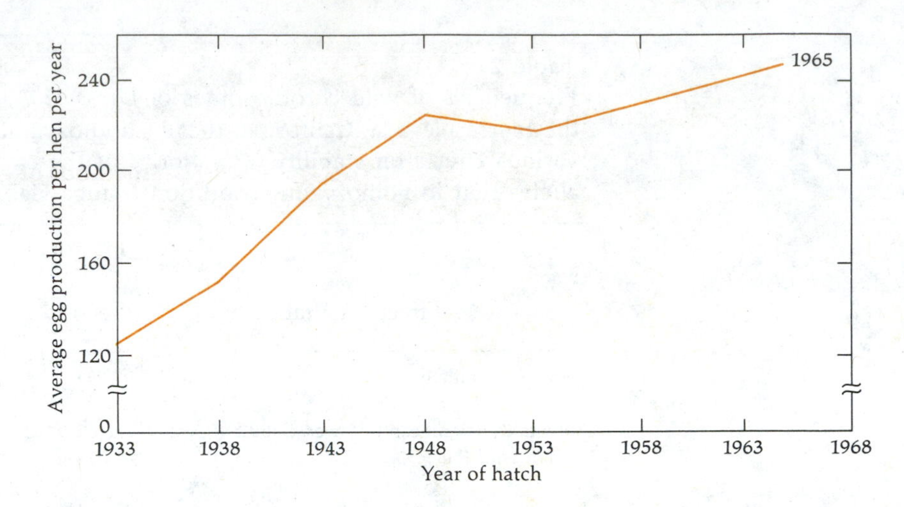

Figure 2.6
An example of artificial selection: egg production per hen per year in a flock of White Leghorn chickens. In the formation stock the average production was 125.6 eggs. Thirty-two years later, selection had increased productivity to 249.6 eggs, double the initial number. The success of selection indicates that the flock had considerable genetic variability with respect to egg production. The economic significance of doubling the number of eggs laid in a year is obvious. (After I. M. Lerner and W. J. Libby, *Heredity, Evolution, and Society*, 2nd ed., W. H. Freeman, San Francisco, 1976.)

The Problem of Measuring Genetic Variation

The evidence mentioned in the previous section may indicate that genetic variation is pervasive in natural populations, and hence that there is ample opportunity for evolutionary change. But in any case we would like to go one step further and find out precisely how much variation there is. For example, what proportion of all gene loci are polymorphic (i.e., variable) in a given population, and what proportion of all gene loci are heterozygous in a typical individual of the population? In trying to solve this problem, we find that the traditional methods of genetic analysis impose a methodological handicap.

Consider what we need to do in order to find out what proportion of the genes are polymorphic in a population. We cannot study every gene locus of an organism, because we do not even know how many loci there are, and because it would be an enormous task. The solution, then, is to look at only a sample of gene loci. If the sample is random, i.e., not biased and thus truly representative of the population, the values observed in the sample can be extrapolated to the whole population. Pollsters do quite well this way; for example, based on a sample of about 2000 individuals, they are able to predict with fair accuracy how millions of Americans will vote in a presidential election.

In order to ascertain how many gene loci are polymorphic in a population, we need to study a few genes that are an unbiased sample of all the gene loci. With the traditional methods of genetics this is impossible, because the existence of a gene is ascertained by examining the progenies of crosses between individuals showing different forms of a given character; from the proportions of individuals in the various classes, we infer whether one or more genes are involved. By such methods, therefore, the only genes known to exist are those that are variable. There is no way of obtaining an unbiased sample of the genome, because invariant genes cannot be included in the sample.

A way of out of this dilemma became possible with the discoveries in molecular genetics. It is now known that the genetic information encoded in the nucleotide sequences of the DNA of a structural gene is translated into a sequence of amino acids making up a polypeptide. We can select for study a series of proteins without previously knowing whether or not they are variable in a population—a series of proteins that, with respect to variation, represents an unbiased sample of all the structural genes in the organism. If a protein is found to be invariant among individuals, it is inferred that the gene coding for that protein is also invariant; if the protein is variable, we know that the gene is variable, and we can measure how variable it is, i.e., how many variant forms of the protein exist, and in what frequencies.

Quantifying Genetic Variation

Biochemists have known since the early 1950s how to obtain the amino acid sequences of proteins. One conceivable way to measure genetic variation in a natural population would therefore be to choose a fair number of proteins, say, 20, without knowing whether or not they were variable in the population, so that they would represent an unbiased sample. Then, each of the 20 proteins could be sequenced in a number of individuals, say, 100 (chosen at random), to find out how much variation, if any, existed for each of the proteins. The average amount of variation per protein found in the 100 individuals for the 20 proteins would be an estimate of the amount of variation in the genome of the population.

Unhappily, obtaining the amino acid sequence of a single protein is so demanding a task that several months, or even years, are usually required to do it. Hence it is hardly practical to sequence 2000 protein specimens in order to estimate the genetic variation in each population we want to study. Fortunately, there is a technique, *gel electrophoresis*, that makes possible the study of protein variation with only a moderate investment of time and money. Since the late 1960s, estimates of genetic variation have

Box 2.1 Gel Electrophoresis

The apparatus and procedures employed in gel electrophoresis for studying genetic variation in natural populations are shown in Figure 2.7. Tissue samples from organisms are individually *homogenized* (ground up) in order to release the enzymes and other proteins from the cells. The homogenate supernatants (liquid fractions) are placed in a gel made of starch, agar, polyacrylamide, or some other jelly-like substance. The gel is then subjected, usually for a few hours, to a direct electric current. Each protein in the gel migrates in a direction and at a rate that depend on the protein's net electric charge and molecular size. After the gel is removed from the electric field, it is treated with a chemical solution containing a substrate that is specific for the enzyme to be assayed, and a salt that reacts with the product of the reaction catalyzed by the enzyme. At the position in the gel to which the specific enzyme has migrated, a reaction takes place that can be written as follows:

$$\text{Substrate} \xrightarrow{\text{Enzyme}} \text{Product} + \text{Salt} \rightarrow \text{Colored spot}$$

The usefulness of the method lies in the fact that the genotype at the gene locus coding for the enzyme can be inferred for each individual in the sample from the number and positions of the spots observed in the gels. Figure 2.8 shows a gel that has been treated to reveal the position of the enzyme phosphoglucomutase; the gel contains the homogenates of 12 *Drosophila* flies. The gene locus coding for this enzyme can be represented as *Pgm.* The first and third individuals in the gel, starting from the left, have enzymes with different electrophoretic mobility, and thus different amino acid sequences; this in turn implies that they are coded by different alleles. Let us represent the alleles coding for the enzymes in the first and third individuals as Pgm^{100} and Pgm^{108}, respectively. (The superscripts indicate that the enzyme coded by allele Pgm^{108} migrates 8 mm farther in the gel than the enzyme coded by Pgm^{100}; this is a common way of representing alleles in electrophoretic studies, although letters—such as *S*, *M*, and *F* for slow, intermediate, and fast, or *a*, *b*, *c*, etc.—are sometimes used.)

Because the first and third individuals in Figure 2.8 each exhibit only one colored spot, we infer that they are homozygotes, with genotypes $Pgm^{100/100}$ and $Pgm^{108/108}$, respectively. The second individual exhibits two colored spots. One of these spots shows the same migration as that of the first individual and is thus coded by allele Pgm^{100}, while the other spot shows the same migration as that of the third individual and is thus coded by allele Pgm^{108}. We conclude that the second individual is heterozygous, with genotype $Pgm^{100/108}$.

Some proteins, such as the enzyme malate dehydrogenase, shown in Figure 2.9, consist of two polypeptides; heterozygotes will then exhibit three colored spots. Let us represent the locus coding for malate dehydrogenase as *Mdh*. The second individual in Figure 2.9 shows only one spot and is thus inferred to be homozygous, with genotype $Mdh^{94/94}$; the first individual is also homozygous, with genotype $Mdh^{104/104}$. A heterozygous individual has two kinds of polypeptides, which we can represent as A and B, coded by alleles Mdh^{94} and Mdh^{104}, respectively. Three associations of two different units are possible, namely, AA, AB, and BB. These correspond to the three colored spots that we see in the fourth individual of Figure 2.9.

There are proteins that consist of four or even more subunits; the electrophoretic patterns of heterozygous individuals will then show five or more colored spots, but the principles used to infer the genotypes from the patterns are similar to those just presented. The patterns shown in Figures 2.8 and 2.9 manifest the existence of two alleles at each locus. An invariant locus will be manifested by a colored spot that is the same for all individuals. On the other hand, more than two alleles are often found, as in Figure 2.10, which shows a pattern of the enzyme acid phosphatase in *Drosophila.*

Protein variants controlled by allelic variants at a single gene locus and detectable by electrophoresis are called *allozymes*, or *electromorphs*. Electromorphs with identical migration in a gel may be the products of more than one allele, because (1) synonymous triplets code for the same amino acid and (2) some amino acid substitutions do not change the electrophoretic mobility of the proteins. Therefore, gel electrophoresis underestimates the amount of genetic variation, although at present it is not known by how much.

---

been obtained for natural populations of many organisms using gel electrophoresis (see Box 2.1).

Electrophoretic techniques show what the genotypes of the individuals in a sample are: how many are homozygous, how many are heterozygous, and for what alleles. In order to obtain an estimate of the amount of variation in a population, about 20 or more gene loci are usually studied. It is desirable to summarize the information obtained for all the loci in a simple way that would express the degree of variability of a population and that would permit comparing one population to another. This can be accomplished in a variety of ways, but two measures of genetic variation are commonly used: polymorphism and heterozygosity.

Polymorphism and Heterozygosity

One measure of genetic variation is the *proportion of polymorphic loci*, or simply the *polymorphism* (P), in a population. Assume that, using electrophoretic techniques, we examine 30 gene loci in *Phoronopsis viridis*, a kind of marine worm that lives on the coast of California, and assume that we find no variation whatsoever at 12 loci, but some variation at the other 18 loci. We can say that $18/30 = 0.60$ of the loci are polymorphic in that population, or that the degree of polymorphism in the population is 0.60. Assume that we examine three other populations of *P. viridis* and that the numbers of polymorphic loci, out of the 30 loci studied, are 15, 16, and 14. The degree of polymorphism in these three populations is 0.50, 0.53, and 0.47, respectively. We can then calculate the average polymorphism in the four populations of *P. viridis* as $(0.60 + 0.50 + 0.53 + 0.47)/4 = 0.525$ (Table 2.7).

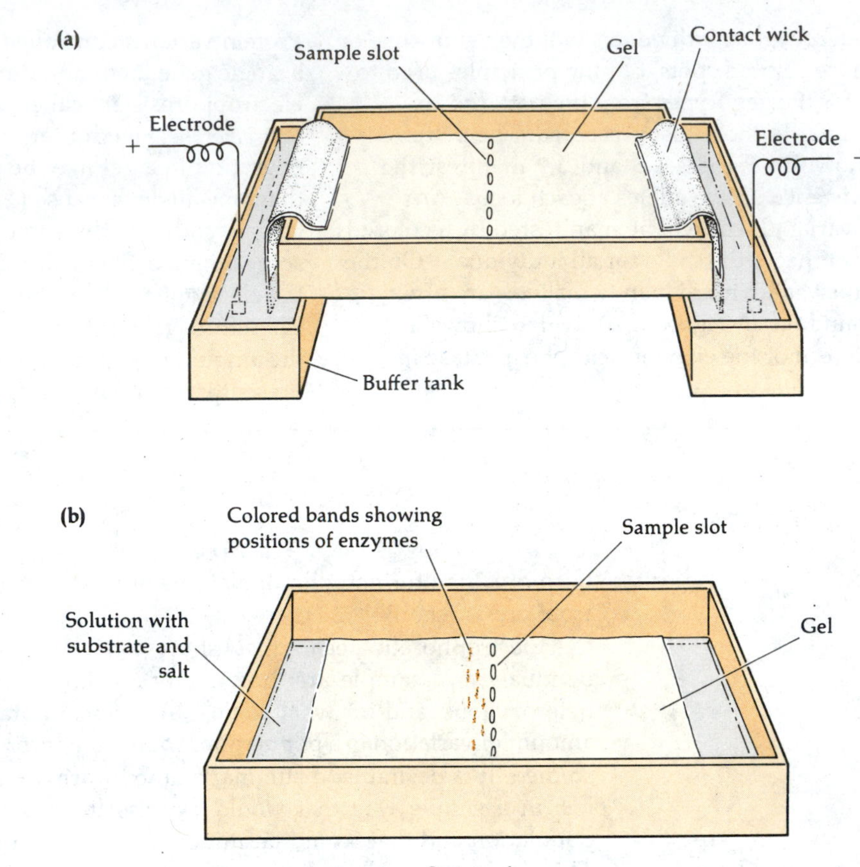

Figure 2.7
The techniques of gel electrophoresis and enzyme assay used to measure genetic variation in natural populations(the techniques are described in somewhat more detail in the text). **(a)**The liquid fractions from homogenized tissue samples are placed in a gel and subjected to a direct electric current. The enzymes and other proteins in the samples migrate to characteristic positions in the gel. **(b)** After the gel is removed from the electric field, it is treated with a specific chemical solution to reveal the positions to which the enzyme being assayed had migrated. The genotype at the gene locus coding for the enzyme can be determined for each individual from the pattern of the spots in the gels.

The amount of polymorphism is a useful measure of variation for certain purposes, but it suffers from two defects: arbitrariness and imprecision.

The number of variable loci observed depends on how many individuals are examined. Assume, for example, that we examined 100 individuals in the first *Phoronopsis* population. If we had examined more individuals, we might have found variation at some of the 12 loci that appeared invariant; if we had examined fewer individuals, some of the 18 polymorphic loci might have appeared invariant. In order to avoid the

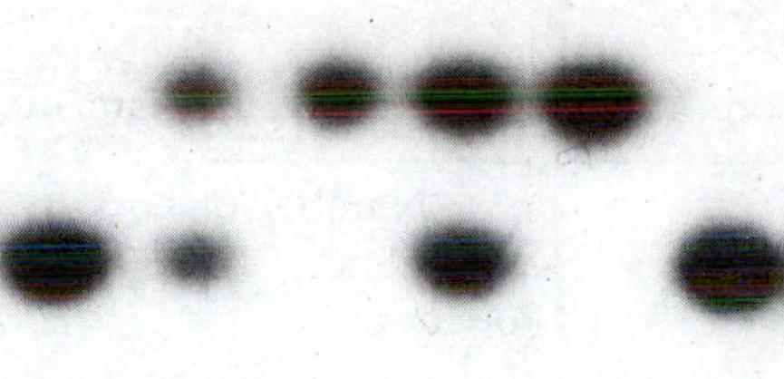

Figure 2.8
An electrophoretic gel stained for the enzyme phosphoglucomutase. The gel contains tissue samples from each of 12 females of *Drosophila pseudoobscura*. Flies with only one colored spot in the gel are inferred to be homozygotes; flies with two spots are heterozygotes. The genotypes of all 12 individuals are, from left to right: $Pgm^{100/100}$, $Pgm^{100/108}$, $Pgm^{108/108}$, $Pgm^{100/108}$, $Pgm^{108/108}$, $Pgm^{100/100}$, $Pgm^{100/100}$, $Pgm^{100/100}$, $Pgm^{108/108}$, $Pgm^{100/108}$, $Pgm^{100/100}$, and $Pgm^{100/100}$.

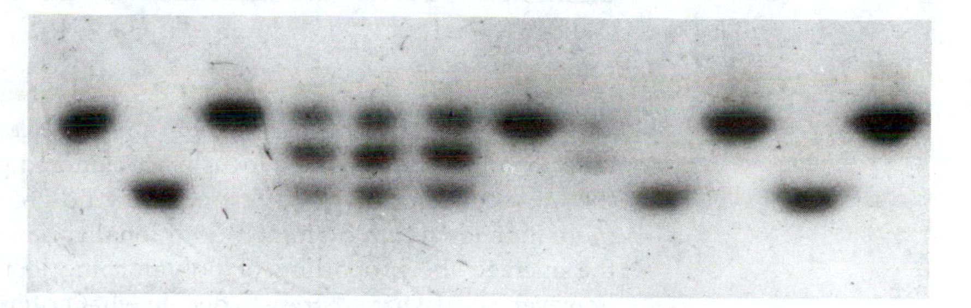

Figure 2.9
An electrophoretic gel stained for the enzyme malate dehydrogenase. The gel contains tissue samples from each of 12 flies of *Drosophila equinoxialis*. As in Figure 2.8, flies with only one colored spot in the gel are inferred to be homozygotes; but the heterozygotes exhibit three bands because malate dehydrogenase is a dimeric enzyme. The genotype of the second and ninth flies is inferred to be $Mdh^{94/94}$; the genotype of the first fly is inferred to be $Mdh^{104/104}$; the fourth, fifth, and sixth flies all have the heterozygous genotype $Mdh^{94/104}$, and so on.

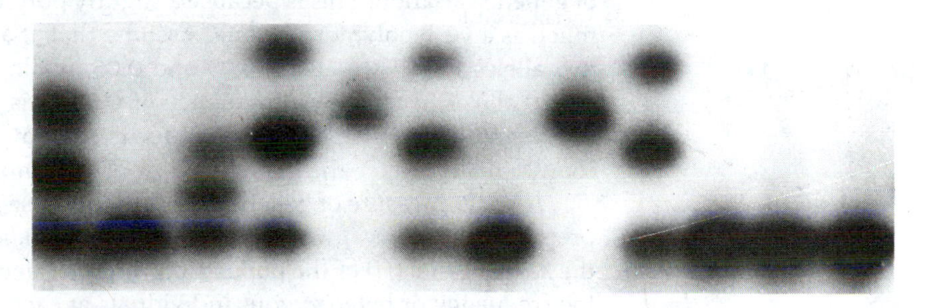

Figure 2.10
An electrophoretic gel stained for the enzyme acid phosphatase. The gel contains tissue samples from each of 12 flies of *Drosophila equinoxialis*. Acid phosphatase is a dimeric enzyme, and hence heterozygotes show three bands. Four different alleles (*88*, *96*, *100*, and *106*) are manifested in the gel. The first fly on the left has the genotype $Acph^{88/100}$, the second has $Acph^{88/88}$, the third has $Acph^{88/96}$, the fourth has $Acph^{88/106}$, the fifth has $Acph^{100/100}$, and so on.

Table 2.7
Calculation of the average polymorphism in four populations.

Population	Number of Loci: Polymorphic	Number of Loci: Total	Polymorphism
1	18	30	18/30 = 0.60
2	15	30	15/30 = 0.50
3	16	30	16/30 = 0.53
4	14	30	14/30 = 0.47
			Average: 0.525

effect of sample size, it is necessary to adopt a *criterion of polymorphism.* One criterion often used is that a locus be considered polymorphic only when the most common allele has a frequency no greater than 0.95. Then, as more individuals are examined, additional variants may be found, but on the average the proportion of polymorphic loci will not change. It is, however, somewhat *arbitrary* to decide what criterion of polymorphism to use. Different values of polymorphism are obtained when different criteria are used. For example, if the criterion of polymorphism is that the frequency of the most common allele be no greater than 0.98, it is possible that some additional loci will be considered polymorphic that are not so with the 0.95 criterion (e.g., a locus with two alleles with the frequencies 0.97 and 0.03).

The polymorphism of a population is, moreover, an *imprecise* measure of genetic variation. This is because a slightly polymorphic locus counts as much as a very polymorphic one. Assume that at a certain locus there are two alleles with frequencies 0.95 and 0.05, while at another locus there are 20 alleles, each with a frequency of 0.05. It is obvious that more genetic variation exists at the second locus than at the first, yet both will count equally under the 0.95 criterion of polymorphism.

A better measure of genetic variation (because it is not arbitrary and is precise) is the *average frequency of heterozygous individuals* per locus, or simply the *heterozygosity* (H) of the population. This is calculated by first obtaining the frequency of heterozygous individuals at each locus and then averaging these frequencies over all loci. Assume that we study four loci in a population and find that the frequencies of heterozygotes at these loci are 0.25, 0.42, 0.09, and 0. The heterozygosity of the population, based on these four loci, is $(0.25 + 0.42 + 0.09 + 0)/4 = 0.19$ (Table 2.8). We

Table 2.8
Calculation of the average heterozygosity at four loci.

Locus	Number of Individuals: Heterozygotes	Number of Individuals: Total	Heterozygosity
1	25	100	25/100 = 0.25
2	42	100	42/100 = 0.42
3	9	100	9/100 = 0.09
4	0	100	0/100 = 0
			Average: 0.19

conclude that the heterozygosity of the population is 19%. Of course, in order for an estimate of heterozygosity to be valid, it must be based on a sample of more than four loci, but the procedure is the same. If several populations of the same species are examined, one can first calculate the heterozygosity in each population and then obtain the average over the various populations. If the heterozygosities in four populations are 0.19, 0.15, 0.13, and 0.17, the average heterozygosity for all four populations is $(0.19 + 0.15 + 0.13 + 0.17)/4 = 0.16$.

The heterozygosity of a population is the measure of genetic variation preferred by most population geneticists. It is a good measure of variation because it estimates the probability that two alleles taken at random from the population are different. (Each gamete from a different individual carries an allele at each locus that can be considered as randomly sampled from the population.) However, the observed heterozygosity does not reflect well the amount of genetic variation in populations of organisms that reproduce by self-fertilization, as some plants do, or organisms in which matings between relatives are common. In a population that always reproduces by self-fertilization, most individuals will be homozygous, even though different individuals may carry different alleles if the locus is variable in the population. There will also be more homozygotes in a population in which matings between relatives are common than in a population where they do not occur, even when the allelic frequencies are identical in both populations.

This difficulty can be overcome by calculating the *expected* heterozygosity, calculated from the allelic frequencies *as if* the individuals in the population were mating with each other at random. Assume that, at a locus, there are four alleles with frequencies f_1, f_2, f_3, and f_4. As we shall see in Chapter 3, the expected frequencies of the four homozygotes if there is random mating are f_1^2, f_2^2, f_3^2, and f_4^2. The expected heterozygosity at the locus will therefore be $H_{\text{expected}} = 1 - (f_1^2 + f_2^2 + f_3^2 + f_4^2)$. For example, if the allelic frequencies at a given locus are 0.50, 0.30, 0.10, and 0.10, the expected heterozygosity will be $H_{\text{expected}} = 1 - (0.50^2 + 0.30^2 + 0.10^2 + 0.10^2) = 1 - (0.25 + 0.09 + 0.01 + 0.01) = 0.64$.

Electrophoretic Estimates of Variation

Electrophoretic techniques were first applied to the estimation of genetic variation in natural populations in 1966, when three studies were published, one dealing with humans and the other two with *Drosophila* flies. Numerous populations of many organisms have been surveyed since that time, and many more are studied every year. Two studies will be reviewed here.

Table 2.9 lists 20 variable loci out of 71 loci sampled in a population of Europeans. The symbol used to represent the locus, the enzyme encoded by the locus, and the frequency of heterozygous individuals at the locus are given for each of the 20 variable loci. The heterozygosity of the population is the sum of the heterozygosities found at the 20 variable loci divided by the total number of loci sampled: $4.78/71 = 0.067$.

A total of 39 gene loci coding for enzymes were studied in a population of the marine worm (phylum Phoronida) *Phoronopsis viridis,* from Bodega Bay, California. Table 2.10 gives the symbols used to represent the 27 loci in which at least two alleles were found. The table shows the observed and expected heterozygosities, as well as the loci that are polymorphic when the criterion of polymorphism is that the frequency of the most common allele be no greater than 0.95. By this criterion, 28.2% of the 39 loci studied are polymorphic. However, using the 0.99 criterion of polymorphism, 20 of the 39 loci, or 51.2%, are polymorphic. The observed heterozygosity is 7.2%, conspicuously less than the expected heterozygosity, 9.4%. This difference may be due to the occurrence of a certain amount of self-fertilization, since *P. viridis* is a hermaphroditic animal.

Table 2.10 also gives the allelic frequencies at the 27 variable loci. The number of alleles per locus ranges from only one (at the 12 invariant loci) to six (at the *Acph-2* and *G3pd-1* loci). Loci with a greater number of alleles do not necessarily have greater heterozygosities than loci with fewer alleles. For example, the observed and expected heterozygosities at the *Acph-2* locus are 0.160 and 0.217, respectively, while at the *Adk-1* locus, which has only two alleles, these heterozygosities are 0.224 and 0.496.

The importance of sampling a fairly large number of loci is apparent in Tables 2.9 and 2.10. For example, if only very few loci had been surveyed in the population of Europeans, the sample might have included a disproportionate number of highly variable loci (such as *Acph*, *Pgm-1*,

Table 2.9
Heterozygosity at 20 variable gene loci out of 71 loci sampled in a population of Europeans, as determined by electrophoresis.

Gene Locus*	Enzyme Encoded	Heterozygosity
Acph	Acid phosphatase	0.52
Pgm-1	Phosphoglucomutase-1	0.36
Pgm-2	Phosphoglucomutase-2	0.38
Adk	Adenylate kinase	0.09
Pept-A	Peptidase-A	0.37
Pept-C	Peptidase-C	0.02
Pept-D	Peptidase-D	0.02
Adn	Adenosine deaminase	0.11
6Pgdh	6-Phosphogluconate dehydrogenase	0.05
Aph	Alkaline phosphatase (placental)	0.53
Amy	Amylase (pancreatic)	0.09
Gpt	Glutamate-pyruvate transaminase	0.50
Got	Glutamate-oxaloacetate transaminase	0.03
Gput	Galactose-1-phosphate uridyl transferase	0.11
Adh-2	Alcohol dehydrogenase-2	0.07
Adh-3	Alcohol dehydrogenase-3	0.48
Peps	Pepsinogen	0.47
Ace	Acetylcholinesterase	0.23
Me	Malic enzyme	0.30
Hk	Hexokinase (white-cell)	0.05
	Average heterozygosity (including 51 invariant loci):	0.067

*Numbers or letters are used to distinguish several related enzymes and the loci coding for them; e.g., *Pgm-1* and *Pgm-2*, or *Pept-A*, *Pept-B*, and *Pept-C*.

After H. Harris and D. A. Hopkinson, *J. Human Genet.* 36:9 (1972).

Table 2.10
Allelic frequencies at 27 variable loci in 120 individuals of the marine worm *Phoronopsis viridis.* The numbers used to represent alleles (*1, 2, 3,* etc.) indicate increasing mobility, in an electric field, of the proteins encoded by the alleles.

Gene Locus	Frequency of Allele: 1	2	3	4	5	6	Heterozygosity: Observed	Expected	Is locus polymorphic by 0.95 criterion?
Acph-1	0.995	0.005					0.010	0.010	No
Acph-2	0.009	0.066	0.882	0.014	0.005	0.024	0.160	0.217	Yes
Adk-1	0.472	0.528					0.224	0.496	Yes
Est-2	0.008	0.992					0.017	0.017	No
Est-3	0.076	0.924					0.151	0.140	Yes
Est-5	0.483	0.396	0.122				0.443	0.596	Yes
Est-6	0.010	0.979	0.012				0.025	0.041	No
Est-7	0.010	0.990					0.021	0.021	No
Fum	0.986	0.014					0.028	0.028	No
αGpd	0.005	0.995					0.010	0.010	No
G3pd-1	0.040	0.915	0.017	0.011	0.011	0.006	0.159	0.161	Yes
G6pd	0.043	0.900	0.057				0.130	0.185	Yes
Hk-1	0.996	0.004					0.008	0.008	No
Hk-2	0.005	0.978	0.016				0.043	0.043	No
Idh	0.992	0.008					0.017	0.017	No
Lap-3	0.038	0.962					0.077	0.074	No
Lap-4	0.014	0.986					0.028	0.027	No
Lap-5	0.004	0.551	0.326	0.119			0.542	0.576	Yes
Mdh	0.008	0.987	0.004				0.025	0.025	No
Me-2	0.979	0.021					0.042	0.041	No
Me-3	0.017	0.824	0.159				0.125	0.296	Yes
Odh-1	0.992	0.008					0.017	0.017	No
Pgi	0.995	0.005					0.010	0.010	No
Pgm-1	0.159	0.827	0.013				0.221	0.290	Yes
Pgm-3	0.038	0.874	0.071	0.017			0.185	0.229	Yes
Tpi-1	0.929	0.071					0.000	0.133	Yes
Tpi-2	0.008	0.004	0.962	0.013	0.013		0.076	0.074	No
Averages (including 12 invariant loci)									
Heterozygosity:							0.072	0.094	
Polymorphism:									11/39 = 0.282

After F. J. Ayala et al., *Biochem. Genet.* *18*:413 (1974).

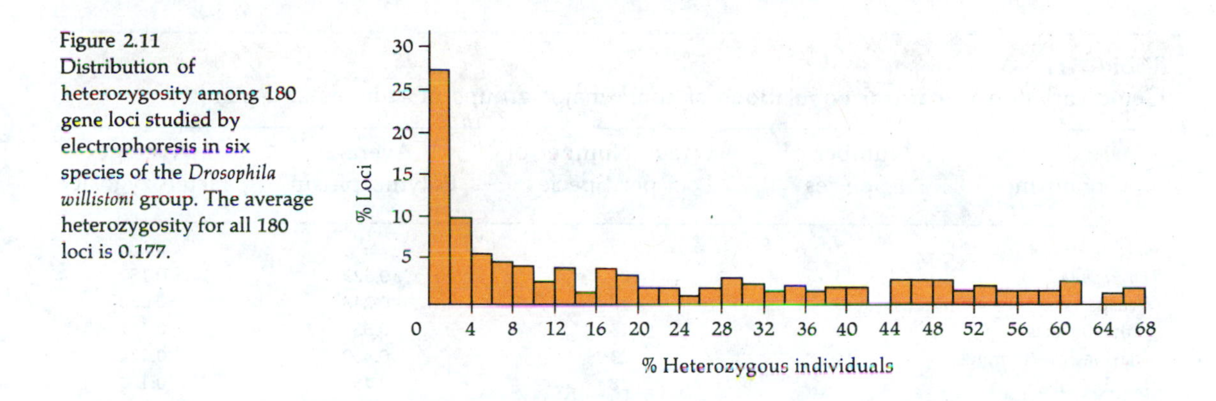

Figure 2.11 Distribution of heterozygosity among 180 gene loci studied by electrophoresis in six species of the *Drosophila willistoni* group. The average heterozygosity for all 180 loci is 0.177.

Pgm-2, and *Pept-A*). A distorted estimate of heterozygosity would then have been obtained. Figure 2.11 shows the distribution of heterozygosity among 180 loci studied in six closely related species of *Drosophila*. Characteristically, the distribution is widely spread (H ranges from 0 to 0.68) and is not at all like a normal distribution.

Experience with electrophoretic studies indicates that a sample of about 20 gene loci is usually sufficient; estimates of heterozygosity usually change little as the number of loci sampled exceeds 20. For example, a value of $H = 0.072$ was obtained for humans, using a sample of 26 gene loci. When the sample was extended to a total of 71 loci, the estimate became $H = 0.067$ (see Table 2.9).

Genetic Variation in Natural Populations

Considerable genetic variation exists in most natural populations. Table 2.11 summarizes the results of electrophoretic surveys obtained for 17 plant species and 125 animal species in which a fair number of loci have been sampled. Among animals it seems that, in general, invertebrates have more genetic variation than vertebrates, although there are exceptions. For the species in Table 2.11, the average heterozygosity is 13.4% for invertebrates and 6.0% for vertebrates. Plants show an average heterozygosity of 4.6%, with outcrossing plants exhibiting more genetic variation than those that reproduce primarily by selfing.

One way to appreciate the large amounts of genetic variation found in natural populations is the following. Consider humans, with a 6.7% heterozygosity detectable by electrophoresis. If we assume that there are 30,000 structural gene loci in a human being, which may be an underestimate, a person would be heterozygous at $30{,}000 \times 0.067 = 2010$ loci. Such an individual can theoretically produce $2^{2010} \approx 10^{605}$ different kinds of gametes. (An individual heterozygous at one locus can produce two

Table 2.11
Genic variation in natural populations of some major groups of animals and plants.

Organisms	Number of Species	Average Number of Loci per Species	Average Polymorphism*	Average Heterozygosity
Invertebrates				
Drosophila	28	24	0.529	0.150
Wasps	6	15	0.243	0.062
Other insects	4	18	0.531	0.151
Marine invertebrates	14	23	0.439	0.124
Land snails	5	18	0.437	0.150
Vertebrates				
Fishes	14	21	0.306	0.078
Amphibians	11	22	0.336	0.082
Reptiles	9	21	0.231	0.047
Birds	4	19	0.145	0.042
Mammals	30	28	0.206	0.051
Plants				
Self-pollinating	12	15	0.231	0.033
Outcrossing	5	17	0.344	0.078
Overall averages				
Invertebrates:	57	22	0.469	0.134
Vertebrates:	68	24	0.247	0.060
Plants:	17	16	0.264	0.046

*The criterion of polymorphism is not the same for all species.

From various sources.

different kinds of gametes, one with each allele; an individual heterozygous at n gene loci has the potential of producing 2^n different gametes.) This number of gametes will never be produced by any individual, however, nor by the whole of mankind; the estimated total number of protons and neutrons in the universe, 10^{76}, is infinitesimal by comparison.

Although not all possible gametic combinations are equally probable, calculations indicate that no two independent human gametes are likely to be identical and that no two human individuals (except those derived from the same zygote, such as identical twins) who exist now, have ever existed in the past, or will ever exist in the future are likely to be genetically identical. And the same can be said, in general, of organisms that reproduce sexually: no two individuals developed from separate zygotes are ever likely to be genetically identical.

Electrophoretic techniques have made it possible to obtain estimates of genetic variation in natural populations. How reliable are these estimates? Two conditions are required for making good estimates of genetic variation: (1) that a random sample of all gene loci be obtained and (2) that all alleles be detected at every locus.

The gene loci studied should represent a random sample of the genome with respect to variation because otherwise the estimates would be biased. The genes studied by gel electrophoresis code for enzymes and other soluble proteins. Such genes represent a considerable portion of the genome, but there are other kinds of gene loci, such as regulatory genes and genes coding for nonsoluble proteins. It is not known whether these genes are as variable as structural genes coding for soluble proteins. The estimates of heterozygosity might be biased for this reason, although we do not know whether they overestimate or underestimate genetic variation in this regard.

Electrophoresis separates proteins on the basis of their differential migration in an electric field. This differential migration is due to differences in molecular configuration and, primarily, to differences in net electric charge. But amino acid substitutions can occur that do not change the net electric charge of a protein or substantially modify its configuration. Electrophoresis therefore detects only a fraction of all differences in amino acid sequence. Several methods have been used to detect protein differences not distinguishable by electrophoresis, but at present we do not have reliable estimates of the proportion of the protein variants that remain undetected. We know that electrophoretic estimates of heterozygosity underestimate genetic variation at the protein level, although we do not know by how much.

Moreover, electrophoretic techniques do not detect nucleotide substitutions in the DNA that do not change the encoded amino acid. However, synonymous mutations that do not change the amino acid sequence of a protein are less likely to be important for evolution than those that do, since the latter usually have larger effects on the biology of the organism.

Problems

1. Three genotypes were observed at the *Pgm-1* locus in a human population. In a sample of 1110 individuals, the three genotypes occurred in the following numbers (*1* and *2* represent two different alleles):

Genotypes:	*1/1*	*1/2*	*2/2*
Numbers:	634	391	85

Calculate the genotypic and allelic frequencies.

2. Two human serum haptoglobins are determined by two alleles at a single locus. In a sample of 219 Egyptians, the three genotypes occurred in the following numbers (*1* and *3* represent the two alleles):

Genotypes:	*1/1*	*1/3*	*3/3*
Numbers:	9	135	75

What are the frequencies of the two alleles?

3. Calculate the expected frequency of heterozygotes, assuming random mating, from the data of problems 1 and 2. Use the chi-square test to determine whether the observed and expected *numbers* of heterozygous individuals are significantly different.

4. The following table gives the number of individuals for each of the M-N blood groups in samples from various human populations. Calculate the genotypic and allelic frequencies as well as the expected *number* of heterozygous individuals for each population. Test whether the observed and expected numbers of heterozygotes agree.

	Number having blood group:			
Population	M	MN	N	Total
Eskimos	475	89	5	569
Pueblo Indians	83	46	11	140
Russians	195	215	79	489
Swedes	433	564	203	1200
Chinese	342	500	187	1029
Japanese	356	519	225	1100
Belgians	896	1559	645	3100
English	121	200	101	422
Egyptians	140	245	117	502
Ainu	90	253	161	504
Fijians	22	89	89	200
Papuans	14	48	138	200

5. Samples from a population of *Drosophila pseudoobscura* taken in 11 successive months during 1973 gave the following numbers for each genotype at the *Hk-1* locus (*96, 100, 104,* and *108* are four different alleles):

	Number having genotype:					
Month	*96/100*	*100/104*	*100/108*	*100/100*	*104/104*	Total
January	0	1	0	20	0	21
February	1	0	0	43	0	44
March	1	0	0	167	0	168
April	1	13	1	363	1	379
May	1	11	0	283	0	295
June	0	20	0	270	1	291
July	0	13	1	257	0	271
August	1	5	0	309	0	315
September	0	3	0	144	0	147
October	0	1	0	177	0	178
November	0	13	0	215	0	228
Total:	5	80	2	2248	2	2337

Calculate the allelic frequencies for each monthly sample. Use the chi-square test to determine whether the observed and expected numbers of heterozygotes are significantly different for the total of all monthly samples.

6. Several chromosomal arrangements, differing by a series of overlapping inversions, are known in the third chromosome of *Drosophila pseudoobscura*. Four arrangements (ST = Standard; AR = Arrowhead; CH = Chiricahua; TL = Tree Line) are found in three natural populations. The observed numbers of each genotype are as follows:

	Number having genotype:									
Locality	*ST/AR*	*ST/CH*	*ST/TL*	*AR/CH*	*AR/TL*	*CH/TL*	*ST/ST*	*AR/AR*	*CH/CH*	Total
Keen Camp	53	66	3	48	3	6	30	11	44	264
Piñon Flat	40	53	5	37	3	7	31	11	21	208
Andreas Canyon	87	47	12	20	4	2	89	18	4	283

Calculate the frequency of each chromosomal arrangement and the expected frequency and number of heterozygotes in each of the three populations.

7. Twenty-two gene loci coding for blood proteins were studied in 23 chimpanzees (*Pan troglodytes*) and in 10 gorillas (*Gorilla gorilla*). The chimpanzees were all homozygotes at 21 loci; at the *Pgm-1* locus, 6 individuals were heterozygotes (*96/100*) and the other 17 were homozygotes (*100/100*). The gorillas were all homozygotes at 19 loci; at 3 loci, the following genotypes were observed (the number of individuals is given in parentheses after each genotype):

Ak:	*98/100* (4)	*100/100* (6)	
Dia:	*85/95* (5)	*85/85* (4)	*95/95* (1)
6-Pgdh:	*97/105* (3)	*105/105* (7)	

Calculate the observed and expected average heterozygosities for all 22 loci in the chimpanzees and in the gorillas. What proportion of the loci in each species are polymorphic by the 0.95 criterion?

3 Processes of Evolutionary Change

Evolution, a Two-Step Process

Biological evolution is the process of change and diversification of organisms through time. Evolutionary change affects all aspects of living things—their morphology, physiology, behavior, and ecology. Underlying those changes, there are genetic changes, i.e., changes in the hereditary materials, which, in interaction with the environment, determine what the organisms are. At the genetic level, evolution consists of changes in the genetic constitution of populations.

Evolution at the genetic level may be seen as a two-step process. We have, first, mutation and recombination, the processes by which hereditary variation arises; then, we have genetic drift and natural selection, the processes by which genetic variants are differentially transmitted from generation to generation.

Evolution can occur only if there is hereditary variability. The ultimate source of all genetic variation is the process of mutation, but the variability is sorted out in new ways by the sexual process, i.e., by the independent assortment of chromosomes and by crossing over. The genetic variants arisen by mutation and recombination are not equally transmitted from one generation to another, but rather some variants may increase in frequency at the expense of others. Besides mutation, the processes by which allelic frequencies change in populations are natural selection, gene flow (migration) from one population to another, and random drift. Genotypic, though not allelic, frequencies may change by assortative mating, i.e., by deviations from random mating.

This chapter and the next two chapters will be dedicated to the processes by which allelic and genotypic frequencies change. Mutation, migration, and drift will be considered in this chapter, and selection and assortative mating in Chapters 4 and 5. Before studying these processes of change, we shall demonstrate that heredity by itself does not change gene frequencies, a principle known as the *Hardy-Weinberg law.*

Random Mating

It might at first appear that individuals exhibiting a dominant genotype should be more common than individuals with a recessive genotype. However, the 3:1 ratio applies to segregation in the offspring of two individuals heterozygous for the same two alleles. Different ratios appear in other mating combinations, and the frequencies of these combinations depend on the frequencies of the genotypes in a population. Mendel's laws do not tell us anything about the frequencies of genotypes in a population; the Hardy-Weinberg law does.

The main statement of the Hardy-Weinberg law is that, in the absence of the evolutionary processes—mutation, migration, drift, and selection—*gene frequencies remain constant from generation to generation.* The law also says that, if matings are random, *the genotypic frequencies are related to the gene frequencies by a simple formula* (the square expansion). A corollary of the Hardy-Weinberg law is that, if the allelic frequencies are initially the same in males and in females, *the equilibrium genotypic frequencies at any given locus are attained in one single generation of random mating.* If the allelic frequencies are initially different in the two sexes, they will become the same in one generation in the case of autosomal loci, because males and females inherit half of their genes from the males of the previous generation and half from the females; then the equilibrium genotypic frequencies will be reached in *two* generations. However, in the case of sex-linked loci, the equilibrium frequencies are attained only gradually (see Box 3.1, page 68). Before demonstrating the Hardy-Weinberg law, we must define random mating.

Random mating occurs when the probability of mating between individuals is independent of their genetic constitution. In a random mating population, therefore, matings between genotypes occur according to the proportions in which the genotypes exist.

The frequencies of the three genotypes for the M-N blood groups in U. S. Caucasians are given in Table 2.3 as $L^M L^M = 0.292$, $L^M L^N = 0.496$, and $L^N L^N = 0.213$. If U. S. Caucasians mate at random with respect to this trait, we expect the various kinds of marriages to occur with the frequencies given in Table 3.1. To obtain the probability of a given type of mating, we simply multiply the frequencies of the two genotypes involved. For example, matings between $L^M L^M$ men and $L^M L^N$ women will occur with a frequency of $0.292 \times 0.496 = 0.145$. We can verify that the frequencies of *all* types of matings add up to one: $0.085 + 0.145 + 0.062$

Table 3.1
Expected frequencies of the various types of marriages among U.S. Caucasians if matings are random with respect to the M-N blood groups.

Men	Women		
	0.292 L^ML^M	0.496 L^ML^N	0.213 L^NL^N
0.292 L^ML^M	♂ *MM* × ♀ *MM* 0.292 × 0.292 = 0.085	♂ *MM* × ♀ *MN* 0.292 × 0.496 = 0.145	♂ *MM* × ♀ *NN* 0.292 × 0.213 = 0.062
0.496 L^ML^N	♂ *MN* × ♀ *MM* 0.496 × 0.292 = 0.145	♂ *MN* × ♀ *MN* 0.496 × 0.496 = 0.246	♂ *MN* × ♀ *NN* 0.496 × 0.213 = 0.106
0.213 L^NL^N	♂ *NN* × ♀ *MM* 0.213 × 0.292 = 0.062	♂ *NN* × ♀ *MN* 0.213 × 0.496 = 0.106	♂ *NN* × ♀ *NN* 0.213 × 0.213 = 0.045

+ 0.145 + 0.246 + 0.106 + 0.062 + 0.106 + 0.045 = 1.002. (The excess over 1.000 is due to rounded numbers.)

Random mating may occur with respect to a given gene locus or a trait, even though matings are not random with respect to some other loci or traits. Indeed, when choosing their mates, people have all sorts of preferences, and are influenced by circumstances such as socioeconomic status, schooling, neighborhood, and the like. But it seems likely that people choose their spouses without particular regard to their M-N blood groups; if so, mating *may* be random with respect to this trait.

Assortative mating occurs when choices of mates are affected by the genotype. For example, marriages in the U.S. are assortative with respect to racial features. Matings between two Caucasians or between two Blacks occur with higher frequency, and matings between one Caucasian and one Black with lesser frequency, than is expected from random mating. Assortative mating also occurs in organisms other than man. An extreme form of assortative mating is self-fertilization, which is the most common form of reproduction in many plants.

The Hardy-Weinberg Law

The Hardy-Weinberg law says that, by itself, the process of heredity does not change either allelic frequencies or (in a random mating population) genotypic frequencies at a given locus. Moreover, the equilibrium genotypic frequencies at any given locus are attained in one single generation of random mating whenever the allelic frequencies are the same in the two sexes.

The equilibrium genotypic frequencies are given by the square of the allelic frequencies. If there are only two alleles, A and a, with frequencies p and q, the frequencies of the three possible genotypes are:

$$\underset{A\ \ a}{(p + q)^2} = \underset{AA}{p^2} + \underset{Aa}{2pq} + \underset{aa}{q^2}$$

where the alleles and genotypes to which the frequencies correspond are written in the second line.

If there are three alleles—say, A_1, A_2, and A_3—with frequencies p, q, and r, the genotypic frequencies are:

$$(\underset{A_1}{p} + \underset{A_2}{q} + \underset{A_3}{r})^2 = \underset{A_1A_1}{p^2} + \underset{A_2A_2}{q^2} + \underset{A_3A_3}{r^2} + \underset{A_1A_2}{2pq} + \underset{A_1A_3}{2pr} + \underset{A_2A_3}{2qr}$$

The square expansion can be used similarly to obtain the equilibrium genotypic frequencies for any number of alleles. Note that the sum of all the allelic frequencies, and of all the genotypic frequencies, must always be 1. If there are only two alleles, with frequencies p and q, then $p + q = 1$, and therefore $p^2 + 2pq + q^2 = (p + q)^2 = 1$; if there are three alleles, with frequencies p, q, and r, then $p + q + r = 1$, and therefore also $(p + q + r)^2 = 1$; and so on.

The Hardy-Weinberg law was formulated in 1908 independently by the mathematician G. H. Hardy in England and by the physician Wilhelm Weinberg in Germany. A simple way of demonstrating the law is the following. Assume that at a given locus there are two alleles, A and a, and that their frequencies, in males as well as in females, are p for A and q for a. Assume also that males and females mate at random; this is equivalent to saying that the male and female *gametes* meet at random in the formation of the zygotes. Then the frequency of a given genotype will simply be the product of the frequencies of the two corresponding alleles (Table 3.2). The probability that one individual will have the AA genotype is the probability (p) of receiving the A allele from the mother multiplied by the probability (p) of receiving the A allele from the father, or $p \times p = p^2$. Similarly, the probability that an individual will have the aa genotype is q^2. The genotype Aa can arise in two ways: A from the mother and a from the father, which will occur with a frequency of pq, or a from the mother and A from the father, which will also occur with a frequency of pq; therefore the total frequency of Aa is $pq + pq = 2pq$. A geometric representation of the Hardy-Weinberg law for two alleles is shown in Figure 3.1, where the allelic frequencies are assumed to be 0.7 and 0.3.

Table 3.2
The Hardy-Weinberg law for two alleles.

Male Gametic Frequencies	Female Gametic Frequencies	
	p (A)	q (a)
p (A)	p^2 (AA)	pq (Aa)
q (a)	pq (Aa)	q^2 (aa)

We can now demonstrate the following three statements contained in the Hardy-Weinberg law:

1. The allelic frequencies do not change from generation to generation. This can easily be shown. The frequency of the A allele among the offspring in Table 3.2 is the frequency of AA plus half the frequency of Aa, or $p^2 + pq = p(p + q) = p$ (remember that $p + q = 1$).
2. The equilibrium genotypic frequencies are given by the square expansion and do not change. Since the allelic frequencies among the offspring are p and q, as they were among the parents, the genotypic frequencies in the following generation will again be p^2, $2pq$, and q^2.
3. The equilibrium genotypic frequencies are attained in one single generation. Notice that in Table 3.2 nothing is assumed with respect to the genotypic frequencies among the parents. Whatever these frequencies may be, if the allelic frequencies are p and q in males as well as in females, the genotypic frequencies among the progeny will be p^2, $2pq$, and q^2.

Table 3.3 uses the M-N blood groups among U.S. Caucasians as an example of the Hardy-Weinberg relationships. From the observed numbers of individuals shown in Table 2.3 for U.S. Caucasians, we can calculate the allelic frequencies. The frequency of L^M is the total number of L^M alleles observed (twice the number of $L^M L^M$ individuals plus the number of $L^M L^N$ individuals) divided by the total number of alleles in the sample (twice the number of individuals), or $[(1787 \times 2) + 3039]/(2 \times 6129) = 0.5395$. Similarly, we obtain the frequency of L^N as 0.4605. The expected equilibrium frequencies calculated according to the Hardy-Weinberg law [0.2911 ($L^M L^M$), 0.4968 ($L^M L^N$), and 0.2121 ($L^N L^N$)] are, in fact, very close to the genotypic frequencies observed in the population (0.292, 0.496, and 0.213).

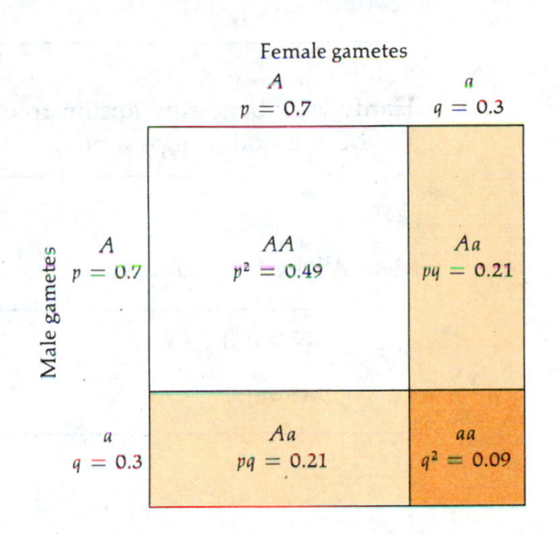

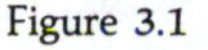

Figure 3.1
Geometric representation of the relationship between allelic and genotypic frequencies according to the Hardy-Weinberg law.

The same procedure used for two alleles can be followed for demonstrating the Hardy-Weinberg law with any number of alleles. Table 3.4 demonstrates the equilibrium genotypic frequencies for a locus with three alleles, with frequencies p, q, and r, so that $p + q + r = 1$. The geometric representation shown in Figure 3.2 uses the ABO blood groups as an example of the three-allele case.

Applications of the Hardy-Weinberg Law

One application of the Hardy-Weinberg law is that it permits the computation of gene and genotypic frequencies in cases where not all genotypes can be distinguished, because of dominance. Albinism in humans is a relatively rare recessive condition. If the allele for normal pigmentation is represented as A and the allele for albinism as a, albinos have aa genotypes, while normally pigmented individuals are either AA or Aa. Assume that in a certain human population the frequency of albino individuals is 1 in 10,000. According to the Hardy-Weinberg law, the frequency of the homozygotes aa is q^2; thus $q^2 = 0.0001$ and $q = \sqrt{0.0001} = 0.01$. It follows that the frequency of the normal allele is 0.99; the frequencies of the other two genotypes are $p^2 = 0.99^2 = 0.98$ for AA, and $2pq = 2 \times 0.99 \times 0.01 \approx 0.02$ for Aa.

Table 3.3
Hardy-Weinberg equilibrium frequencies for the three genotypes of the M-N blood groups among U.S. Caucasians.

	Female Allelic Frequencies	
Male Allelic Frequencies	0.5395 (L^M)	0.4605 (L^N)
0.5395 (L^M)	0.2911 ($L^M L^M$)	0.2484 ($L^M L^N$)
0.4605 (L^N)	0.2484 ($L^M L^N$)	0.2121 ($L^N L^N$)

The ABO blood groups can serve as an example for three alleles. Assume that in a certain population the frequencies of the four blood groups are:

A (genotypes $I^A I^A$ and $I^A i$) = 0.45

B (genotypes $I^B I^B$ and $I^B i$) = 0.13

AB (genotype $I^A I^B$) = 0.06

O (genotype ii) = 0.36

Let us represent the frequencies of I^A, I^B, and i as p, q, and r, respectively. According to the Hardy-Weinberg law, the frequency of the genotype ii is r^2; therefore $r = \sqrt{0.36} = 0.60$, which is the frequency of the i allele. We now note that the joint frequency of blood groups B and O is $(q + r)^2$ (see Figure 3.2). Therefore, $(q + r)^2 = 0.13 + 0.36 = 0.49$, and $q + r = \sqrt{0.49} = 0.70$. Since we already know that $r = 0.60$, the frequency of allele I^B is $0.70 - 0.60 = 0.10$. Finally, the frequency of allele I^A is $p = 1 - (q + r) = 1 - 0.70 = 0.30$.

One interesting implication of the Hardy-Weinberg law is that rare alleles exist in populations, for the most part, in heterozygous genotypes, not in the homozygotes. Consider the albinism example given above. The frequency of albinos (aa) is 0.0001. The frequency of heterozygotes (Aa) is 0.02. Only half the alleles of the heterozygotes are a. Therefore the frequency of a in the population is 0.01 in heterozygous individuals and 0.0001 in homozygous individuals. There are about 100 times more a alleles in heterozygotes than in albinos.

In general, if the frequency of a recessive allele in a population is q, there will be pq recessive alleles (half the alleles in $2pq$ individuals) in the heterozygotes and q^2 recessive alleles in the homozygotes. The ratio of one to the other is $pq/q^2 = p/q$, which, if q is very small, will be approximately

Table 3.4
The Hardy-Weinberg law for three alleles.

Male Gametic Frequencies	Female Gametic Frequencies		
	$p\ (A_1)$	$q\ (A_2)$	$r\ (A_3)$
$p\ (A_1)$	$p^2\ (A_1A_1)$	$pq\ (A_1A_2)$	$pr\ (A_1A_3)$
$q\ (A_2)$	$pq\ (A_1A_2)$	$q^2\ (A_2A_2)$	$qr\ (A_2A_3)$
$r\ (A_3)$	$pr\ (A_1A_3)$	$qr\ (A_2A_3)$	$r^2\ (A_3A_3)$

$1/q$. Thus, the lower the frequency of an allele, the greater the proportion of that allele that exists in the heterozygotes. The frequency of the recessive gene for alkaptonuria is about 0.001. The frequency of alkaptonurics is $q^2 = 0.000001$, or 1 in 1 million, while the frequency of the heterozygotes is $2pq$, or about 0.002. The number of alkaptonuria genes is about 1000 times greater in heterozygotes than in homozygotes.

Imagine now that a misguided dictator, with eugenic ideals of "improving the race," wants to eliminate albinism from the population. If the heterozygotes cannot be identified, his program will be based on the elimination, or sterilization, of the recessive homozygotes. This will

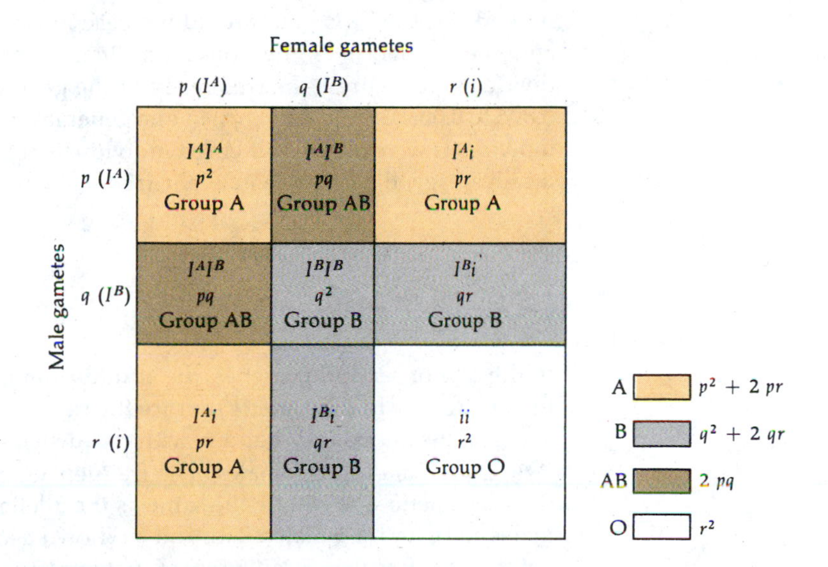

Figure 3.2
Geometric representation of the relationships between allelic and genotypic frequencies for the ABO blood groups.

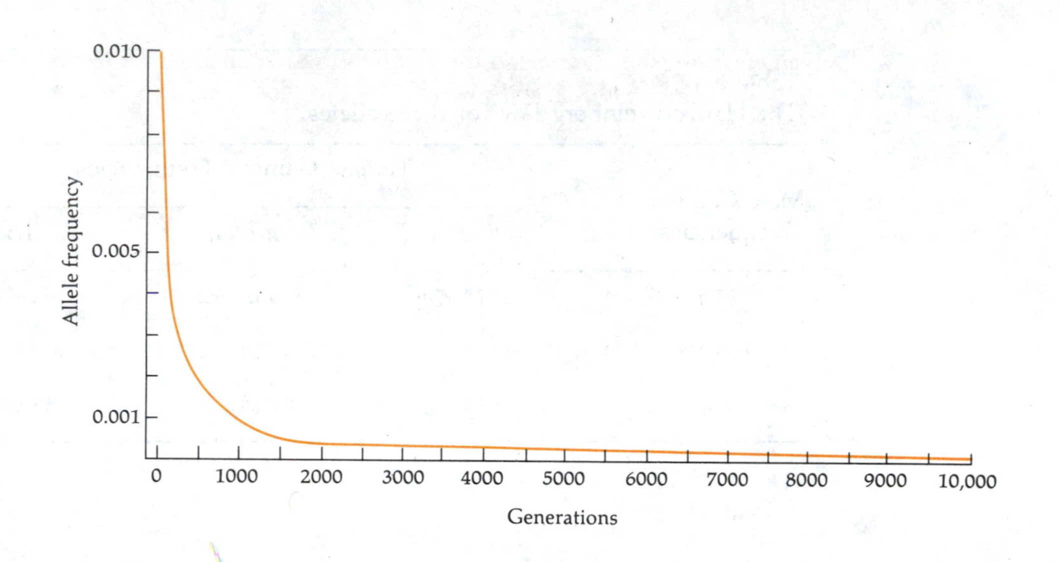

Figure 3.3
Change in allele frequency when the recessive homozygotes are eliminated from a population. If the initial allele frequency is 0.01, it takes 900 generations to reduce it to 0.001, and 9900 generations to reduce it to 0.0001. In general, the number of generations, t, required to change the allele frequency from q_0 to q_t is $t = 1/q_t - 1/q_0$ (see Chapter 4, page 99).

reduce the frequency of the allele in the population only very slightly, however, since most albino alleles go undetected in the heterozygotes. Albinos will appear in the next generation in very nearly the same frequency as they did previously. Many generations will be required to reduce greatly the frequency of the recessive allele (Figure 3.3).

The converse of this imaginary situation occurs in human populations for lethal recessive conditions that can now be cured. An example is phenylketonuria (PKU). The frequency of the allele is about 0.006. Even if all homozygotes were cured and they reproduced as effectively as normal individuals, the PKU gene would increase in frequency only very slowly, and the incidence of the condition even more slowly. If all PKU individuals were cured, the frequency of the gene would change only from 0.006 to 0.006036 ($q_1 = q + q^2$) in one generation. And, of course, if not all individuals were cured or if cured individuals retained some reproductive disadvantage, the increase in allelic frequency would be even slower.

Sex-Linked Genes

In the case of sex-linked genes, the equilibrium genotypic frequencies for females (or the homogametic sex) are the same as in the case of autosomal genes. If the alleles are A and a, with frequencies p and q, there will be p^2 AA, $2pq$ Aa, and q^2 aa females. The frequencies of the hemizygous males (heterogametic sex) will be the same as the allelic frequencies: p for the A males and q for the a males. This can be shown using the same reasoning as before. AA females receive one A gamete from the father and one A

gamete from the mother; if the frequency of A in males as well as in females is p, then AA females will be produced with a frequency p^2. Similarly, the frequency of aa females will be q^2 and the frequency of Aa females will be $2pq$. Males, however, receive their only X chromosome from the mother. Thus, the frequencies of the two hemizygous genotypes are the same as the frequencies of the two alleles among the females in the previous generation.

It follows that phenotypes determined by recessive genes will be more frequent among males than among females. If the frequency of a sex-linked recessive allele is q, the incidence of the phenotype will be q among males and q^2 among females. The ratio of one to the other will be $q/q^2 = 1/q$; the smaller the value of q, the greater the ratio of males to females exhibiting the recessive phenotype. The frequency of the allele for red-green color blindness is 0.08; therefore there are $1/0.08 = 12.5$ times more color-blind men than women. The frequency of the recessive gene for the most common form of hemophilia is 0.0001; according to the Hardy-Weinberg law, we expect $1/0.0001 = 10{,}000$ times more victims of this kind of hemophilia among men than among women (but very few in either case—1 in 10,000 men and 1 in 100 million women).

Mutation

The Hardy-Weinberg law in genetics is analogous to Newton's first law in mechanics, which says that a body remains at rest or maintains a constant velocity when not acted upon by a net external force. Bodies are always acted upon by external forces, but the first law is the point of departure for applying other laws. The Hardy-Weinberg law says that, in the absence of disturbing processes, gene frequencies do not change. But processes that change gene frequencies are always present, and without them there would be no evolution. The Hardy-Weinberg law is the point of departure from which we can calculate the effects of the processes of change.

The first process we shall consider is mutation. Although gene and chromosome mutations are the ultimate source of all genetic variability, they occur with very low frequency. Mutation is a very slow process that, by itself, changes the genetic constitution of populations at a very low rate. If mutation were the *only* process of genetic change, evolution would occur at an impossibly low rate. This is the main lesson to be learned from the calculations that follow.

Assume that there are two alleles, A_1 and A_2, at a locus, and that mutation from A_1 to A_2 occurs at a rate u per gamete per generation. Assume also that at a given time the frequency of A_1 is p_0. In the next generation, a fraction u of all A_1 alleles become A_2 by mutation. The

Box 3.1 Calculation of Allele Frequencies and Approach to Equilibrium for Sex-Linked Genes

In the case of sex-linked genes, the homogametic sex carries two-thirds of all genes in the population, while the heterogametic sex carries only one-third. Assume that there are two alleles, *A* and *a*, in a population and that the frequency of *A* is p_f among the females and p_m among the males. The frequency of *A* in the whole population will be

$$p = \tfrac{2}{3}p_f + \tfrac{1}{3}p_m$$

Similarly,

$$q = \tfrac{2}{3}q_f + \tfrac{1}{3}q_m$$

where q, q_f, and q_m are the frequencies of the *a* allele in the whole population, in females, and in males, respectively.

In humans, the *ma* (macroglobulin a) locus coding for the α_2 macroglobulin portion of blood serum is sex-linked. Presence of the antigen (ma^+) is dominant over absence (ma^-). The numbers of the two phenotypes in a sample of Norwegians were 57 ma^+ and 44 ma^- among females, but 23 ma^+ and 77 ma^- among males. Therefore the frequency of the ma^- allele among females was

$$q_f = \sqrt{44/101} = 0.66$$

and, among males,

$$q_m = 77/100 = 0.77$$

Hence the frequency of the ma^- allele in the population was

$$q = \tfrac{2}{3}(0.66) + \tfrac{1}{3}(0.77) = 0.70$$

When the allele frequencies at sex-linked loci are different between females and males, the population does not reach the equilibrium frequency in a single generation. Rather, the frequency of a given allele among the males in a given generation is the frequency of that allele among the females in the previous generation (because the males inherit their only X chromosome from their mothers), while the frequency among the females in a given generation is the average of its frequency in the females and the males of the previous generation (because the females inherit one X chromosome from their fathers and the other from their mothers). Consequently, the distribution of the allele frequencies of the two sexes oscillates, but the difference between the sexes is halved each generation; therefore the population rapidly approaches an equilibrium in which both sexes have the same frequency (Figure 3.4).

---

frequency (p_1) of A_1 after mutation will be the initial frequency (p_0) minus the frequency of mutated alleles (up_0), or

$$p_1 = p_0 - up_0 = p_0(1 - u)$$

In the following generation, a fraction u of the remaining A_1 alleles (p_1) will mutate to A_2, and the frequency of A_1 will become

$$p_2 = p_1 - up_1 = p_1(1 - u)$$

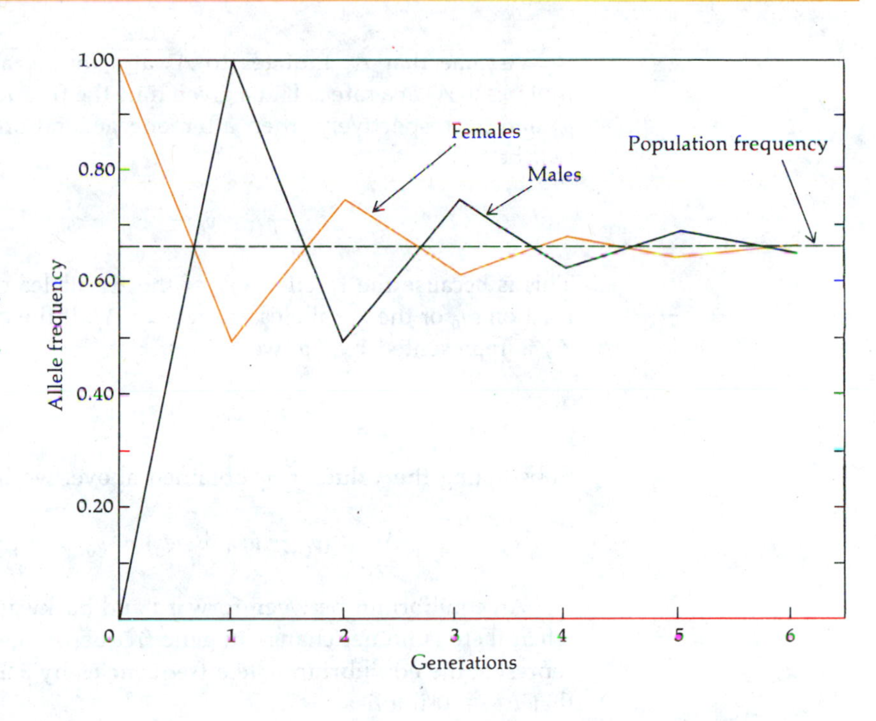

Figure 3.4
Changes in allele frequencies through the generations at a sex-linked locus when the two sexes have different allele frequencies. The case shown is the most extreme possible, namely, when the initial frequency of an allele is 1 in females and 0 in males. The initial frequency of the allele in the population (which is also the ultimate equilibrium frequency in males as well as in females) is therefore 0.67.

Substituting the value of p_1 obtained above, we have

$$p_2 = p_1(1 - u) = p_0(1 - u)(1 - u) = p_0(1 - u)^2$$

After t generations, the frequency of A_1 alleles will be

$$p_t = p_0(1 - u)^t$$

Because $1 - u$ is less than 1, it follows that, as t increases, p_t becomes ever smaller; if the process continues indefinitely, the frequency of A_1 will eventually decrease to zero. This result is intuitively obvious—the frequency of A_1 gradually decreases because a fraction of A_1 alleles changes to A_2 in every generation.

The rate of change is nevertheless very slow. For example, if the mutation rate is $u = 10^{-5}$ per gamete per generation, which is a typical gene mutation rate for eukaryotes, it will take 1000 generations to change the frequency of A_1 from 1.00 to 0.99, 2000 generations to change it from 0.50 to 0.49, and 10,000 generations to change it from 0.10 to 0.09. In general, as the frequency of A_1 decreases, it takes ever longer to accomplish a given amount of change (0.01 in the examples just given).

The model of mutation from one genetic variant to another without reverse mutation may apply to certain cases: for example, chromosomal inversions—a given chromosomal arrangement may yield inversions at a certain rate, but it is extremely unlikely that an inverted arrangement will revert to exactly the original sequence. Mutations, however, are often reversible: allele A_2 may mutate back to A_1.

Assume that A_1 mutates to A_2 at a rate u, as before, and that A_2 mutates to A_1 at a rate v. If at a given time the frequencies of A_1 and A_2 are p_0 and q_0, respectively, then after one generation the frequency of A_1 will be

$$p_1 = p_0 - up_0 + vq_0$$

This is because the fraction up_0 of the A_1 alleles changes to A_2, but the fraction vq_0 of the A_2 alleles changes to A_1. If the change in frequency of A_1 is represented by Δp, we have

$$\Delta p = p_1 - p_0$$

Substituting the value of p_1 obtained above, we have

$$\Delta p = (p_0 - up_0 + vq_0) - p_0 = vq_0 - up_0$$

An equilibrium between forward and backward mutations will exist when there is no net change in gene frequency, i.e., when $\Delta p = 0$. If we represent the equilibrium allele frequencies by $\hat{p}$ and $\hat{q}$, the requirement that $\Delta p = 0$ implies

$$u\hat{p} = v\hat{q}$$

This result says that the allelic frequencies will be at equilibrium when the number of A_1 alleles changing to A_2 alleles is the same as the number of A_2 changing to A_1. Since $p + q = 1$, and therefore $q = 1 - p$, there will be equilibrium when

$$u\hat{p} = v(1 - \hat{p})$$

$$u\hat{p} + v\hat{p} = v$$

$$\hat{p} = \frac{v}{u + v}$$

and, because $\hat{p} + \hat{q} = 1$,

$$\hat{q} = \frac{u}{u + v}$$

Assume that the mutation rates are $u = 10^{-5}$ and $v = 10^{-6}$; then

$$\hat{p} = \frac{10^{-6}}{10^{-5} + 10^{-6}} = \frac{1}{11} = 0.09$$

$$\hat{q} = \frac{10^{-5}}{10^{-5} + 10^{-6}} = \frac{10}{11} = 0.91$$

Two points need to be added. The first point is that allelic frequencies are usually not in mutational equilibrium, because other processes affect them. In particular, natural selection may favor one allele over the other; the equilibrium frequencies are then decided by the interaction between mutation and selection, as we shall see in Chapter 4. The second point is that allelic frequencies change at a slower rate when there is forward and backward mutation than when mutation occurs in only one direction, because backward mutation partially counteracts the effects of forward mutation. This emphasizes what has already been said—that mutation by itself takes a very long time in order to effect any substantial change in allele frequencies.

Migration

Migration, or *gene flow,* occurs when individuals move from one population to another and interbreed with the latter. Gene flow does not change allele frequencies for the whole species, but may change them locally when the allele frequencies in the migrants are different from those in resident individuals.

Assume that individuals from surrounding populations migrate at a certain rate into a local population and there interbreed with the residents. The proportion of migrants is m, so that in the next generation $(1 - m)$ of the genes are descendants of residents, and m are descendants of migrants. Assume that in the surrounding population a certain allele, A_1, has an average frequency P, while in the local population it has the frequency p_0. In the next generation, the frequency of A_1 in the local population will be

$$p_1 = (1 - m)p_0 + mP$$

$$= p_0 - m(p_0 - P)$$

That is, the new allelic frequency will be the original allelic frequency (p_0) multiplied by the proportion of reproducing individuals that are residents $(1 - m)$, plus the proportion of reproducing migrant individuals (m) multiplied by their gene frequency (P). Reorganizing the terms as shown, we see that the new allelic frequency will be the original allelic frequency (p_0) minus the proportion of migrant individuals (m) multiplied by the difference in allelic frequency between residents and migrants $(p_0 - P)$.

The change Δp in allele frequency is

$$\Delta p = p_1 - p_0$$

Substituting the value of p_1 obtained above, we have

$$\Delta p = p_0 - m(p_0 - P) - p_0 = -m(p_0 - P)$$

That is, the greater the proportion of migrant individuals and the greater the difference between the two allele frequencies, the greater Δp becomes. Notice that Δp will be zero only when either m or $p_0 - P$ is zero. Therefore, unless migration stops ($m = 0$), the allelic frequency will continue to change until it becomes the same in the local population as in the surrounding populations ($p - P = 0$).

It is worthwhile to look at the difference in allele frequency between the local and the surrounding populations. After the first generation, it will be

$$p_1 - P = p_0 - m(p_0 - P) - P$$

$$= p_0 - mp_0 - P + mP$$

$$= (1 - m)p_0 - (1 - m)P$$

$$= (1 - m)(p_0 - P)$$

After the second generation, the difference in allele frequencies will be

$$p_2 - P = (1 - m)^2(p_0 - P)$$

and after t generations of migration, we have

$$p_t - P = (1 - m)^t(p_0 - P)$$

This formula allows one to calculate the effect of t generations of migration at a certain rate (m) if the initial allele frequencies (p_0 and P) are known:

$$p_t = (1 - m)^t(p_0 - P) + P$$

The formula is also helpful in investigating other interesting questions. For example, if we know the initial allelic frequencies (p_0 and P), the allelic frequency (p_t) in the resident population at a certain time, as well as the number of generations (t), we can calculate the rate of gene flow (m).

In the United States, people of mixed Caucasian and Black descent are considered members of the Black population. Racial admixture can hence be seen as a process of gene flow from the Caucasian to the Black population. The frequency of the R^0 allele at the locus determining the Rhesus blood groups is $P = 0.028$ in U.S. Caucasian populations. Among the African populations from which the ancestors of U.S. Blacks came, the frequency of R^0 is 0.630; this can be considered the initial allele frequency of the U.S. Black population, $p_0 = 0.630$. The African ancestors of the U.S. Blacks came to the United States about 300 years, or some 10 generations, ago; hence $t = 10$. The frequency of R^0 among U.S. Blacks is at present $p_t = 0.446$.

Rearranging the terms in the last equation shown above, we have

$$(1 - m)^t = \frac{p_t - P}{p_0 - P}$$

Substituting the values of the various parameters, we have

$$(1 - m)^{10} = \frac{0.446 - 0.028}{0.630 - 0.028} = 0.694$$

$$1 - m = \sqrt[10]{0.694} = 0.964$$

$$m = 0.036$$

Therefore the gene flow from U.S. Caucasians into the U.S. Black population has occurred at a rate equivalent to an average of 3.6% per generation. Ten generations of gene flow have left $(1 - m)^{10} = 0.694$ of all genes in U.S. Blacks derived from their African ancestors, and $1 - 0.694 = 0.306$, or somewhat more than 30%, of their genes derived from Caucasian ancestors.

The calculations above are only approximations, but they give us an idea of the extent of racial admixture in the United States. Using allele frequencies at other loci, calculations similar to the one just made lead to somewhat different results. Moreover, the degree of racial admixture may be different in various parts of the United States (Table 3.5). But it is clear that considerable gene flow has taken place.

Random Genetic Drift

Random genetic drift (*genetic drift*, for short, or simply *drift*) refers to changes in gene frequencies due to sampling variation from generation to generation. Assume that in a certain population two alleles, *A* and *a*, exist in frequencies 0.40 and 0.60; the frequency of *A* in the following generation may be less (or greater) than 0.40 simply because, by chance, allele *A* is present less often (or more often) than expected among the gametes that form the zygotes of that generation.

Genetic drift is a process of pure chance and represents a special case of the general phenomenon known as *sampling errors*, or *sampling variation*. The general principle is that the magnitude of the "errors" due to sampling is inversely related to the size of the sample—the smaller the sample, the larger the effects. With respect to organisms, this principle means that the smaller the number of breeding individuals in a population, the larger the allele frequency changes due to genetic drift are likely to be.

Table 3.5
Allele frequencies at several loci in African Blacks, U.S. Blacks, and U.S. Caucasians. The Africans are from a region from which slaves were shipped to the United States. The U.S. Blacks are from one southern city and one western city.

Allele	Blacks (Africa)	Blacks (Claxton, Georgia)	Blacks (Oakland, California)	Caucasians (Claxton, Georgia)
R^0	0.630	0.533	0.486	0.022
R^1	0.066	0.109	0.161	0.429
R^2	0.061	0.109	0.071	0.137
r	0.248	0.230	0.253	0.374
A	0.156	0.145	0.175	0.241
B	0.136	0.113	0.125	0.038
M	0.474	0.484	0.486	0.507
S	0.172	0.157	0.161	0.279
Fy^a	0.000	0.045	0.094	0.422
P	0.723	0.757	0.737	0.525
Jk^a	0.693	0.743		0.536
Js^a	0.117	0.123		0.002
T	0.631	0.670		0.527
Hp^1	0.684	0.518		0.413
$G6PD$	0.176	0.118		0.000
Hb^S	0.090	0.043		0.000

After J. Adams and R. H. Ward, *Science 180*:1137 (1973).

It is simple to see why there should be an inverse relation between sample size and sampling errors. Assume that we toss a coin and that the probability of getting heads in any one toss is 0.5. If we toss the coin only once, we can get heads or tails, but not both; although the probability of getting heads is 0.5, we get heads either once or not at all, but not half the time. If we toss the coin ten times, we are likely to get several heads and several tails; we would be surprised (and suspicious of the coin) if we got only heads, but not if we got, say, six heads and four tails. In the latter case the frequency of heads would be 0.6 rather than the expected 0.5, but we would attribute such deviation from expectation to chance. Assume now that we toss the coin 1000 times. We would be extremely suspicious of the coin if we got only heads, or even if we got 600 heads and 400 tails, although the frequency of heads in the latter case would be 0.6, the same frequency we observed without surprise when we threw the coin ten times. When the coin is tossed 1000 times, however, we would not be surprised to get 504 heads and 496 tails, which means a frequency of heads of 0.504, although the expected frequency is still 0.500.

The point of the coin example is that the larger the sample, the more likely it is to show a close agreement between the expected frequency of heads (0.5) and the observed frequency (1, 0.6, and 0.504 for one, ten, and 1000 throws in the example). With populations, we also expect that the larger the number of individuals producing the next generation, the closer the agreement between the expected allelic frequency (which is that in the parental generation) and the observed allelic frequency (which is that in the progeny).

Note that the relevant figure is not the total number of individuals in the population, but the *effective population size*, which is determined by the number of parents producing the following generation. This is so because the genes sampled to make up the following generation are from the individuals that become parents, not from the rest of the population.

There is an important difference between the coin examples and genetic drift. With a coin, the probability of obtaining heads in a run remains 0.5 even though in the previous run we may have obtained heads more or less often than expected. In populations, however, the frequency of an allele in a given sample (i.e., a generation) becomes the probability of obtaining the allele in the following sample (generation). If the frequency of an allele in one generation changes from, say, 0.5 to 0.6, the probability of obtaining the allele in the following generation becomes 0.6. Thus changes in allele frequencies become cumulative through the generations. However, because chance changes are random in direction, changes may be reversed, so long as the frequency of an allele does not become 0 or 1 (Figure 3.5). An allele that has increased in frequency in one generation may, in the next generation, increase again or decrease instead, with equal probability. If the allele becomes "lost" or "fixed" (i.e., if it reaches a frequency of either 0 or 1), the process stops. The allele cannot change in frequency again, unless a new allele arises by mutation.

Consider the following example. Assume that we have a large number of pea plants, *Pisum sativum*, like those used by Mendel, and that the frequency of the allele responsible for yellow peas, Y, is 0.5, the same as the frequency of the y allele, which produces green peas in homozygous condition. Assume also that the three genotypes occur in the expected frequencies of $\frac{1}{4}YY$, $\frac{1}{2}Yy$, and $\frac{1}{4}yy$. Suppose now that we pick up one seed (pea), without observing the phenotype, and grow a plant from it. What is the frequency of the Y allele among the peas produced by this plant through self-fertilization? Clearly, there are three possibilities: the frequency of Y will be 1, ½, or 0, depending on the genotype of the pea used

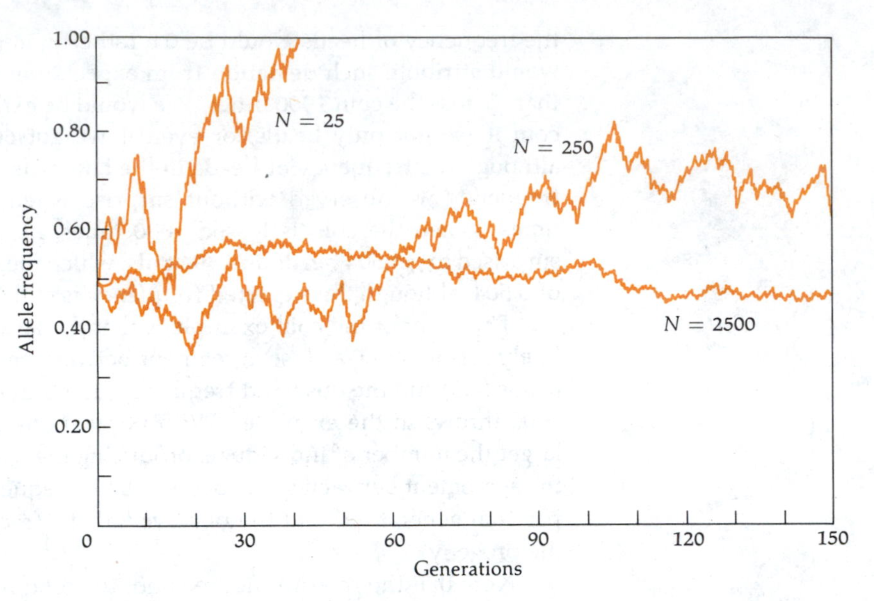

Figure 3.5
Population size and drift. The graphs show the results of computer experiments that simulate chance effects in three populations of different size, each starting with an allele frequency of 0.50. N is the effective population size. (After W. F. Bodmer and L. L. Cavalli-Sforza, *Genetics, Evolution, and Man*, W. H. Freeman, San Francisco, 1976.)

to produce the plant. The probability is ¼ that the pea was YY, and also ¼ that it was yy; thus the frequency of Y is likely to change to either 1 or 0 with a probability of ½. Now suppose that we collect 1000 peas from the original population and grow 1000 plants from them; the frequency of the Y allele among the peas produced by these plants is likely to be very nearly ½, although it may be slightly higher or lower.

Whenever we know the number of parents used to produce the following generation, and the allelic frequencies, as in the previous examples, it is possible to calculate the probability of obtaining a given allelic frequency in the following generation. In order to do this we need to know the *variance* of the allelic frequencies in the following generation, which is a measure of the amount of variation that would be found among different samples (see Appendix A.III). If there are two alleles, with frequencies p and q, and the number of parents is N (so that the number of genes in the sample used to produce the next generation is $2N$), the *variance* (s^2) of the allelic frequency in the following generation is

$$s^2 = \frac{pq}{2N}$$

and the *standard deviation* is

$$s = \sqrt{\frac{pq}{2N}}$$

These equations show the inverse relationship between sample size, $2N$, and the expected variation in allelic frequencies.

Table 3.6
Effects of random genetic drift from one generation to the next.

Population Size (N)	Gametes ($2N$)	Variance ($pq/2N$)	Standard Deviation ($\sqrt{pq/2N}$)	Expected Range in p in 95% of Populations ($p \pm 2$ s.d.)
Case 1: $p = q = 0.5$				
5	10	0.025	0.16	0.18–0.82
50	100	0.0025	0.05	0.40–0.60
500	1000	0.00025	0.016	0.468–0.532
Case 2: $p = 0.3$, $q = 0.7$				
5	10	0.021	0.145	0.01–0.59
50	100	0.0021	0.046	0.208–0.392
500	1000	0.00021	0.0145	0.271–0.329

Table 3.6 shows the likely effects of drift from one generation to the next in two cases: (1) when $p = q = 0.5$ and (2) when $p = 0.3$ and $q = 0.7$. In each case three different effective population sizes are considered: $N =$ 5, 50, and 500. The expected range of variation in p in 95% of the populations is given by $p \pm 2$ standard deviations (see Appendix A.III). In small populations with an effective size of 5 individuals, the range in the frequency p extends in one single generation from 0.18 to 0.82; in larger populations the range in allelic frequency is lower. Notice that the extent of the range decreases as the square root of the ratio of population sizes. For example, the range for the populations with 5 individuals is 0.64 (from 0.18 to 0.82), while in the populations 100 times greater (500 individuals), the range is $\sqrt{100} = 10$ times smaller ($0.532 - 0.468 = 0.064$).

Table 3.6 shows the range within which 95% of the populations (or the genes) are expected to fall. Within this range, the intermediate frequencies have a greater probability than the extreme frequencies. This is illustrated in Figures 3.6 and 3.7 for two cases with initial "frequencies" of 0.5 and 0.4, respectively, and each having two "population sizes," 10 and 100.

The *cumulative* effects of random genetic drift over the generations are shown in Figure 3.8. The experiment, performed by Peter F. Buri, consisted of 107 different populations, reproduced each generation by selecting at random 8 females and 8 males from the offspring of the preceding

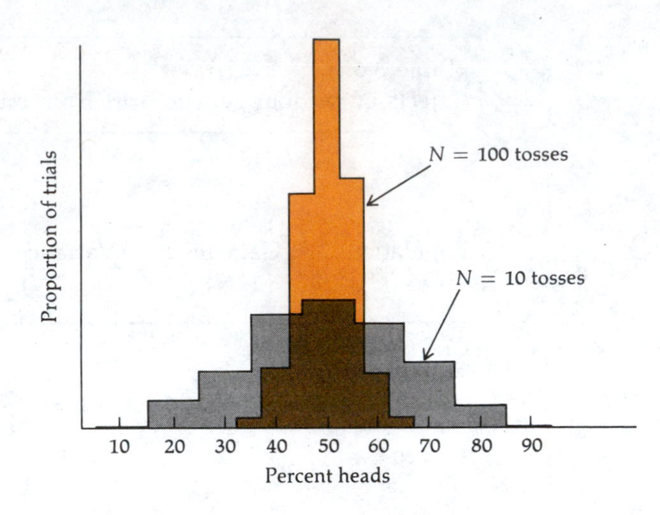

Figure 3.6
Experiments in coin tossing. Each of the two graphs represents 1000 independent trials of either 10 or 100 tosses. The ordinate represents the proportion of times a given number of heads was obtained. The distributions are approximately normal, with the mode at the expected probability of 0.50. The variance of the distribution is larger in the experiments with 10 tosses (equivalent to an effective population size of 5 diploid individuals) than in those with 100 tosses (effective population size of 50). (After W. F. Bodmer and L. L. Cavalli-Sforza, *Genetics, Evolution, and Man*, W. H. Freeman, San Francisco, 1976.)

generation (the approximate effective population size was 16 individuals, or 32 genes). The initial frequency of the two alleles, *bw* and bw^{75}, was 0.5 (all individuals were heterozygotes for these two alleles). In the first generation, the allele frequencies are spread around the mean value of 0.5, but with a greater number of populations in the intermediate frequencies. The frequencies obtained in the first generation become the initial frequencies for the second generation, etc. The spread in gene frequencies increases through the generations. The first "fixed" population appears in the fourth generation (frequency of allele $bw^{75} = 1$). The number of fixed populations gradually increases until generation 19, when the experiment was terminated (30 populations fixed for allele *bw* and 28 fixed for allele bw^{75}). If the experiment had continued indefinitely, all populations would have become fixed, approximately half for each allele.

Founder Effect and Bottlenecks

Unless a population is very small, changes in allelic frequencies due to genetic drift will be small from one generation to another, but the effects

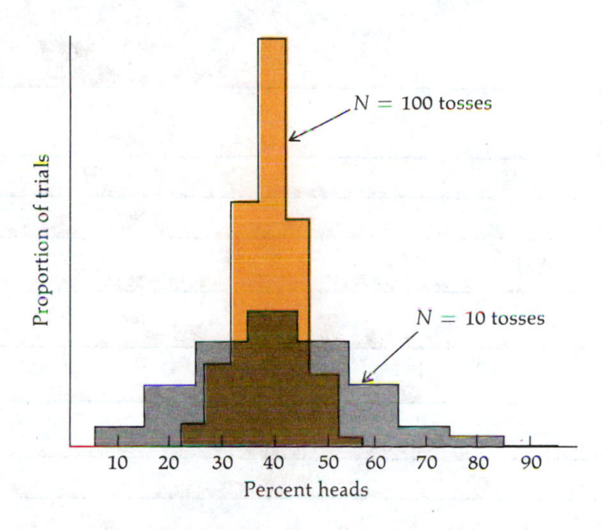

Figure 3.7
Computer-simulated experiments with coins that are "loaded" so as to give heads only 40% of the time. The experiments are analogous to those in Figure 3.6. The mode of the distribution is now at the expected probability of 0.40. (After W. F. Bodmer and L. L. Cavalli-Sforza, *Genetics, Evolution, and Man*, W. H. Freeman, San Francisco, 1976.)

over many generations may be large. If no other processes (mutation, migration, or selection) affect allelic frequencies at a gene locus, evolution will ultimately result in the fixation of one allele and the elimination of all others. When only drift is operating, the probability that a given allele will ultimately be fixed is precisely its frequency; e.g., if a certain allele has a frequency of 0.2 at a given time, it has a probability of 0.2 of ultimately being the only allele in the population. But this may require a very long time; the average number of generations required for fixation is about four times as large as the number of parents per generation.

Assume that a new allele arises by mutation in a population with effective size N. Since there are $2N$ alleles in the population, the frequency of the new mutant is $1/2N$, which is also the probability that the population will become fixed for that mutant by drift alone, since all alleles have the same probability of becoming fixed. The population will eventually be made up of genes all descended from that new mutant (or of one of the other $2N$ alleles in the population), but this can be shown to require, on the average, $4N$ generations when drift alone is operating. If the effective size of the population is 1 million individuals, the process will require about 4 million generations.

It is unlikely that random drift *alone* will affect allelic frequencies at any locus during long periods of time; mutation, migration, and selection are likely to take place at one time or another. These last three processes

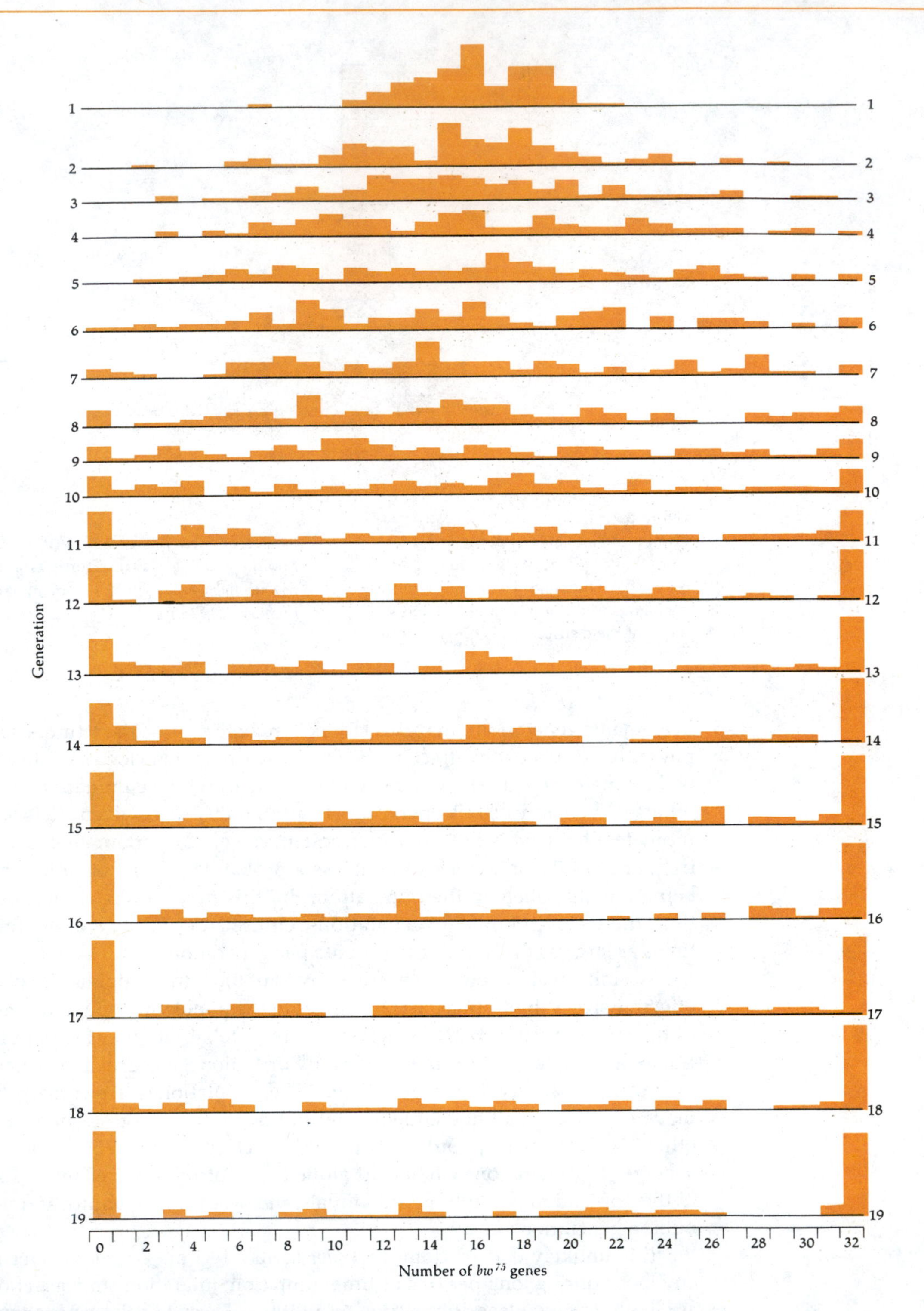

Figure 3.8
Distribution of allele frequencies in 19 consecutive generations among 107 lines of *Drosophila melanogaster,* each with 16 individuals. [After P. Buri, *Evolution 10*:367 (1956).]

are the *deterministic* processes of evolutionary change. Let us use x to represent the mutation rate (u) or the migration rate (m), or the selection coefficient (s, to be defined in Chapter 4); then gene frequency changes will be governed primarily by random genetic drift if, and only if,

$$4Nx << 1$$

where $<<$ means "much less than." If $4Nx \approx 1$ or greater, then gene frequency changes will be determined for the most part by the deterministic processes.

For example, assume that the mutation rate from allele A to allele a is $u = 10^{-5}$ (and assume there is no migration or selection). In a population of 100 breeding individuals, this rate will have little effect on gene frequencies relative to the effects of drift, because $4Nu = 4 \times 10^2 \times 10^{-5} = 4 \times 10^{-3} << 1$. In a population of one million breeding individuals, however, it would have a greater effect than drift, because then $4Nu = 4 \times 10^6 \times 10^{-5} = 40 > 1$. If the migration rate is 0.02 (or two individuals for every hundred) per generation and there is no mutation or selection, gene frequencies will change toward the frequencies in the population from which the migrants come, even in a small population with only 100 individuals, because in such case $4Nm = 4 \times 100 \times 0.02 = 8 > 1$.

Extreme cases of random genetic drift occur when a new population is established by only very few individuals; this has been called by Ernst Mayr the *founder effect.* Populations of many species living on oceanic islands, although they may now consist of millions of individuals, are descendants of one or very few colonizers arrived long ago by accidental dispersal. The situation is similar in lakes, isolated forests, and other ecological isolates. Because of sampling errors, gene frequencies at various loci are likely to be different in the few colonizers than in the population from which they came, which may have lasting effects on the evolution of such isolated populations.

An experimental demonstration of the founder effect is shown in Figure 3.9. Laboratory populations of *Drosophila pseudoobscura* were begun with samples from a population in which a certain chromosomal arrangement, represented as *PP,* had a frequency of 0.50. There were two types of populations; some ("large") were started with 5000 individuals, and the others ("small"), with 20 individuals. After 1½ years, or about 18 generations, the mean frequency of *PP* was about 0.30 in the large as well as in the small populations, but the range of frequencies was considerably greater in the small populations.

Chance variations in allelic frequencies similar to those due to the founder effect occur when populations go through *bottlenecks.* When climatic or other conditions are unfavorable, populations may be drastically reduced in numbers and run the risk of extinction. Such populations may later recover their typical size, but random drift may considerably alter their allelic frequencies during the bottleneck and, therefore, in the following generations. In primitive mankind, many tribes were decimated owing to various disasters; some of these tribes undoubtedly

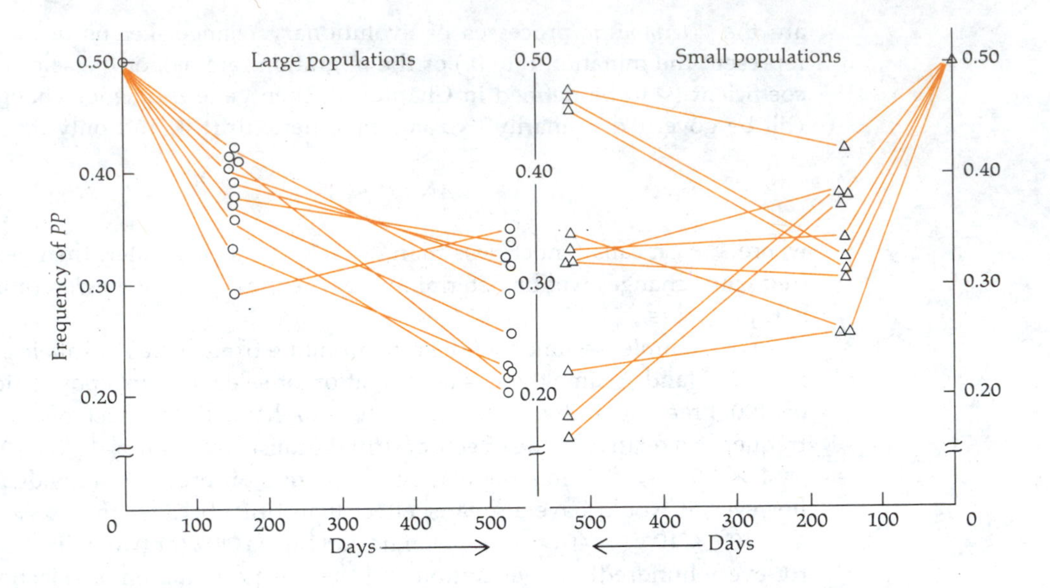

Figure 3.9
Founder effect in laboratory populations of *Drosophila pseudoobscura*. The graphs show the changes in the frequency of a chromosomal inversion, *PP*. Note that time proceeds from left to right for the 10 large populations, but from right to left for the 10 small populations. [After Th. Dobzhansky and O. Pavlovsky, *Evolution* 11:311 (1957).]

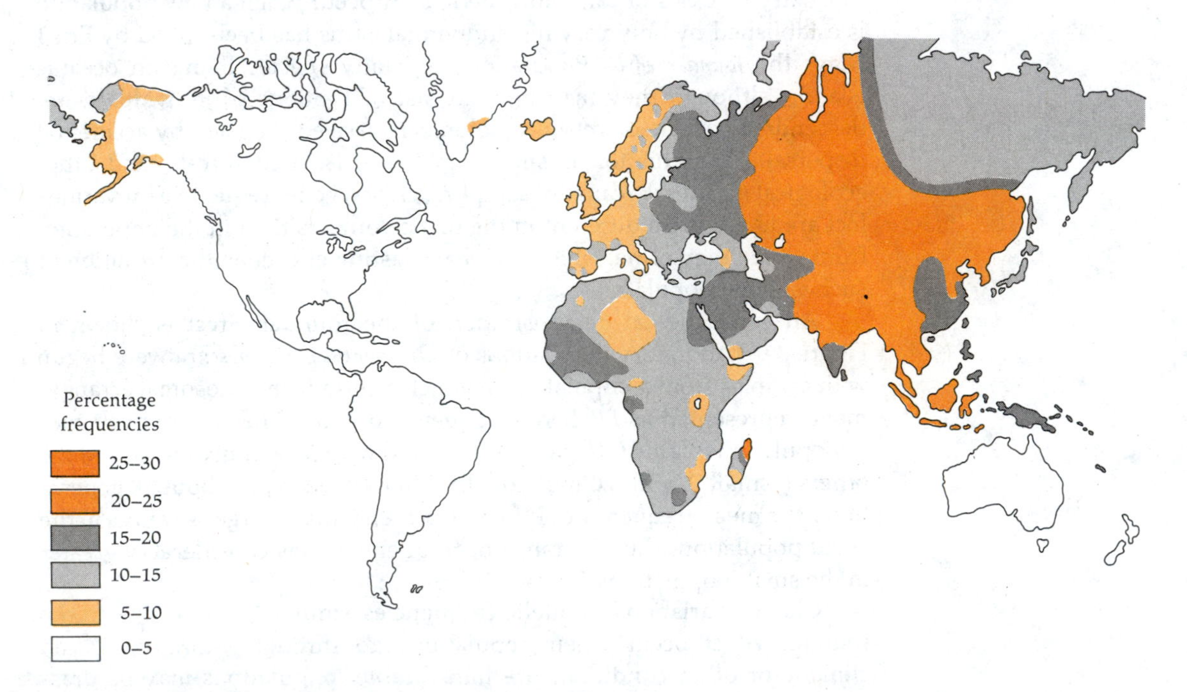

Figure 3.10
Frequency of the blood-group allele I^B in aboriginal populations of the world. The I^B allele is nearly or altogether absent among American Indians and aboriginal Australians unmixed with Europeans, but it is present in all Old World populations. The frequency of I^B is highest in northern India, Mongolia, central Asia, and in some aboriginal peoples from Siberia.

suffered rapid extinction as a result, but presumably most recovered from the survivors, possibly augmented by migrants from other tribes. Differences among human populations in the frequency of the ABO blood-group alleles may have resulted, at least in part, from population bottlenecks and founder effects (Figure 3.10).

A recent example of the founder effect in a human population is the Dunker communities in Pennsylvania. The Dunkers are a religious sect established by 27 families that emigrated from Germany in the mid-eighteenth century. Since then they have lived in small communities, intermarrying little with the surrounding populations. The effects of drift can be noted in several gene loci. The frequency of blood group A (genotypes I^AI^A and I^Ai) is 40–45% in German populations and in U.S. populations of German descent; among the Dunkers, however, it is 60%, while the I^B allele has gone nearly to extinction (frequency 2.5%). At the locus for the M-N blood groups, the frequency of the M allele is about 54% in both German and U.S. populations, but 65% among the Dunkers.

Problems

1. In a sample of 1100 Japanese from Tokyo, it was found that 356, 519, and 225 individuals had blood groups M, MN, and N, respectively. Calculate the allele frequencies and the expected Hardy-Weinberg genotypic frequencies. Use the chi-square test to determine whether the expected and observed numbers are in statistical agreement.

2. For the data in the previous problem, assume that the genotypic frequencies are the same in men and women. Assuming random mating, calculate the probabilities of all possible types of mating (see Table 3.1) and the probability of the progeny produced by each mating. When the progenies of all matings are taken into account, the expected genotypic frequencies should be the same as in the previous problem (i.e., those predicted by the Hardy-Weinberg law).

3. The frequency of red-green color blindness in the men in a certain population is 0.08. This form of color blindness is caused by a sex-linked recessive allele. What are the expected frequencies of the three genotypes among women?

4. The most common form of hemophilia is due to a sex-linked recessive allele with a frequency of 0.0001. What are the expected frequencies of the two male genotypes and the three female genotypes in the population?

5. Tay-Sachs disease is caused by an autosomal recessive allele. The disease is characterized by mental deficiency and blindness, with death occurring

by four years of age. The incidence of the disease among newborns is about 10 per million births. Assuming Hardy-Weinberg equilibrium, estimate the frequency of the allele and of the heterozygotes.

6. Cystic fibrosis is a recessive disease characterized by malabsorption of food and obstruction of the bronchial tubes and other tissues. Death usually occurs by the late teens. About 4 in 10,000 newborns will suffer from cystic fibrosis. Assuming Hardy-Weinberg equilibrium, what are the frequencies of the three genotypes among newborns?

7. Acatalasia is a recessive condition first discovered in Japan. Heterozygotes can be identified by the intermediate level of catalase in their blood. The frequency of heterozygotes is 0.09% in Hiroshima and Nagasaki, but 1.4% in the rest of Japan. Assuming Hardy-Weinberg equilibrium, calculate the allelic frequencies in (1) Hiroshima and Nagasaki and (2) the rest of Japan.

8. An experimental population of *Drosophila melanogaster* is started with 100 bw/bw females and 100 bw^+/bw^+ males. What will be the genotypic frequencies in the F_1, the F_2, and the following generations, assuming random mating and assuming that all genotypes reproduce equally effectively?

9. A population begins with the following composition at a sex-linked locus:

Males:	400 *A*	600 *a*	
Females:	640 *AA*	320 *Aa*	40 *aa*

Assuming random mating, what will be the equilibrium genotypic frequencies?

10. The ratio between the recessive alleles (a) that exist in heterozygotes and in homozygotes is approximately $1/q$ when q is small. What proportion of *all a* alleles are found in homozygotes? (Note that in this case the approximation used in the text of Chapter 3 is not required.)

11. Calculate the proportion of all recessive alleles that are found in homozygotes for Tay-Sachs disease and for cystic fibrosis (see problems 5 and 6).

12. The procedure given on page 64 for calculating the allelic frequencies is not the most efficient, because it does not use the AB blood group for the calculations and therefore ignores part of the available information. The maximum likelihood method is the best for estimating the allelic frequencies, but it is quite complicated. A better method than the one used on page 64 starts by calculating p, q, and r, as is done there. Let $D = 1 -$

$p - q - r$. The corrected allelic frequencies are then given by $p^* = p(1 + D/2)$, $q^* = q(1 + D/2)$, and $r^* = (r + D/2)(1 + D/2)$. Use this method to estimate the allelic frequencies in the following populations:

Population	Number of individuals having blood group: AB	B	A	O	Total
English	5,782	16,279	79,341	88,774	190,177
Chinese	606	1,626	1,920	1,848	6,000
Pygmies	103	300	313	316	1,032
Eskimos	7	17	260	200	484

(Note that $D = 0$ for the frequencies given on page 64, because those assumed frequencies exactly fit the Hardy-Weinberg expectations.)

13. In *Escherichia coli* the rate of mutation from histidine independence (his^+) to histidine requirement (his^-) and the rate of the reverse mutation have been estimated as

$$his^+ \rightarrow his^- \quad 2 \times 10^{-6}$$

$$his^- \rightarrow his^+ \quad 4 \times 10^{-8}$$

Assuming that no other processes are involved, what will be the equilibrium frequencies of the two alleles?

14. Assume that the forward and backward mutation rates at a certain locus of *Drosophila melanogaster* are

$$A \rightarrow a \quad 2 \times 10^{-5}$$

$$a \rightarrow A \quad 6 \times 10^{-7}$$

What are the expected allelic equilibrium frequencies if no other processes are involved? Is the calculation of mutation equilibrium frequencies of alleles different for diploid and haploid organisms?

15. Assume that at a certain locus the mutation rate of $A \rightarrow a$ is 10^{-6} and that there is no back mutation. What will be the frequency of A after 10, 1000, and 100,000 generations of mutation?

16. Assuming that ten generations have elapsed since the African ancestors came to the U.S., calculate the average rate of gene flow per generation between U.S. Blacks and Caucasians from Claxton, Georgia, using the frequencies of the Fy^a allele given in Table 3.5.

17. A population of *Drosophila melanogaster* is polymorphic for two alleles A_1 and A_2. One thousand populations are derived, and each is maintained by selecting at random ten females and ten males in each generation as parents of the following generation. After many generations, it is observed that 220 populations are fixed for allele A_1, and 780 for allele A_2. Estimate the allele frequencies in the original population, assuming that only genetic drift is involved.

18. What will be the range of allele frequencies (standard deviation) in the first generation among the 1000 populations of problem 17?

19. The effective size of a population, N_e, can be estimated by the equation

$$N_e = \frac{4N_m N_f}{N_m + N_f}$$

where N_m and N_f are the number of males and females producing the following generation. If $N_m = N_f$, then $N_e = N_m + N_f$; otherwise N_e is smaller than the sum of the two parents. Assume that in a cattle ranch 100 bulls and 400 cows are used to produce the following generation. What is the effective population size? A neighboring rancher maintains 500 cows, all artificially inseminated with the sperm of a single bull; what is the effective population size in this case?

4

Natural Selection

The Concept of Natural Selection

In Chapter 3 we considered three of the four processes that change gene frequencies—mutation, migration, and drift. Now we introduce the fourth and most critical one—natural selection. But first let us recall some essential features of the other three. We can predict the *direction* and *rate* of change in allele frequencies due to mutation or to migration whenever the appropriate parameters are known (i.e., the mutation or migration rate and the allele frequencies). With respect to drift, a knowledge of the relevant parameters (effective population size and allele frequencies) makes it possible to calculate the expected magnitude of allele frequency changes—i.e., the expected *rate* of change—but not the *direction* of change, because this is random.

There is, however, an important attribute that mutation, migration, and drift have in common: none of them is oriented with respect to adaptation. These processes change gene frequencies independently of whether or not such changes increase or decrease the adaptation of organisms to their environments. Therefore, because these processes are random with respect to adaptation, they would, by themselves, destroy the organization and adaptation characteristic of living beings. Natural selection is the process that promotes adaptation and keeps the disorganizing effects of the other processes in check. In this sense, natural selection is the most critical evolutionary process, because only natural selection accounts for the *adaptive* and highly organized nature of living creatures. Natural selection also explains the *diversity* of organisms because it promotes their adaptation to different ways of life.

The idea of natural selection as the fundamental process of evolutionary change was reached independently by Charles Darwin and Alfred Russel Wallace. In 1858 they made a joint presentation of their discovery to the Linnean Society of London. The argument for evolution by natural selection was developed fully, with considerable supporting evidence, in *The Origin of Species*, published by Darwin in 1859. He proposed that carriers of hereditary variations that are useful as adaptations to the environment are likely to survive better and produce more progeny than organisms possessing less useful variations. As a consequence, adaptive variations will gradually increase in frequency over the generations, at the expense of less adaptive ones. This process of differential multiplication of hereditary variations was called *natural selection.* The outcome of the process is organisms well adapted to their environments.

Natural selection can be defined simply as *the differential reproduction of alternative genetic variants*, determined by the fact that some variants increase the chances of survival and reproduction of their carriers relative to the carriers of other variants. Natural selection may be due to differential survival or differential fertility, or both.

Darwin emphasized that competition for limited resources, a common situation in nature, results in natural selection of the most effective competitors. He wrote, for example: "As more individuals are produced than can possibly survive, there must in every case be a struggle for existence, either one individual with another of the same species, or with the individuals of distinct species." But Darwin also noted that natural selection may occur without competition, as a result of inclement weather and other aspects of "the physical conditions of life." Populations of all sorts of organisms are often depleted during unfavorable seasons; some organisms are better equipped than others to withstand inclement weather. Moreover, natural selection may occur even if no organism would die before completing its reproductive period, simply because some organisms produce more offspring than others.

Darwinian Fitness

The parameters used to measure mutation, migration, and drift are the mutation rate, the migration rate, and the variance of allelic frequencies, respectively. The parameter commonly used to measure natural selection is *Darwinian fitness*, or *relative fitness* (also called *selective value* and *adaptive value*). Fitness is a measure of the reproductive efficiency of a genotype.

Natural selection operates by *differential* reproduction; accordingly, fitness is often expressed as a relative, not absolute, measure of reproductive efficiency. For mathematical convenience, geneticists usually

Table 4.1
Computation of the fitnesses of three genotypes when the number of progeny produced by each genotype is known.

	Genotype			
	A_1A_1	A_1A_2	A_2A_2	Total
(*a*) Number of zygotes in one generation	40	50	10	100
(*b*) Number of zygotes produced by each genotype in next generation	80	90	10	180
Computation				
1. Average number of progeny per individual in next generation (b/a)	80/40 = 2	90/50 = 1.8	10/10 = 1	
2. Fitness (*relative* reproductive efficiency)	2/2 = 1	1.8/2 = 0.9	1/2 = 0.5	

assign the fitness value 1 to the genotype with the highest reproductive efficiency. Assume that at a certain locus there are three genotypes and that, for every one progeny produced by an A_1A_1 homozygote, the heterozygote A_1A_2 produces, on the average, one progeny also, but the A_2A_2 homozygote produces only 0.8 progeny. Then the fitnesses of the three genotypes are 1, 1, and 0.8, respectively.

Table 4.1 shows how fitnesses can be computed when the number of progeny produced by each genotype is known. There are two steps. First, one calculates the average number of progeny produced per individual by each genotype in the next generation. Second, one divides the average number of progeny of each genotype by that of the *best* genotype.

If we know the genotypic fitnesses, we can predict the rate of change in the frequency of the genotypes. The converse is also true, and geneticists often compute fitnesses based on the changes in genotypic frequencies. A simple example is the following. Assume that in a haploid organism, such as *Escherichia coli*, the frequency of the two genotypes, A and a, in a large population is 0.50 at a given time, but the frequencies change to 0.667 A and 0.333 a in one generation; we infer that the fitnesses of A and a are 1 and 0.50, respectively. Note that a large population is assumed so that drift can be ignored; it is also assumed that there is neither mutation nor migration, or that they are so low that they can be ignored.

Table 4.2
Fitness differences due to differences in survival rate.

	Genotype		
Component	A_1A_1	A_1A_2	A_2A_2
Survival fitness	1	0.9	0.5
Fertility fitness	1	1	1
Net fitness	$1 \times 1 = 1$	$0.9 \times 1 = 0.9$	$0.5 \times 1 = 0.5$

Fitness is often represented by the letter w. A related measure is the *selection coefficient* (not to be confused with *selective value*, which is the same as fitness); it is usually represented by s, and is defined as $s = 1 - w$ (therefore, $w = 1 - s$). The selection coefficient measures the reduction in fitness of a genotype. In Table 4.1, the selection coefficient is 0 for the genotype A_1A_1, 0.1 for A_1A_2, and 0.5 for A_2A_2.

Relative fitness predicts the course of selection, i.e., it predicts how gene frequencies will change, but not how well the population will do. Because relative fitness values are relative measures, they do not say whether the population will increase or decrease in numbers. Assume, for example, that in Table 4.1 the numbers of zygotes produced by the three genotypes in the next generation are 40, 45, and 5. The relative fitnesses would be the same as those shown in the table now, although the total number of zygotes in the population would have decreased from 100 to 90 rather than increased from 100 to 180.

Aspects of the life of an individual that may affect its reproductive success contribute to natural selection and, therefore, to the fitness of genotypes. These various aspects—survival, rate of development, mating success, fertility, etc.—are called *fitness components.* The main ones are survival (often called *viability*) and fertility. Other components can be treated separately or incorporated into these two. For example, rate of development, mating success, and age of reproduction can be incorporated into fertility if this is expressed as a function of age.

Fitness differences may be due to one or several fitness components. From the point of view of natural selection, what counts is the overall, or net, fitness, not which components are involved (although this may be interesting in other regards). Tables 4.2 and 4.3 show three genotypes that have identical fitnesses in the two examples, although the components involved are different. Tay-Sachs disease is a human condition caused by the accumulation of complex fatty substances, called gangliosides, in the central nervous system, leading to complete mental degeneration, blind-

Table 4.3
Fitness differences due to differences in fertility.

	Genotype		
Component	A_1A_1	A_1A_2	A_2A_2
Survival fitness	1	1	1
Fertility fitness	1	0.9	0.5
Net fitness	$1 \times 1 = 1$	$1 \times 0.9 = 0.9$	$1 \times 0.5 = 0.5$

ness, and early death. Achondroplastic dwarfs reproduce, on the average, only 20% as efficiently as normal individuals. The fitness of Tay-Sachs patients is 0, owing to poor viability; that of achondroplastics is 0.20, owing to reduced fertility.

The ultimate outcome of natural selection may be either the elimination of one or another allele (although mutation may keep deleterious alleles at low frequencies, as shown below) or a stable polymorphism with two or more alleles. The effects of natural selection can be simply treated when there are only two alleles, and therefore three genotypes, at a single

Box 4.1 Kinds of Equilibrium

A physical system is in equilibrium when its state will not change if left undisturbed by external forces. The equilibrium may be stable, unstable, or neutral, depending on the behavior of the system when it is disturbed. A *stable* equilibrium exists when the system returns to its original state after the disturbance. An equilibrium is *unstable* when, after the disturbing force has ceased, the state of the system continues to change in the direction of the disturbance (until it finds a natural boundary or limit). An equilibrium is *neutral* when the state of the system does not change any more (does not react) after the disturbing force ceases.

Model examples of these three kinds of equilibria are (see Figure 4.1): a ball at the lowest point of a smooth concave surface (stable equilibrium), a ball at the highest point of a smooth convex surface (unstable equilibrium), and a ball on a perfectly horizontal surface (neutral equilibrium).

All three types of equilibrium may occur with respect to allele frequencies. The equilibrium may involve only one allele (*monomorphic* equilibrium) or more than one allele (*polymorphic* equilibrium).

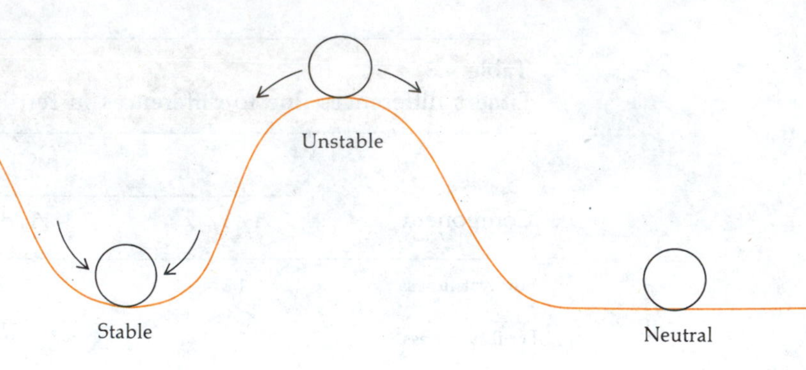

Figure 4.1
Three kinds of equilibrium. When displaced from a stable or unstable equilibrium position, the ball will move in the direction indicated by the arrows; a ball in neutral equilibrium, however, will remain at the position to which it has been displaced.

gene locus. In the following sections we will consider five cases: (1) selection against a recessive allele; (2) selection against a dominant allele; (3) selection against an allele without dominance; (4) selection in favor of the heterozygotes; and (5) selection against the heterozygotes. The first three cases lead to the eventual elimination of the disfavored allele. Case 4 leads to a stable polymorphism with both alleles present at frequencies determined by the selection coefficients against the two homozygotes. Case 5 has a point of polymorphic equilibrium, but the equilibrium is unstable, so that selection leads to fixation for one or the other allele. In the models considered, it is assumed that genotypic fitnesses are constant, i.e., independent of the frequency of the alleles themselves, of the density of the population, and of any other factors. Frequency-dependent selection—when the fitnesses are a function of the genotypic frequencies—will be considered later. Like selection favoring the heterozygotes, frequency-dependent selection may lead to stable polymorphic equilibrium.

Selection Against Recessive Homozygotes

Recessive alleles—such as those for colorless seeds (*c*) in corn, vestigial wings (*vg*) in *Drosophila*, or phenylketonuria in humans—produce heterozygotes that have phenotype and fitness identical to those of the dominant homozygotes. However, the recessive homozygotes may have considerably reduced fitness. Selection will, then, occur against the recessive homozygotes. We can study the effects of selection using the following general model:

Genotype:	*AA*	*Aa*	*aa*
Fitness (w):	1	1	$1 - s$

Table 4.4
Allele frequency changes after one generation of selection against recessive homozygotes.

	Genotype				
	AA	Aa	aa	Total	Frequency of a
1. Initial zygote frequency	p^2	$2pq$	q^2	1	q
2. Fitness (w)	1	1	$1 - s$		
3. Contribution of each genotype to next generation	p^2	$2pq$	$q^2(1 - s)$	$1 - sq^2$	
4. Normalized frequency	$\frac{p^2}{1 - sq^2}$	$\frac{2pq}{1 - sq^2}$	$\frac{q^2(1 - s)}{1 - sq^2}$	1	$q_1 = \frac{q - sq^2}{1 - sq^2}$
5. Change in allele frequency					$\Delta q = \frac{-spq^2}{1 - sq^2}$

The procedure used to calculate gene frequency changes from one generation to the next is shown in Table 4.4; additional details are shown in Box 4.2. The initial zygote frequencies are assumed to be in Hardy-Weinberg equilibrium, owing to the random combination of gametes from the previous generation. The basic step in the calculation is shown in the third row of Table 4.4: multiplication of the initial zygote frequencies (first row) by their relative fitnesses (second row), which gives the contribution of each genotype to the next generation. However, the values in the third row do not add up to 1. In order to convert them to frequencies that add up to 1, we must divide them by their sum total, an operation called *normalization*, which is done in the fourth row. From the frequencies of the progenies produced by each genotype, we can calculate the allele frequencies after selection (by the procedure explained in Chapter 2, page 35); the *change* in allele frequency due to selection is obtained by subtracting the original allele frequency from the frequency after selection. In the first, fourth, and fifth rows of Table 4.4 are shown the initial frequency q of the a allele, the frequency q_1 after one generation of selection, and the change in the frequency due to selection, $\Delta q = q_1 - q$.

The effect of selection against recessive homozygotes is a decrease in the frequency of the recessive allele, a result we would expect, because the recessive homozygotes are reproducing less effectively than the genotypes carrying the dominant allele.

Box 4.2 Selection Against Recessive Homozygotes

The model for selection against recessive homozygotes is given in the text. The basic steps are shown in Table 4.4. The frequencies of the two alleles at the beginning (zygote stage) of a certain generation are p and q, for A and a, respectively. The genotypes are assumed to be in Hardy-Weinberg equilibrium and will thus have the frequencies given in the first row. Since there are only two alleles, $p + q = 1$, and therefore $p^2 + 2pq + q^2 = (p + q)^2 = 1$. The basic step is shown in the third row of the table: multiplication of the initial zygote frequencies (first row) by their fitnesses (second row), which represent the relative rates of reproduction of the genotypes.

When we sum the values given in the third row, we see that they do not add up to 1:

$$p^2 + 2pq + q^2(1 - s) = p^2 + 2pq + q^2 - sq^2 = (p^2 + 2pq + q^2) - sq^2 = 1 - sq^2 \neq 1$$

In order to convert the third-row values to frequencies, we divide each value by their sum total, as shown in the fourth row; these now add up to 1:

$$\frac{p^2}{1 - sq^2} + \frac{2pq}{1 - sq^2} + \frac{q^2(1 - s)}{1 - sq^2} = \frac{p^2 + 2pq + q^2 - sq^2}{1 - sq^2} = \frac{1 - sq^2}{1 - sq^2} = 1$$

The frequency of the a allele after one generation of selection, q_1, is calculated by adding the frequency of the aa genotype and half the frequency of the Aa genotype:

$$q_1 = \frac{q^2(1 - s)}{1 - sq^2} + \frac{pq}{1 - sq^2} = \frac{pq + q^2 - sq^2}{1 - sq^2} = \frac{q(p + q) - sq^2}{1 - sq^2} = \frac{q - sq^2}{1 - sq^2}$$

The frequency of the A allele after one generation of selection, p_1, can be obtained by adding the frequency of the AA homozygotes and half the frequency of the Aa heterozygotes or, alternatively, by subtracting from 1 the frequency of a after selection. Using the first procedure, we obtain

$$p_1 = \frac{p^2}{1 - sq^2} + \frac{pq}{1 - sq^2} = \frac{p^2 + pq}{1 - sq^2} = \frac{p(p + q)}{1 - sq^2} = \frac{p}{1 - sq^2}$$

The allele frequency change is shown in the fifth row. The initial frequency of the a allele was q, and the frequency after selection is $q_1 = (q - sq^2)/(1 - sq^2)$. Therefore, the change in allele frequency per generation is

$$\Delta q = q_1 - q = \frac{q - sq^2}{1 - sq^2} - q = \frac{q - sq^2 - q(1 - sq^2)}{1 - sq^2} = \frac{q - sq^2 - q + sq^3}{1 - sq^2} = \frac{-sq^2(1 - q)}{1 - sq^2}$$

And, because $1 - q = p$, we obtain

$$\Delta q = \frac{-spq^2}{1 - sq^2}$$

Because the values of s, p, and q are either positive (but less than one) or zero, the numerator in the expression for Δq will be either negative or zero. And, because both s and q have values less than one, the denominator will be a positive number. Therefore, the value of Δq will be negative (unless it is zero), indicating that the value of q will decrease as a result of selection.

What will be the ultimate outcome of selection? By definition, there will be no further change in allele frequencies when

$$\Delta q = \frac{-spq^2}{1 - sq^2} = 0$$

Δq will be zero when the numerator is zero; this will happen only when $q = 0$ (or when s or p is zero, but this implies either that there is no selection or that the dominant allele is absent from the population). Therefore the value of q will gradually decrease (as it progresses through the values of q_1, q_2, q_3, etc.) to zero. The ultimate outcome of selection against recessive homozygotes is the elimination of the recessive allele.

Another interesting question concerns the size of Δq, i.e., the amount of change in allele frequency per generation. For a given value of s, the product pq^2 is greatest when $p = 0.33$ and $q = 0.67$; it becomes ever smaller as q decreases from that frequency to lower frequencies. This is so because, although p increases as q decreases, the decrease in q^2 is greater than the increase in p (the square of a number less than 1 is smaller than the number itself). Furthermore, the denominator increases as q^2 becomes smaller. Therefore, the rate of selection becomes extremely slow as q approaches zero (Figure 4.2 and Table 4.5).

The phenomenon of *industrial melanism* has been best studied in the moth *Biston betularia* in England. Until the middle of the nineteenth century, these moths were a uniformly peppered light gray color. Then, darkly pigmented variants started to appear in industrial regions, where the vegetation had gradually blackened owing to pollution from soot and other wastes. In some localities the dark varieties have almost completely replaced the light ones. The light gray moths are homozygous (*dd*) for a recessive allele; the dark moths are either heterozygous (*Dd*) or homozygous (*DD*) for the dominant allele.

The replacement of dark forms for light forms of *B. betularia* in industrial regions is due to differential predation by birds: on tree trunks darkened by pollution, the light gray forms are conspicuous, while the dark forms are well camouflaged. H. B. D. Kettlewell released marked dark and light moths near Birmingham, a heavily industrialized area. The proportions of recaptured moths were 53% for the dark form and 25% for the light form. Since the fertility of the two forms is about the same, we may assume that the relative fitnesses are determined exclusively by the difference in survival rate, mostly due to predation (Table 4.6). The procedure used to estimate fitnesses is essentially that shown in Table

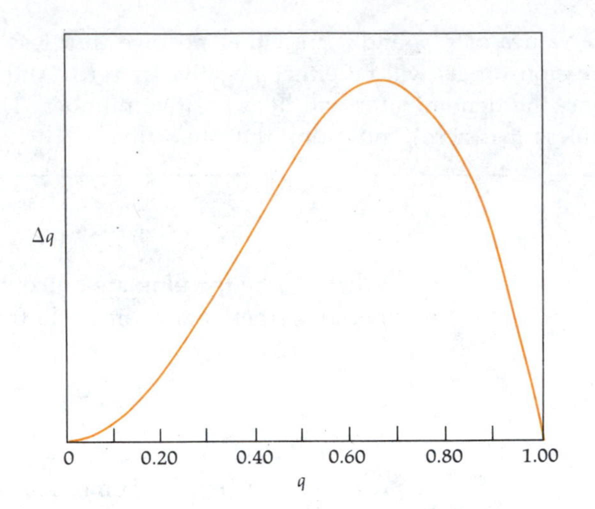

Figure 4.2
The change in allele frequency per generation (Δq) as a function of allele frequency (q) in the case of selection against recessive homozygotes.

4.1, except that the dominant homozygotes and the heterozygotes are treated jointly, since they are phenotypically indistinguishable.

Assume now that, at a certain time, the frequency of the *d* allele is $q = 0.50$ (frequency of the *dd* homozygotes $= 0.25$). Using the fitness values obtained in Table 4.6, we can calculate the selection coefficient against the recessive homozygotes (light form) as $s = 1 - w = 1 - 0.47 = 0.53$. With the formulas given in Table 4.4, we can calculate the changes that would occur owing to selection in a single generation. The frequency of the *d* allele is

$$q_1 = \frac{q - sq^2}{1 - sq^2} = \frac{0.50 - (0.53 \times 0.50^2)}{1 - (0.53 \times 0.50^2)} = \frac{0.3675}{0.8675} = 0.424$$

The change in allele frequency is

$$\Delta q = \frac{-spq^2}{1 - sq^2} = \frac{-0.53 \times 0.50 \times 0.50^2}{0.8675} = \frac{-0.06625}{0.8675} = -0.076$$

We can check that $0.424 + 0.076 = 0.50$, the initial frequency of the *d* allele. Note that the frequency of the *dd* homozygotes in the following generation will be $q^2 = 0.424^2 = 0.180$. This is *not* the same as the value for the zygotes *produced* by *dd* individuals, which is 0.135, as determined by the expression for *aa* in the fourth row of Table 4.4, namely, $q^2(1 - s)/(1 - sq^2)$. Recessive homozygotes are produced by matings involving the heterozygotes as well as the recessive homozygotes of the previous generation. Similarly, not all offspring produced by recessive homozygotes are

Table 4.5
Number of generations required to effect a given reduction in allele frequency (q) for various values of the selection coefficient (s) against recessive homozygotes.

	Generations Required				
Reduction in q	$s = 1$	$s = 0.50$	$s = 0.10$	$s = 0.01$	$s = 0.001$
0.99 to 0.50	1	11	56	559	5,585
0.50 to 0.10	8	20	102	1,020	10,198
0.10 to 0.01	90	185	924	9,240	92,398
0.01 to 0.001	900	1,805	9,023	90,231	902,314
0.001 to 0.0001	9,000	18,005	90,023	900,230	9,002,304

Table 4.6
Relative fitness of dark and light *Biston betularia* moths in Birmingham, England.

	Dark	Light
Genotype	*DD* and *Dd*	*dd*
(*a*) Number released	154	64
(*b*) Number recaptured	82	16
Survival rate (b/a)	0.53	0.25
Relative fitness (w)	$0.53/0.53 = 1$	$0.25/0.53 = 0.47$

also recessive homozygotes; matings with dominant homozygotes or with heterozygotes produce heterozygotes.

Biston betularia illustrates an important point concerning natural selection: the fitness of a genotype may be different in different environments. In unpolluted areas, the light-gray moths survive better than dark moths because, on lichen-covered trees, they are quite inconspicuous to bird predators. When light and dark moths were released in the unpolluted region of Dorset, a greater proportion of light moths than of dark moths was recovered. The data and the fitness calculations are shown in

Table 4.7
Relative fitness of dark and light *Biston betularia* moths in Dorset, England.

	Dark	Light
Genotype	*DD* and *Dd*	*dd*
(*a*) Number released	406	393
(*b*) Number recaptured	19	54
Survival rate (*b*/*a*)	0.047	0.137
Relative fitness (*w*)	0.047/0.137 = 0.343	0.137/0.137 = 1

Table 4.7. The reversal of fitness values between polluted and unpolluted areas is impressive. The fitness of light moths relative to dark moths is 0.47:1 in Birmingham, but 1:0.34 in Dorset.

Recessive Lethals

The limiting case of selection against recessive alleles occurs when the recessive homozygotes have zero fitness, i.e., when they die before attaining reproductive age or are sterile. A familiar example is phenylketonuria. The fitness of the *aa* homozygotes, if left untreated, is $w = 0$; therefore $s = 1$. Substituting 1 for s simplifies the formulas in Table 4.4. In particular, the frequency of allele a after one generation of selection becomes

$$q_1 = \frac{q - sq^2}{1 - sq^2} = \frac{q - q^2}{1 - q^2} = \frac{q(1 - q)}{(1 + q)(1 - q)} = \frac{q}{1 + q}$$

and the change in allele frequency becomes

$$\Delta q = \frac{-spq^2}{1 - sq^2} = \frac{-pq^2}{1 - q^2} = \frac{-(1 - q)q^2}{(1 + q)(1 - q)} = \frac{-q^2}{1 + q}$$

The amount of change in allele frequency after a given number of generations can now easily be calculated. Representing the frequency of a in the initial and following generations by $q_0, q_1, q_2, \ldots, q_t$, we have

$$q_1 = \frac{q_0}{1 + q_0} \quad \text{and} \quad q_2 = \frac{q_1}{1 + q_1}$$

Substituting the value of q_1 in the equation for q_2, we get

$$q_2 = \frac{\dfrac{q_0}{1 + q_0}}{1 + \dfrac{q_0}{1 + q_0}} = \frac{\dfrac{q_0}{1 + q_0}}{\dfrac{1 + q_0 + q_0}{1 + q_0}} = \frac{q_0}{1 + 2q_0}$$

Similarly, after t generations of selection,

$$q_t = \frac{q_0}{1 + tq_0}$$

The number of generations, t, required to change the allele frequency from a certain value, q_0, to another, q_t, can be derived from the last formula:

$$q_t(1 + tq_0) = q_0$$

$$q_t + tq_0 q_t = q_0$$

$$t = \frac{q_0 - q_t}{q_0 q_t} = \frac{1}{q_t} - \frac{1}{q_0}$$

This formula was used to calculate the values for $s = 1$ in Table 4.5. For the particular case when q_t is one-half the value of q_0, the formula becomes

$$t = \frac{1}{q_0/2} - \frac{1}{q_0} = \frac{2}{q_0} - \frac{1}{q_0} = \frac{1}{q_0}$$

That is, the number of generations required to reduce a certain gene frequency to half its initial value is 1 divided by the gene frequency. Thus it takes 10 generations to change q from 0.1 to 0.05, 100 generations to change it from 0.01 to 0.005, and 1000 generations to change it from 0.001 to 0.0005.

The frequency of the recessive allele for albinism in Norway is about 0.01. Assume that a eugenic goal is established to eliminate the allele from the population by sterilizing all albino individuals (see Chapter 3, page 65). It would take 100 generations to reduce the allele frequency to half its present value, and 9900 generations (see Table 4.5 and Figure 3.3) to reduce it to 0.0001. Eugenic measures are inefficient in the case of recessive alleles.

Table 4.8
Allele frequency changes after one generation of selection against genotypes carrying the dominant allele.

	Genotype				
	AA	Aa	aa	Total	Frequency of A
1. Initial zygote frequency	p^2	$2pq$	q^2	1	p
2. Fitness (w)	$1 - s$	$1 - s$	1		
3. Contribution of each genotype to next generation	$p^2(1 - s)$	$2pq(1 - s)$	q^2	$1 - s + sq^2$	
4. Normalized frequency	$\frac{p^2(1 - s)}{1 - s + sq^2}$	$\frac{2pq(1 - s)}{1 - s + sq^2}$	$\frac{q^2}{1 - s + sq^2}$	1	$p_1 = \frac{p(1 - s)}{1 - s + sq^2}$
5. Change in allele frequency					$\Delta p = \frac{-spq^2}{1 - s + sq^2}$

Selection Against Dominants and Selection Without Dominance

Selection is more effective against dominant alleles than against recessive alleles because a dominant allele is expressed in the heterozygotes as well as in the homozygotes. Assume that dominance with respect to fitness is complete, so that dominant homozygotes and heterozygotes have equal fitness. The model is

Genotype:	AA	Aa	aa
Fitness (w):	$1 - s$	$1 - s$	1

The effects of one generation of selection are shown in Table 4.8. The change in allele frequency due to selection is

$$\Delta p = \frac{-spq^2}{1 - s + sq^2}$$

As long as both alleles are in the population (p and q are positive) and there is selection (s is positive), the values of spq^2 and of $1 - s + sq^2$ will be positive. Therefore Δp will have a negative value and the frequency of A will gradually decrease to zero. If a dominant allele is sterile or le-

Table 4.9
Allele frequency changes after one generation of selection when there is no dominance

	Genotype				
	A_1A_1	A_1A_2	A_2A_2	Total	Frequency of A_2
1. Initial zygote frequency	p^2	$2pq$	q^2	1	q
2. Fitness (w)	1	$1 - s/2$	$1 - s$		
3. Contribution of each genotype to next generation	p^2	$2pq(1 - s/2)$	$q^2(1 - s)$	$1 - sq$	
4. Normalized frequency	$\frac{p^2}{1 - sq}$	$\frac{2pq(1 - s/2)}{1 - sq}$	$\frac{q^2(1 - s)}{1 - sq}$	1	$q_1 = \frac{q - sq(1 + q)/2}{1 - sq}$
5. Change in allele frequency					$\Delta q = \frac{-spq/2}{1 - sq}$

thal ($s = 1$), the change in allele frequency is

$$\Delta p = \frac{-pq^2}{1 - 1 + q^2} = \frac{-pq^2}{q^2} = -p$$

The frequency of the dominant allele then becomes zero in one single generation of selection—a result that is obvious, since neither the homozygotes nor the heterozygotes for the dominant allele leave any progeny.

The frequency of the allele selected against is represented by p in the case of selection against dominants (Table 4.8), and by q in the case of selection against recessives (Table 4.4). For any given frequency of the allele selected against, if s is the same, the change in allele frequency is greater in the case of selection against dominants than in the case of selection against recessives. This is as expected, because in the former case the heterozygotes are also selected against, but not in the latter case.

In some instances, the fitness of the heterozygotes is intermediate between the fitnesses of the two homozygotes. We shall consider only the general case when the selection coefficient against the heterozygotes is exactly half the selection coefficient against the disfavored homozygotes, i.e., when there is no dominance. The model is

Genotype:	A_1A_1	A_1A_2	A_2A_2
Fitness (w):	1	$1 - s/2$	$1 - s$

The effects of one generation of selection are shown in Table 4.9.

The change in allele frequency is

$$\Delta q = \frac{-spq/2}{1 - sq}$$

It will be negative as long as both alleles persist in the population and there is selection. The equilibrium condition, $\Delta q = 0$, will be reached only when $q = 0$, i.e., when the allele selected against, A_2, is completely eliminated.

Selection and Mutation

The ultimate outcome of selection in all three cases considered so far (selection against recessive homozygotes, against dominants, and without dominance) is the elimination of the deleterious allele. Such alleles nevertheless remain in populations owing to mutation. Selection and mutation have opposite effects with respect to deleterious alleles. The net effect of the two processes will be nil when the number of deleterious alleles eliminated by selection is the same as that produced by mutation.

Let us first consider the case of recessive alleles. The frequency q of the recessive allele a will decrease, owing to selection, by the amount (Table 4.4)

$$\Delta q = \frac{-spq^2}{1 - sq^2}$$

Because the a allele will be at low frequency, the denominator will be nearly 1, and the approximate change in allele frequency due to selection will be

$$\Delta q \approx -spq^2$$

The a allele will, however, increase in frequency by the amount up, owing to mutation from A to a (Chapter 3, page 70). (We may ignore back mutation from a to A because the frequency of a is low.) There will be equilibrium between mutation and selection when

$$spq^2 \approx up$$

Canceling out p in both sides, we have

$$q \approx \sqrt{\frac{u}{s}}$$

When $s = 1$, this equation reduces to

$$q \approx \sqrt{u}$$

That is, the equilibrium frequency of an allele that causes lethality or sterility in homozygous condition is, approximately, the square root of the mutation rate. If $u = 10^{-5}$, the approximate equilibrium frequency of a recessive lethal allele will be $q = \sqrt{10^{-5}} = 0.003$. On the other hand, if $u = 10^{-5}$ but $s = 0.1$, the equilibrim frequency of a recessive deleterious allele will be $q = \sqrt{10^{-5}/10^{-1}} = \sqrt{10^{-4}} = 0.01$, or about three times higher than that of a lethal allele.

In the case of dominance, the frequency p of the dominant allele A will decrease, owing to selection, by the amount (Table 4.8)

$$\Delta p = \frac{-spq^2}{1 - s + sq^2}$$

But the A allele will increase in frequency by the amount uq, owing to mutation pressure. Making the same approximations as before (i.e., ignoring back mutation as well as the denominator in the expression for Δp), there will be an equilibrium when

$$spq^2 \approx uq$$

Because p is small, q, will be nearly 1. Replacing q by 1, we obtain

$$p \approx \frac{u}{s}$$

And, if $s = 1$ (i.e., the allele is lethal),

$$p \approx u$$

That is, in the case of a lethal dominant allele, the equilibrium frequency of the allele is simply the mutation rate. This we should expect. Individuals carrying a lethal dominant allele fail to reproduce; hence, the only such alleles found in a population will be those that have newly arisen by mutation in that generation.

Assuming the same mutation rates and the same selection coefficients, the equilibrium frequency is much higher for a recessive allele than for a dominant allele. (Note that the square root of a positive number less than one is greater than the number.) We expect this result because recessive alleles are hidden from selection in the heterozygotes.

If $u = 10^{-5}$, the equilibrium frequency of a lethal dominant allele is about 10^{-5}, 300 times smaller than 0.003, the approximate equilibrium frequency obtained for a lethal recessive allele. If $u = 10^{-5}$ and $s = 0.1$, the equilibrium frequency of a deleterious dominant allele is about $10^{-5}/10^{-1} = 10^{-4}$, 100 times smaller than the frequency obtained for a deleterious recessive allele. However, the number of *individuals* exhibiting the deleterious phenotype will be about twice as high for a dominant allele as for a recessive deleterious allele. In the former case, the frequency of individuals with the deleterious phenotype will be $2pq$ (the homozygotes with a frequency of p^2 will be very few if p is negligibly small). Because q will be nearly 1, we have $2pq \approx 2p$, which is approximately equal to $2u/s$. In the case of a recessive allele, the deleterious phenotype is manifested only in the homozygotes, which have a frequency of $q^2 = (\sqrt{u/s})^2 = u/s$.

The allele equilibrium frequencies are, for dominant as well as recessive alleles, directly related to u and inversely related to s. Thus the equilibrium frequencies will be higher when u is greater or when s is smaller.

Achondroplasia is a deleterious condition caused by a dominant allele present at low frequencies in human populations. Because of abnormal growth of the long bones, achondroplastics have short, squat, and often deformed limbs, as well as bulging skulls (Figure 4.3). The mutation rate from the normal allele to the achondroplasia allele is 5×10^{-5}. Achondroplastics reproduce only 20% as efficiently as normal individuals; hence $s = 0.8$. The equilibrium frequency of the allele can therefore be calculated as

$$p = \frac{u}{s} = \frac{5 \times 10^{-5}}{0.8} = 6.25 \times 10^{-5}$$

Because q is nearly 1, the heterozygote frequency, $2pq$, reduces approximately to $2p = 2 \times 6.25 \times 10^{-5} = 1.25 \times 10^{-4}$, or 1.25 individuals per 10,000 births, which is indeed the observed frequency of achondroplastic dwarfs in the population. The expected frequency of homozygotes is $(6.25 \times 10^{-5})^2 = 39 \times 10^{-10}$, or about 4 per billion individuals. Homozygotes have an extreme form of the syndrome, and the few known cases died in the fetal stage.

Chromosomal abnormalities may be treated as dominant mutations. Since people suffering from Down's syndrome do not reproduce, $s = 1$, and therefore $p \approx u/s \approx u$. The frequency of Down's syndrome trisomy is thus simply the frequency with which it arises in human populations owing to chromosomal nondisjunction. However, as with all dominant mutations, the frequency of individuals suffering from Down's syndrome is about twice the mutation rate (the frequency of heterozygotes is $2pq \approx 2p \approx 2u$). The incidence of Down's syndrome is about 1 per 700 births; the "mutation" rate to Down's trisomy is about 1 per 1400 gametes.

Figure 4.3
An achondroplastic dwarf. Detail from *Las Meninas* (the ladies-in-waiting) by the Spanish painter Diego Velázquez. (Prado Museum, Madrid.)

Estimation of Mutation Rates

Mutation rates from recessive to dominant alleles can be estimated directly by counting the number of dominant individuals born to recessive parents. In human populations, for example, the number of achondroplastics born to normal parents amounts to about 1 per 10,000 births. Hence the mutation rate to achondroplasia is 1 per 20,000 gametes, or 5×10^{-5} per gamete per generation.

In the case of recessive alleles, this simple method of estimation is not possible, because mutants in heterozygous condition are not expressed in the phenotype. The equations determining the allele equilibrium frequencies under mutation and selection can be used for estimating the mutation rates to recessive alleles (and, of course, to dominant alleles as well). If the selection coefficients and the equilibrium frequencies are known, the mutation rates can be calculated. For dominant alleles,

$$p = \frac{u}{s} \quad \text{or} \quad u = sp$$

For recessive alleles,

$$q = \sqrt{\frac{u}{s}} \quad \text{or} \quad u = sq^2$$

As pointed out in the previous section, these equations are only approximations. Moreover, selection coefficients may change in populations, and, consequently, the allelic frequencies observed may not always represent equilibrium frequencies. Despite these and other possible pitfalls, the equations for mutation-selection equilibrium are the best method for estimating recessive mutation rates in human populations, in which other methods (requiring, for example, inbreeding) cannot be used.

The frequency at birth of the recessive defect phenylketonuria (PKU) is approximately 4 per 100,000; therefore $q^2 = 4 \times 10^{-5}$. The reproductive efficiency of untreated PKU patients is zero, or $s = 1$. Then

$$u = sq^2 = 4 \times 10^{-5}$$

The frequency of this allele in human populations is

$$q = \sqrt{4 \times 10^{-5}} = 6.3 \times 10^{-3}$$

and the frequency of heterozygotes is

$$2pq \approx 2q = 2 \times 6.3 \times 10^{-3} = 1.26 \times 10^{-2}$$

That is, about 1.3 per 100 humans carry the allele, although only 4 per 100,000 suffer from PKU. The frequency of PKU alleles carried by heterozygotes is one-half of 1.26×10^{-2}, or 6.3×10^{-3}; the frequency of the allele carried by homozygotes is 4×10^{-5}. Therefore $(6.3 \times 10^{-3})/(4 \times 10^{-5}) = 158$ times more PKU alleles are present in PKU heterozygotes than in homozygotes. As pointed out earlier, rare alleles exist in populations mostly in heterozygous combinations.

Heterozygote Advantage

Selection in favor of the heterozygote over both homozygotes is known as *overdominance*, or *heterosis*. The model is

Genotype:	AA	Aa	aa
Fitness (w):	$1 - s$	1	$1 - t$

Table 4.10
Allele frequency changes after one generation of selection when there is overdominance.

	Genotype				
	AA	Aa	aa	Total	Frequency of a
1. Initial zygote frequency	p^2	$2pq$	q^2	1	q
2. Fitness (w)	$1 - s$	1	$1 - t$		
3. Contribution of each genotype to next generation	$p^2(1 - s)$	$2pq$	$q^2(1 - t)$	$1 - sp^2 - tq^2$	
4. Normalized frequency	$\frac{p^2(1 - s)}{1 - sp^2 - tq^2}$	$\frac{2pq}{1 - sp^2 - tq^2}$	$\frac{q^2(1 - t)}{1 - sp^2 - tq^2}$	1	$q_1 = \frac{q - tq^2}{1 - sp^2 - tq^2}$
5. Change in allele frequency					$\Delta q = \frac{pq(sp - tq)}{1 - sp^2 - tq^2}$

The effects of one generation of selection are summarized in Table 4.10.

Selection in favor of the heterozygotes is different from the other modes of selection considered thus far, in a very significant way: overdominance leads to a stable polymorphic equilibrium, with frequencies determined by the coefficients of selection against the two homozygotes. The change in allele frequency due to selection is

$$\Delta q = \frac{pq(sp - tq)}{1 - sp^2 - tq^2}$$

The equilibrium condition, $\Delta q = 0$, will be satisfied only when the numerator is zero. If both alleles are present in a population (p and q are greater than zero), this will occur only when

$$sp = tq$$

$$s(1 - q) = tq$$

$$q = \frac{s}{s + t}$$

Correspondingly, the equilibrium frequency of the A allele is

$$p = \frac{t}{s + t}$$

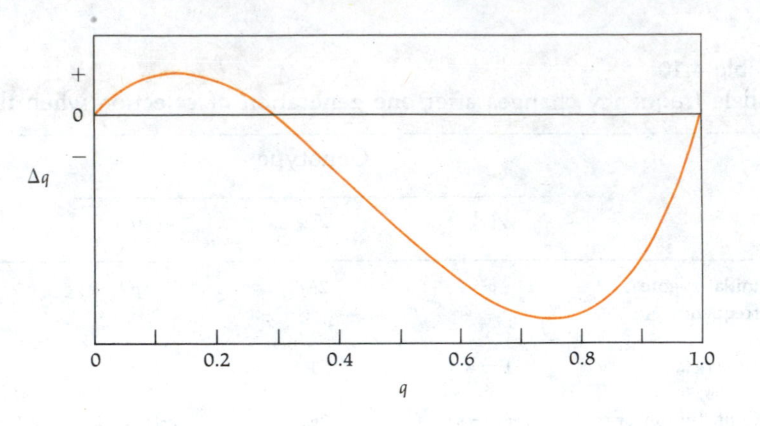

Figure 4.4
The change in allele frequency when selection favors the heterozygotes. The curve shown is the equation $\Delta q = pq(sp - tq) / (1 - sp^2 - tq^2)$, where the equilibrium frequency $q = s/(s + t) = 0.3$. The change in q is positive when $q < 0.3$ and negative when $q > 0.3$.

These two equilibrium frequencies are stable, because selection will change the two allele frequencies until the equilibrium values are reached. If p is greater than its equilibrium value, i.e., if $p > t/(s + t)$, then $sp > tq$, and Δq will be positive. Consequently, q will increase at the expense of p until $sp = qt$. On the other hand, if $p < t/(s + t)$, then $sp < tq$, and Δq will be negative, leading to a decrease in the value of q until the equilibrium frequencies are reached (Figure 4.4).

The equilibrium frequencies in overdominance depend on the relative magnitudes of the two selection coefficients, not on their actual values. Thus, an equilibrium frequency of $q = 0.25$ is obtained, for example, both when $s = 0.1$ and $t = 0.3$ and when $s = 0.02$ and $t = 0.06$. It follows that knowing the equilibrium allele frequencies does *not* allow one to calculate the actual values of s and t, but only their relative magnitudes.

A well-known example of overdominance in human populations is sickle-cell anemia, a disease that is fairly common in some African and Asian populations. The anemia is due to homozygosis for an allele, Hb^S, that produces an abnormal hemoglobin instead of the normal hemoglobin produced by allele Hb^A. Most of the homozygotes Hb^SHb^S die before reaching sexual maturity, so their fitness is only slightly greater than zero. In spite of this, the Hb^S allele has fairly high frequencies in certain regions of the world, precisely in those regions where a certain form of malaria, caused by the parasite *Plasmodium falciparum*, is common (Figure 4.5).

The reason for the high frequency of the Hb^S allele in malarial regions is that heterozygotes Hb^AHb^S are resistant to malarial infections, whereas homozygotes Hb^AHb^A are not. Where *falciparum* malaria is rife, the het-

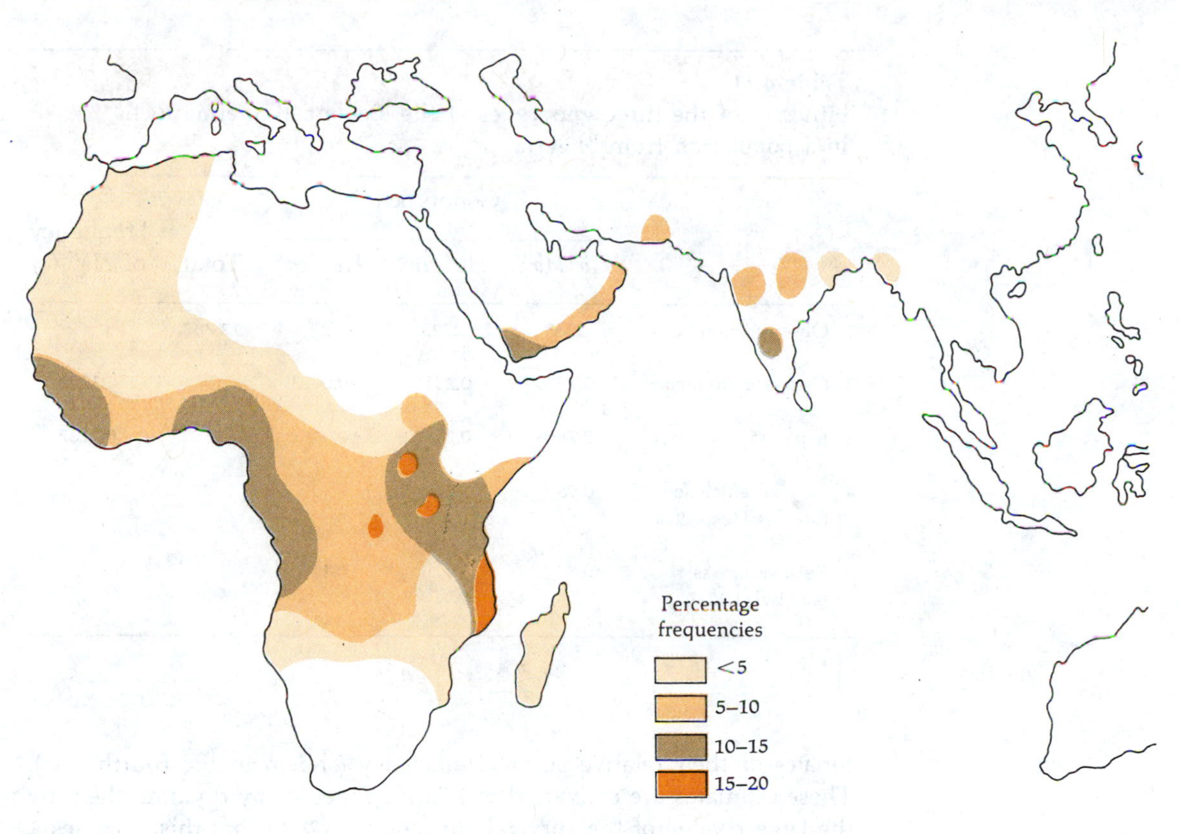

Figure 4.5
Distribution of the allele Hb^S, which in homozygous condition is responsible for sickle-cell anemia. The frequency of Hb^S is high in those regions of the world where *falciparum* malaria is endemic, because Hb^AHb^S individuals, heterozygous for the Hb^S and the "normal" allele, are highly resistant to malarial infection.

erozygotes have selective advantage over both homozygotes, which have a fairly high probability of dying from either anemia (Hb^SHb^S homozygotes) or malaria (Hb^AHb^A homozygotes).

Among 12,387 adult individuals examined in Nigeria, 29 were Hb^SHb^S homozygotes, 2993 were Hb^AHb^S heterozygotes, and 9365 were Hb^AHb^A homozygotes. The fitnesses of the three genotypes are computed in Table 4.11, using a slightly different procedure from that illustrated in Table 4.1. First, the allele frequencies are calculated from the observed genotypic frequencies: the frequency of Hb^S is $q = 0.1232$. If we assume that the population is at equilibrium with respect to this locus, zygotes will be produced in the frequencies p^2, $2pq$, and q^2. If selection had operated completely by the time the adult genotypes were observed, then the ratios of the observed to the expected genotypic frequencies would give es-

Table 4.11
Fitnesses of the three genotypes at the sickle-cell anemia locus in a population from Nigeria.

	Genotype				
	Hb^AHb^A	Hb^AHb^S	Hb^SHb^S	Total	Frequency of Hb^S (q)
1. Observed number	9365	2993	29	12,387	
2. Observed frequency	0.7560	0.2416	0.0023	1	0.1232
3. Expected frequency	0.7688	0.2160	0.0152	1	0.1232
4. Survival efficiency (observed/expected)	0.98	1.12	0.15		
5. Relative fitness (survival/1.12)	0.88	1	0.13		

timates of their relative survival efficiency (shown in the fourth row). These estimates are converted to relative fitnesses by dividing them by the largest value of the survival efficiency, 1.12, so that this becomes 1 (fifth row).

Using the formula developed earlier, we can estimate the allelic equilibrium frequency. The selection coefficients are $s = 1 - 0.88 = 0.12$ against the Hb^AHb^A homozygotes and $t = 1 - 0.13 = 0.87$ against the Hb^SHb^S homozygotes. The expected equilibrium frequency of the Hb^S allele is $0.12/(0.12 + 0.87) = 0.121$ (the frequency of Hb^S calculated from the observed genotypic frequencies is 0.123). The severity of the anemia is manifest in Table 4.11: the homozygotes for the sickle-cell allele survive only 13% as effectively as the Hb^AHb^S heterozygotes. On the other hand, owing to malarial mortality, the homozygotes for the "normal" allele survive only 88% as effectively as the heterozygotes.

Sickle-cell anemia is an additional example of how fitness depends on the environmental conditions. In places where malaria has been eradicated or never existed, the Hb^AHb^A homozygotes have the same fitness as the Hb^AHb^S heterozygotes. The selection mode is no longer overdominance, but selection against recessive homozygotes, which leads to the elimination of the recessive allele. A gradual reduction of Hb^S has occurred in U.S. Blacks, who have a much lower frequency of the sickle-cell allele than their African ancestors (even after taking into account the admixture with Caucasian ancestry, discussed in Chapter 3, page 72). Many more examples of the dependence of fitness on the environmental conditions could be given, such as industrial melanism discussed earlier.

Table 4.12
Allele frequency changes after one generation of selection against heterozygotes.

	Genotype				
	AA	Aa	aa	Total	Frequency of a
1. Initial zygote frequency	p^2	$2pq$	q^2	1	q
2. Fitness (w)	1	$1 - s$	1		
3. Contribution of each genotype to next generation	p^2	$2pq(1 - s)$	q^2	$1 - 2spq$	
4. Normalized frequency	$\frac{p^2}{1 - 2spq}$	$\frac{2pq(1 - s)}{1 - 2spq}$	$\frac{q^2}{1 - 2spq}$	1	$q_1 = \frac{q - spq}{1 - 2spq}$
5. Change in allele frequency					$\Delta q = \frac{spq(q - p)}{1 - 2spq}$

Selection Against Heterozygotes

There are situations in which the heterozygotes have lower fitness than either homozygote. Translocation polymorphisms are examples—the heterozygotes usually have lower fitness owing to reduced fertility. We shall consider the simplest case, when the two types of homozygotes have equal fitnesses. The model is

Genotype:	AA	Aa	aa
Fitness (w):	1	$1 - s$	1

The effects of one generation of selection are shown in Table 4.12.

The change in allele frequencies will be zero when $\Delta q = 0$. This will occur when $p = q$ (but the equilibrium values of p and q will be different when the two types of homozygotes have different fitnesses). However, the equilibrium frequencies are unstable. If $q > p$, then Δq is positive, and q will increase until the A allele is eliminated from the population. If $q < p$, then Δq is negative, and q will decrease further until the a allele is eliminated. Thus a population that is not at equilibrium will depart ever further from the equilibrium frequencies until the allele that initially had a higher-than-equilibrium frequency becomes fixed. The unlikely situation of a population that is initially at equilibrium will not persist; deviations

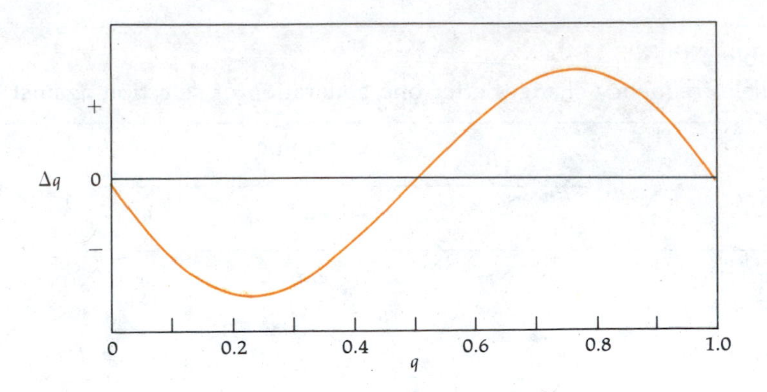

Figure 4.6
The change in allele frequency in the case of selection against the heterozygotes when the two homozygotes are assumed to have equal fitnesses. The equilibrium value ($\Delta q = 0$), when $q = 0.5$, is unstable: if the population is not at equilibrium, the allele frequencies will change until the allele with the lower-than-equilibrium frequency is eliminated from the population.

from equilibrium caused by drift and other factors will tend to eliminate one or another allele (Figure 4.6).

The dynamics of selection against heterozygotes can be used for such practical purposes as pest control. Assume that a population has some undesirable property, e.g., mosquitoes carrying malarial parasites. In the laboratory we might be able to obtain a strain that lacks the undesirable trait, e.g., one in which a mutant allele makes the mosquitoes unsuitable hosts for the parasites. A translocation can be induced and the desirable strain made homozygous for the translocation. Mosquitoes of the desirable strain can then be released in sufficiently large numbers that their translocated genotype will occur in the population with greater frequency than the equilibrium value. The translocation genotype will then run to fixation, carrying with it the desirable allele.

Box 4.3 General Model of Selection at a Single Locus

Various modes of selection (against recessive alleles, against dominant alleles, without dominance, in favor of heterozygotes, and against heterozygotes) have been discussed in the text. They are all special cases of a more general model of selection at a single locus. The general model is

Genotype:	A_1A_1	A_1A_2	A_2A_2
Fitness:	w_1	w_2	w_3

The effects of one generation of selection are shown in Table 4.13. The frequency of A_2 after selection is

$$q_1 = \frac{pqw_2 + q^2w_3}{\overline{w}} = \frac{q(pw_2 + qw_3)}{\overline{w}}$$

The change in the frequency of A_2 is

$$\Delta q = \frac{q(pw_2 + qw_3)}{p^2w_1 + 2pqw_2 + q^2w_3} - q = \frac{pqw_2 + q^2w_3 - p^2qw_1 - 2pq^2w_2 - q^3w_3}{\overline{w}}$$

$$= \frac{pqw_2(1 - 2q) + q^2w_3(1 - q) - p^2qw_1}{\overline{w}}$$

$$= \frac{pqw_2(p - q) + pq^2w_3 - p^2qw_1}{\overline{w}}$$

$$= \frac{pq(w_2p - w_2q + qw_3 - pw_1)}{\overline{w}}$$

$$= pq\frac{p(w_2 - w_1) + q(w_3 - w_2)}{\overline{w}}$$

The allele frequency after selection and the change in allele frequency for the special cases presented in Tables 4.4, 4.8, 4.9, 4.10 and 4.12 can be obtained from the formulas for the general case by substituting the appropriate fitness values (w).

The change in allele frequency due to one generation of selection, Δq, can be analyzed in terms of three component parts—two in the numerator, and the third the denominator. The first component of the numerator is the product pq, which will always be positive (or zero) but will be small when either p or q is small, and larger when p and q have intermediate values. Indeed, the effects of selection are generally greatest when the two alleles have intermediate frequencies.

The second component of the Δq expression is

$$p(w_2 - w_1) + q(w_3 - w_2)$$

which shows that the sign and magnitude of Δq are a function of the differences between the fitness values weighted by the allele frequencies.

The third component is the denominator,

$$p^2w_1 + 2pqw_2 + q^2w_3$$

often called the *mean fitness* of the population and represented by $\overline{w}$. This will always have a positive value; therefore, the sign of Δq is always the same as that of the second component of the expression for Δq. The magnitude of Δq is, of course, inversely related to that of the denominator, and it can be shown that natural selection tends to increase $\overline{w}$. Therefore, as the allele frequencies approach the equilibrium value determined by selection, the magnitude of Δq will tend to decrease, and the approach to equilibrium will be slower.

Table 4.13
General model of selection at a single gene locus.

	Genotype				
	A_1A_1	A_1A_2	A_2A_2	Total	Frequency of A_2
1. Initial zygote frequency	p^2	$2pq$	q^2	1	q
2. Fitness	w_1	w_2	w_3		
3. Contribution of each genotype to next generation	p^2w_1	$2pqw_2$	q^2w_3	$\overline{w} = p^2w_1 + 2pqw_2 + q^2w_3$	
4. Normalized frequency	$\frac{p^2w_1}{\overline{w}}$	$\frac{2pqw_2}{\overline{w}}$	$\frac{q^2w_3}{\overline{w}}$	1	$q_1 = \frac{q(pw_2 + qw_3)}{\overline{w}}$
5. Change in allele frequency					$\Delta q = pq \frac{p(w_2 - w_1) + q(w_3 - w_2)}{\overline{w}}$

Frequency-Dependent Selection

Other forms of selection besides heterozygote advantage may lead to balanced (i.e., stable-equilibrium) genetic polymorphisms. One is *frequency-dependent selection,* which is probably common in nature. Selection is frequency-dependent when the genotypic fitnesses vary with their frequency. In the examples of selection discussed so far, it was assumed that the fitnesses were constant, no matter what the frequencies of the genotypes. This simplifies the mathematical treatment of selection but is often unrealistic. Assume that the fitnesses of two genotypes, *AA* and *aa*, are inversely related to their frequencies—high fitness when a genotype is rare and low fitness when it is common. If a genotype is rare at a given time, natural selection will enhance its frequency; but as its frequency

Table 4.14
Numbers of matings of two strains of *Drosophila pseudoobscura* placed together in variable proportions. Each line summarizes the results of several replicate experiments conducted in observation chambers.

Number of Flies in Chamber*	Males Mated			Females Mated		
	C	T	C:T	C	T	C:T
23C, 2T	77	24	3.2:1	93	8	11.6:1
20C, 5T	70	39	1.8:1	84	25	3.4:1
12C, 12T	55	49	1.1:1	50	54	1:1.1
5C, 20T	39	65	1:1.7	30	74	1:2.5
2C, 23T	30	70	1:2.3	12	88	1:7.3

*C = California; T = Texas.

After C. Petit and L. Ehrman, *Evol. Biol.* 3:177 (1969).

increases, its fitness diminishes, while the fitness of the alternative genotype increases. If there is a frequency at which the two genotypes have equal fitness, a stable polymorphic equilibrium will occur, even without heterosis.

In heterogeneous environments, a genotype may have high fitness when it is rare, because the subenvironments in which it is favored are relatively abundant. But when the genotype is common, its fitness may be low, because its favorable subenvironments are saturated. Frequency-dependent selection has been extensively demonstrated in experimental populations of *Drosophila* and in cultivated plants. For example, in the lima bean, *Phaseolus lunatus*, the fitnesses of three genotypes, *SS*, *Ss*, and *ss*, change over the generations as their frequencies change. The fitness of the heterozygotes is equal to that of the homozygotes when the heterozygotes represent about 17% of the population, but is nearly three times as high when the heterozygotes represent only 2% of the population.

Frequency-dependent *sexual* selection occurs when the probability of mating of a genotype depends on its frequency. Often, the mates preferred are those that happen to be rare, a phenomenon not surprising, perhaps, to people who have experienced the exotic appeal of blondes in Mediterranean countries or of brunettes in Scandinavia. This phenomenon, known as the *rare-mate advantage*, has been thoroughly studied in *Drosophila*, where it commonly involves the males. The results of an experiment are shown in Table 4.14. *Drosophila pseudoobscura* males and females from

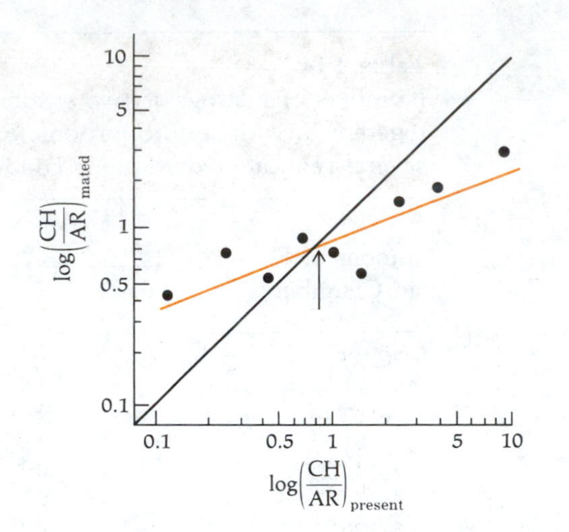

Figure 4.7
Frequency-dependent selection due to mating preferences. Two strains of *Drosophila pseudoobscura* (CH and AR) were combined in different ratios and the numbers of matings of each kind were recorded (in a way similar to that shown in Table 4.14). The graph plots the logarithm of the ratio (CH/AR) of the numbers of males that mated versus the logarithm of the ratio of the numbers of males that were present. Points above the diagonal indicate that males of the CH strain mated disproportionately more often than males of the AR strain; points below the diagonal indicate the opposite. It can be seen that the mating fitness of the CH males decreases as their frequency increases. If male mating differences are the only factor contributing to fitness, there will be a stable equilibrium between the CH and AR strains at the frequency determined by the point at which the diagonal is crossed by the curve representing mating success. [After F. J. Ayala, *Behav. Genet.* 2:85 (1972).]

California (C) and from Texas (T) were placed together in variable proportions. When flies from the C and T localities occur in equal frequencies (12C:12T), they mate with about equal frequencies (55:49 for males, 50:54 for females). But when the two localities are unequally represented, the less common males mate disproportionately more often than the more common males. For example, when the C and T flies exist in the ratio 23:2 (11.5:1), the ratio of matings in the males is 77:24 (3.2:1); that is, each T male is mating nearly four times as often as each C male (11.5/3.2 = 3.6). When the ratio of flies is reversed (2C:23T), the now rare C males mate five times as often as the common T males (11.5/2.3 = 5.0).

Frequency-dependent selection in favor of rare genotypes is a mechanism contributing to the maintenance of genetic polymorphisms, since the fitness of a genotype increases as it becomes rarer (Figure 4.7). Frequency-dependent sexual selection may be particularly important in cases of migration. Immigrants may have a mating advantage because they are rare, thus making it more likely that their genes will become established in the population they have joined.

Problems

1. Using the generalized expression for Δq given in Table 4.13 and the appropriate relative fitness values given in Tables 4.4, 4.8, 4.9, 4.10, and 4.12, derive the Δq expressions given in these tables.

2. In an industrialized region, the fitness of *Biston betularia* moths is 1 for the dark form (*DD* and *Dd*) and 0.47 for the light form (*dd*). The allele frequencies at a certain time are $p = 0.40$ (*D* allele) and $q = 0.60$ (*d* allele). Place the appropriate values in the first and second rows of Table 4.4 and obtain the numerical values for all the expressions given in rows 3, 4, and 5 of that table. Assume now that the allele frequencies are (1) $p = 0.10$, $q = 0.90$ and (2) $p = 0.90$, $q = 0.10$. Calculate the corresponding values of Δq and compare them with each other and with the one obtained in the first part of this problem.

3. For a given gene locus, assume that the heterozygotes have fitness intermediate between the two homozygotes, but not exactly halfway between them, as it is in Table 4.9. That is, the fitness of the heterozygotes is $1 - hs$, where h is some positive number between 0 and 1. Derive the expression for Δq, the change in allele frequency after one generation of selection.

4. Retinoblastoma is a disease, due to a dominant allele, that leads to early death if left untreated. Assume that the mutation rate from the normal allele to the retinoblastoma allele is 10^{-5}. What is the equilibrium frequency of the allele in a population where the condition is not treated?

5. Assume that the mutation rate to a lethal recessive allele, such as that for Tay-Sachs disease, is 10^{-5}. What is the equilibrium frequency of the allele? Compare the answers to this problem and the previous one.

6. A certain allele in homozygous condition causes sterility in both male and female rats, but has no detectable effect in heterozygotes. The frequency of homozygotes in a wild population is 1 per 1000. Assuming Hardy-Weinberg equilibrium, what is the frequency of the heterozygotes? If the mutation rate is doubled, what will be the *equilibrium* frequency of sterile individuals and of the heterozygotes?

7. The equilibrium frequency of a lethal recessive allele in a random-mating population of mice is 0.333. What are the fitnesses of the three genotypes?

8. The marine copepod *Tisbe reticulata* can be reared in seawater cultures in the laboratory, although larval mortality is considerable, particularly at

high population densities. Heterozygotes V^VV^M for two codominant alleles responsible for color differences were bred at low and high population densities. The numbers of adult F_1 progeny were as follows:

Density	V^VV^V	V^VV^M	V^MV^M	Total
Low	904	2023	912	3839
High	353	1069	329	1751

What are the relative fitnesses (viabilities) of the three genotypes at the two densities?

9. At a locus (*Est-6*) coding for an esterase, there are two alleles, *Est-6*^{F} and *Est-6*^{S}, in an experimental population of *Drosophila.* First-instar larvae of the three genotypes are placed in cultures, and the numbers of emerging adults are recorded. The results of two experiments are as follows:

	Number of larvae			Number of adults		
Experiment	*FF*	*FS*	*SS*	*FF*	*FS*	*SS*
1	160	480	360	80	240	90
2	360	480	160	90	240	80

Assuming that relative fitness depends only on larval viability, what are the fitnesses of the three genotypes in each experiment? Do you think that a stable polymorphic equilibrium might occur in the population and, if so, at what frequency?

10. In a region where industrial pollution has been under control for a number of years, the fitness of *Biston betularia* moths is 0.47 for the dark form and 1 for the light form. Calculate the change in allele frequency, Δp, after one generation of selection when (1) $p = 0.40$, (2) $p = 0.10$, and (3) $p = 0.90$.

11. Individuals carrying the sickle-cell allele, Hb^S, can be identified through appropriate tests, because their red blood cells take on a characteristic sickle-like shape when exposed to low oxygen tension. This happens in heterozygotes, $Hb^A Hb^S$, as well as in homozygotes, $Hb^S Hb^S$, although to a lesser extent in the former. Is the Hb^S allele dominant with respect to sickling? In regions free of *falciparum* malaria, the fitness of the heterozygotes is similar to that of normal homozygotes, $Hb^A Hb^A$, while homozygotes for the Hb^S allele have very low fitness. Is the Hb^S allele dominant with respect to fitness in malaria-free regions? What about those regions of the world infested with *falciparum* malaria?

5

Inbreeding, Coadaptation, and Geographic Differentiation

The Inbreeding Coefficient

The Hardy-Weinberg law applies only when mating is random, i.e., when the probability of mating between two genotypes is the product of their frequencies. Random mating was generally assumed throughout the two previous chapters. *Assortative mating* (see Chapter 3, page 60) prevails when matings do not occur at random: individuals with certain genotypes are more likely to mate with individuals of certain other genotypes than would be expected from their frequencies. Assortative mating does not, by itself, change gene frequencies, but it does change *genotypic* frequencies. If the probability of mating between like genotypes is greater than would be expected from randomness, the frequency of homozygotes will increase; if the probability is smaller, the frequency of homozygotes will decrease. In general, if the *mating system* is known (i.e., if we know the probabilities of the various types of matings), then the expected genotypic frequencies can be calculated from the genotypic frequencies in the previous generation.

A particularly interesting form of assortative mating is *inbreeding*—when matings between relatives are more frequent than would be expected from randomness. Because relatives are genetically more similar than unrelated individuals, inbreeding increases the frequency of homozygotes and decreases the frequency of heterozygotes relative to the expectations from random mating, although it does not change the allele frequencies. The most extreme kind of inbreeding is *self-fertilization*, or *selfing*, a common form of reproduction in some plant groups. Inbreeding is

Table 5.1
Results of selfing in a population started entirely with heterozygotes, *Aa*.

Generation	Genotypic Frequency			F	Frequency of a
	AA	Aa	aa		
0	0	1	0	0	0.5
1	$\frac{1}{4}$	$\frac{1}{2}$	$\frac{1}{4}$	$\frac{1}{2}$	0.5
2	$\frac{3}{8}$	$\frac{1}{4}$	$\frac{3}{8}$	$\frac{3}{4}$	0.5
3	$\frac{7}{16}$	$\frac{1}{8}$	$\frac{7}{16}$	$\frac{7}{8}$	0.5
⋮	⋮	⋮	⋮	⋮	⋮
n	$\frac{1-(\frac{1}{2})^n}{2}$	$(\frac{1}{2})^n$	$\frac{1-(\frac{1}{2})^n}{2}$	$1-(\frac{1}{2})^n$	0.5
∞	$\frac{1}{2}$	0	$\frac{1}{2}$	1	0.5

often practiced in horticulture and animal husbandry. In human populations (as well as in those of other organisms), inbreeding increases the frequency of recessive hereditary infirmities.

The genetic consequences of inbreeding are measured by the *coefficient of inbreeding*, which is the probability that an individual receives, at a given locus, two alleles that are *identical by descent*, i.e., that are both copied from one single allele carried by an ancestor, belonging to a generation that must be specified. Two alleles with the same DNA sequence are identical *in structure* (or *in state*), but they will not be identical by descent if they have been inherited from unrelated ancestors. The coefficient of inbreeding is usually represented as F.

The results of inbreeding in the case of self-fertilization were worked out by Mendel, who calculated that, after n generations of selfing, the progeny of a heterozygote, *Aa*, consists of homozygotes and heterozygotes in the ratio $2^n - 1$ to 1 (Table 5.1). The progeny of a selfed heterozygote (*Aa*) consist of one-half heterozygotes (*Aa*) and one-half homozygotes (either *AA* or *aa*). The two alleles in each heterozygote are obviously not identical by descent. But in the homozygotes, both alleles are identical by descent because both are copies of the only allele of that type (either *A* or *a*) present in the selfed heterozygous parent. Thus, the proportion of individuals carrying two alleles identical by descent in the first generation of selfing is the same as the total frequency (one-half) of homozygotes, or $F = \frac{1}{2}$.

A selfed homozygote produces only homozygous offspring. These homozygotes all have two alleles that are identical by descent. Therefore the one-half inbreeding acquired in the first generation will remain thereafter and will accumulate with any additional inbreeding acquired in the following generations. In the second generation of selfing, one-half the progeny of the heterozygotes will again consist of homozygotes each having two alleles that are identical by descent. Thus the inbreeding coefficient in the progeny of the heterozygotes is again ½, and, since the heterozygotes represent one-half the population, the increment in the inbreeding coefficient is $\frac{1}{2} \times \frac{1}{2} = \frac{1}{4}$; this, added to the pre-existing ½, becomes $F = \frac{3}{4}$. In each of the following generations, the value of F will increase by one-half multiplied by the frequency of the heterozygotes in the previous generation.

Measuring Inbreeding

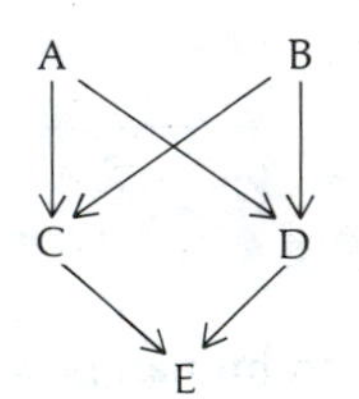

Figure 5.1
Pedigree of an offspring of a brother × sister mating.

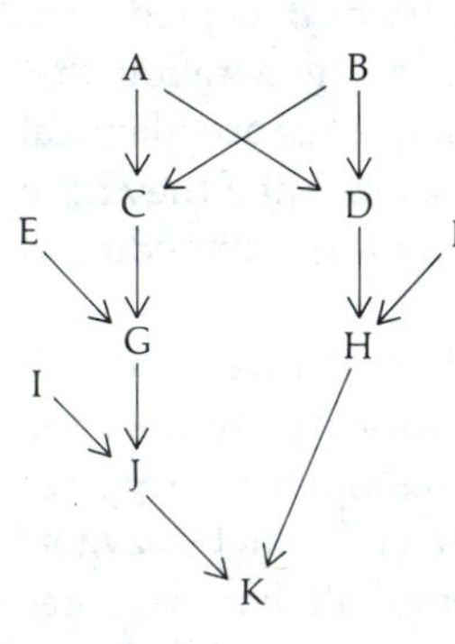

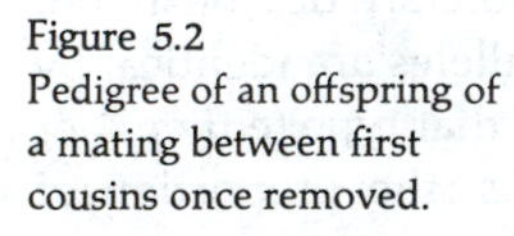

Figure 5.2
Pedigree of an offspring of a mating between first cousins once removed.

Let us calculate the F value in the offspring of full-sibs, i.e., of individuals having the same father and the same mother. Figure 5.1 represents the pedigree of a brother × sister mating; each arrow represents the transmission of one gamete. A and B are two unrelated parents, each of which contributes one gamete to C and one to D; E receives one gamete each from C and D, which are the full-sibs. Because A and B are unrelated, we assume that their alleles at a certain locus are not identical by descent. The two alleles in A can be represented as a_1a_2, and in B as a_3a_4. (Here, different subscripts indicate that the alleles are not identical by descent; however, this does not necessarily imply that the alleles are not identical in structure.) The probabilities of the four types of progeny from the mating A × B are $\frac{1}{4}(a_1a_3)$, $\frac{1}{4}(a_1a_4)$, $\frac{1}{4}(a_2a_3)$, and $\frac{1}{4}(a_2a_4)$. We are interested in the probability that offspring of the sibs are homozygous for any one allele, i.e., the probability that an offspring is homozygous a_1a_1 or a_2a_2 or a_3a_3 or a_4a_4. The answer is ¼.

Parent A produces two kinds of alleles, a_1 and a_2, each with probability ½. Therefore, the probability that C receives allele a_1 from A is ½, and the probability that C passes allele a_1 (if it carries it) to E is also ½. The probability that allele a_1 has passed from A to C to E is thus $\frac{1}{2} \times \frac{1}{2} = \frac{1}{4}$. The probability that A passes allele a_1 to D and that D passes it to E is also $\frac{1}{2} \times \frac{1}{2} = \frac{1}{4}$. Hence E has a probability of ¼ of receiving allele a_1 from C and a probability of ¼ of receiving a_1 from D; the probability that E receives a_1 from both C *and* D is $\frac{1}{4} \times \frac{1}{4} = \frac{1}{16}$.

We can repeat the reasoning of the previous paragraph for each of the other alleles. The probability that E is homozygous a_2a_2 is also $\frac{1}{16}$, and

Table 5.2
Coefficient of inbreeding, F, in the offspring of matings between various kinds of relatives.

Type of Mating	F
Selfing	$1/2$
Full-sibs	$1/4$
Uncle × niece, aunt × nephew, or double first cousins	$1/8$
First cousins	$1/16$
First cousins once removed	$1/32$
Second cousins	$1/64$
Second cousins once removed	$1/128$
Third cousins	$1/256$

the same is true for a_3a_3 and a_4a_4. The probability that E is homozygous for *any one* of the four alleles present in its two grandparents is therefore $1/16 + 1/16 + 1/16 + 1/16 = 1/4$.

There is a simple method, called *path analysis,* based on the kind of reasoning just used for full-sib progeny, that allows one to compute the coefficient of inbreeding for an individual with known pedigree. The method consists of tracing the arrows in a pedigree from one individual back to itself through *each* ancestor common to *both* parents. Figure 5.2 shows the pedigree of an individual, K, whose parents are first cousins once removed. A and B are the two ancestors common to both parents, H and J. Hence, there are two paths: one is K–J–G–C–A–D–H–K, with seven steps; the other is K–J–G–C–B–D–H–K, also with seven steps. Because K appears twice in each path, the number of steps is reduced by one in each path. The contribution of each path to the inbreeding coefficient is $(1/2)^n$, where n is the number of steps in the path minus one (or just the number of steps, if the individual under consideration appears only once in each path). The value of F is obtained by adding the contributions of the various paths. In the pedigree in Figure 5.2, the contribution of each path is $(1/2)^6 = 1/64$, and the sum of the two paths is $F = 1/64 + 1/64 = 1/32$.

The coefficient of inbreeding in the progeny of matings between various kinds of relatives is given in Table 5.2. Systematic inbreeding is sometimes practiced in plant or animal breeding to obtain a certain degree of homozygosity. If the same type of inbred mating is practiced every generation, the coefficient of inbreeding will increase every generation (Figure 5.3).

The population consequence of inbreeding is to increase the frequency of homozygotes at the expense of heterozygotes. In a randomly mating population with two alleles having frequencies p and q, the frequency of heterozygotes is $2pq$. In a population with a coefficient of inbreeding F, the frequency of heterozygotes will be reduced by a fraction F of their total. The genotypic frequencies in an inbred population are

Genotype:	AA	Aa	aa
Frequency:	$p^2 + pqF$	$2pq - 2pqF$	$q^2 + pqF$

When there is no inbreeding, $F = 0$, and the genotypic frequencies reduce to the familiar Hardy-Weinberg equilibrium values.

The coefficient of inbreeding, F, measures the increase in the frequency of homozygous *individuals* at a locus; it also measures the increase in the proportion of homozygous *loci* per individual.

Inbreeding Depression and Heterosis

Plant and animal breeders try to improve their stocks with respect to certain traits (grain yield, egg production, etc.) by using the "best" individuals in each generation as parents of the next generation (artificial selection). Breeders also want homogeneity; they try to achieve this by systematic inbreeding, which increases homozygosity. However, breeders have long known that inbreeding usually leads to a reduction in fitness, owing to deterioration in important attributes, such as fertility, vigor, and resistance to disease. This phenomenon is known as *inbreeding depression*.

Inbreeding depression results from homozygosis for deleterious recessive alleles. Consider a recessive lethal allele, with mutation rate $u = 10^{-5}$. The equilibrium frequency of the allele is $q = \sqrt{u} = 0.0032$. In a random-mating population, the frequency of homozygotes is $q^2 = 10^{-5}$. Assume now that the coefficient of inbreeding in a certain stock is $F = 1/16$, similar to what is achieved in a single generation by matings between first cousins. Then the frequency of homozygotes for the allele will be

$$q^2 + pqF = 10^{-5} + (0.9968 \times 0.0032 \times 0.0625)$$

$$\approx 10^{-5} + (2 \times 10^{-4}) \approx 2 \times 10^{-4}$$

The frequency of homozygotes is thus about 20 times greater than in a random-mating population. A comparable increase in the frequency of

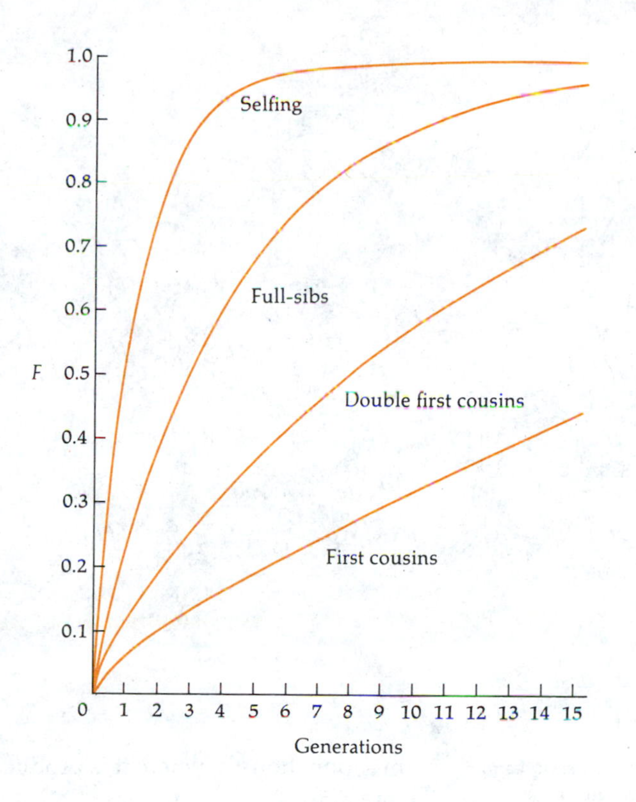

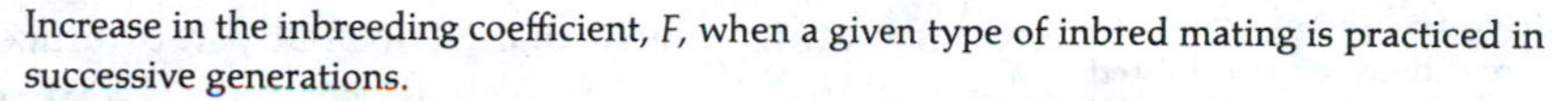

Figure 5.3
Increase in the inbreeding coefficient, F, when a given type of inbred mating is practiced in successive generations.

homozygotes will also occur with respect to other recessive deleterious alleles in the population.

Note that the increase in the proportion of homozygotes at any one locus is directly proportional to the value of F, since it is determined by pqF; thus if $F = \frac{1}{4}$, the frequency of homozygotes in the previous example will be $10^{-5} + (8 \times 10^{-4})$, or about 80 times greater than in a random-mating population.

The effects of inbreeding depression can be counteracted by crossing independent inbred lines. The hybrids usually show a marked increase in fitness—size, fertility, vigor, etc. (Figure 5.4). This is called *hybrid vigor*, or *heterosis*. Independent inbred lines are likely to become homozygous for different deleterious recessive alleles. Intercrossing two inbred lines may retain homogeneity for the artificially selected traits while making the deleterious alleles heterozygous.

Hybrid vigor as a technique for crop improvement was first exploited in corn, with great success. The increase in productivity obtained with hybrid corn is indeed very large. The practice has been extended to other plants and to animals. It requires, however, that hybrid seed be obtained

Figure 5.4
Inbreeding depression and hybrid vigor in corn. Heterosis arises in the F_1 of a cross between two inbred lines (P_1 and P_2). Selfing leads to increased inbreeding depression in the following generations (F_2 to F_8). [After D. F. Jones, *Genetics* 9:405 (1924).]

from supply houses, where it is produced using appropriate inbred stocks (Figure 5.5).

In nature, many plants normally reproduce by selfing. Inbreeding depression does not occur in these plants, because natural selection keeps deleterious recessive alleles at much lower frequencies than in random-mating populations. In normally self-fertilized organisms, homozygosis is very high; deleterious recessive alleles are eliminated by natural selection as they become homozygous. However, inbreeding depression follows the inbreeding of normally outbred animals and plants, because deleterious recessive alleles, mostly present in heterozygotes, become homozygous.

Inbreeding in Human Populations

Mating between parents and their children or between brothers and sisters is known as *incest.* There is an "incest taboo" in most human cultures, although many Egyptian pharaohs had incestuous marriages. Matings between close relatives, such as first cousins, are often forbidden by law or by religion. In the United States, for example, about half the states have laws prohibiting uncle-niece, aunt-nephew, and first-cousin marriages; in the other states, such marriages are legal.

Most religions and countries forbid marriages between close relatives, but exceptions are sometimes granted by the authorities. In the Roman

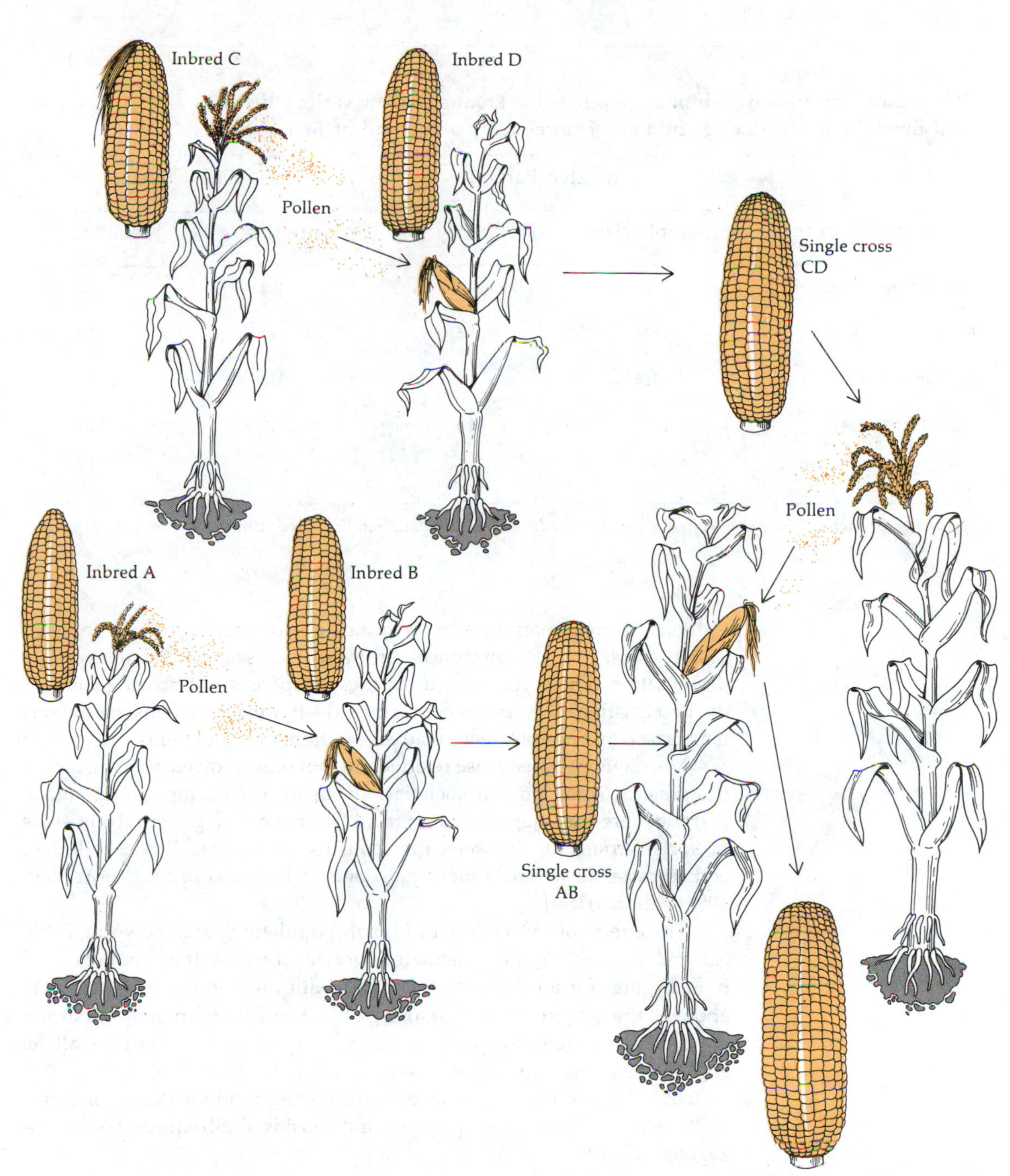

Figure 5.5
Production of hybrid corn from four inbred lines, A, B, C, and D. Paired crosses between the original inbred lines produce two vigorous hybrid plants, AB and CD, which are then intercrossed to yield the double-cross hybrid ABCD.

Table 5.3
Inbreeding depression in human populations. Frequencies of various diseases and of physical and mental defects among children of unrelated parents and of first cousins.

Population	Unrelated Parents		First Cousins	
	Sample size	Frequency (%)	Sample size	Frequency (%)
United States (1920–1956)	163	9.8	192	16.2
France (1919–1925)	833	3.5	144	12.8
Sweden (1947)	165	4	218	16
Japan (1948–1954)	3570	8.5	1817	11.7
Average:		6.5		14.2

After C. Stern, *Principles of Human Genetics*, 3rd ed., W. H. Freeman, San Francisco, 1973.

Catholic Church, marriages between uncle and niece, aunt and nephew, first cousins, first cousins once removed, and second cousins require dispensation. The parish records of these dispensations provide some of the best existing information concerning the frequency of consanguineous marriages—matings between relatives—in human populations.

Marriages between close relatives are not only allowed but considered desirable in some human societies. In Japan, for example, first-cousin marriages are encouraged, and up to 10% of the marriages in certain areas or social groups are between first cousins. In Andhra Pradesh (India), certain castes favor uncle-niece marriages, which account for more than 10% of all marriages.

The effects of inbreeding in human populations are shown in Table 5.3 and Figure 5.6. We calculated previously that the frequency of homozygotes for a *lethal* recessive allele with mutation rate $u = 10^{-5}$ is about 20 times greater in the offspring of first cousins than in the offspring of random-mating individuals. In the case of a *deleterious* recessive allele, with $s = 0.1$, the equilibrium frequency is $q = \sqrt{10^{-5}/10^{-1}} = \sqrt{10^{-4}} = 0.01$. The frequency of homozygotes among random-mating individuals is $q^2 = 10^{-4}$. In the offspring of first cousins, the frequency of homozygotes will be

$$q^2 + pqF = 10^{-4} + (0.99 \times 0.01 \times 0.0625)$$

$$\approx 10^{-4} + (6 \times 10^{-4}) = 7 \times 10^{-4}$$

or about seven times greater than in the offspring of random-mating individuals. If $s = 0.01$, the incidence of homozygotes is about three times

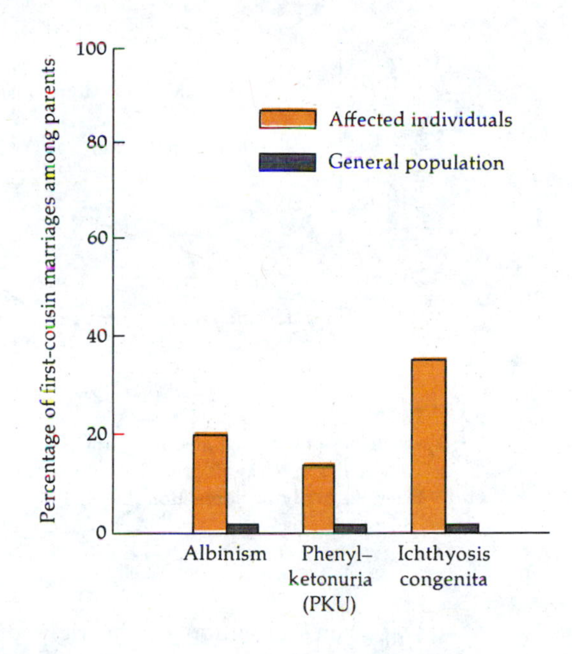

Figure 5.6
Consanguinity among parents of recessive homozygotes. The bars represent the frequencies, in European populations, of first-cousin marriages among the parents of affected individuals and in the general population. Ichthyosis congenita is a severe disease of the skin. (After W. F. Bodmer and L. L. Cavalli-Sforza, *Genetics, Evolution, and Man*, W. H. Freeman, San Francisco, 1976.)

greater in the offspring of first cousins than in the offspring of random-mating individuals.

Table 5.3 shows that, on the average, the incidence of defective newborn children is about twice as high when the parents are first cousins as when the parents are unrelated. This increase is considerably less than might be expected from the previous calculations. The calculations, however, apply to *recessive* infirmities. With respect to dominant alleles, the incidence of hereditary conditions is no greater, on the average, in consanguineous marriages than in unrelated marriages. Moreover, the data in Table 5.3 include nonhereditary defects. Figure 5.6 shows that deleterious recessive conditions increase by a large factor in the offspring of first cousins.

Cultural attitudes often have genetic consequences. Figure 5.7 shows the frequency of consanguineous marriages in some European populations. Until the year 1700, the Roman Catholic Church rarely granted dispensation from the prohibition against consanguineous marriages. Such marriages in Catholic Europe increased during the eighteenth century until the first half of the nineteenth century, when they started to decrease. The high frequency of consanguineous marriages during the first half of the nineteenth century seems to have been due in part to

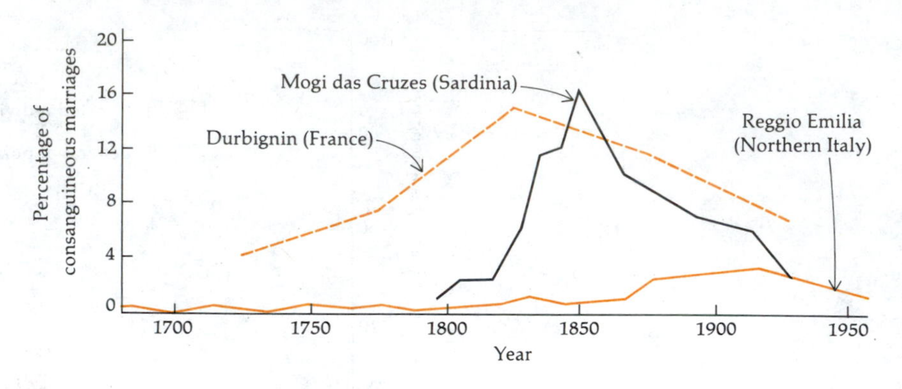

Figure 5.7
Frequency of consanguineous marriages in three European populations. (After A. Moroni, *Historical Demography, Human Ecology, and Consanguinity,* International Union for the Scientific Study of Population, Liège, 1969.)

Napoleon's abolition of the right of primogeniture, which caused the splitting of land property. This could be counteracted to some extent by marriages between close relatives. The Industrial Revolution, which greatly increased geographic mobility, may have been partially responsible for the decrease in consanguineous marriages observed since the nineteenth century. Whatever the reasons may have been for these changes, they had important genetic and health consequences because of their effects on the incidence of deleterious recessive infirmities.

Genetic Coadaptation

The mechanisms of evolutionary change—mutation, migration, drift, and selection—as well as inbreeding, have so far been considered primarily as they affect individual gene loci. However, genes survive and reproduce in whole organisms; a functional allele may fail to be passed on to the following generation if it occurs in an organism that fails to reproduce. Genes interact with the environment but also with other genes. At any one locus, natural selection favors alleles that interact well with the alleles present at other loci. The term *genetic coadaptation* refers to the adaptive interaction between the genes that make up a gene pool.

Imagine a zygote made up of human genes, whale genes, and corn genes in equal numbers: such a chimera could not develop into a functional organism. Most living species cannot be intercrossed with each other; interspecific fertilization is sometimes possible between closely related species, but more often than not the hybrid zygotes fail to develop, or they develop into sterile organisms, such as the mule. The inviability or

Table 5.4
Allelic frequencies at the *Mdh-2* locus in two *Drosophila* species in Tame, Colombia.

Species	Frequency of Allele: *86*	*94*	Other*
D. equinoxialis	0.005	0.992	0.003
D. tropicalis	0.995	0.004	0.001

*Refers to several alleles with very low frequencies in both species.

sterility of interspecific hybrids is conspicuous evidence of genetic coadaptation. The horse and donkey genotypes are not mutually coadapted.

Whenever a new genic or chromosomal mutation arises that does not interact well with the rest of the genome, it is eliminated or kept at low frequency by natural selection, even though it might be functional in a different genetic background. The role of coadaptation between alleles at different loci can be illustrated by an analogy. A successful performance by a symphony orchestra requires not only that each player know *how* to play his instrument (a gene must be *able* to function), but also that he master his *part* in the piece being performed (a gene must *interact* well with the other genes). A violinist playing his part for Beethoven's Sixth Symphony while the rest of the orchestra was playing Ravel's *Bolero* would be cacophonic.

Owing to genetic coadaptation, a certain allele or set of alleles may be favorably selected in one species but be unfavorably selected in a different species. *Drosophila equinoxialis* has, at the *Mdh-2* locus (which codes for the cytoplasmic enzyme malate dehydrogenase), allele *94* with a frequency about 0.99, while *D. tropicalis* has allele *86* with a frequency about 0.99. Table 5.4 gives the allelic frequencies in both species in a certain South American locality. These allelic frequencies are relatively constant throughout the distributions of both species, which are abundant in the rain forests of Central and South America.

The polypeptides coded by alleles *86* and *94* differ by at least one amino acid, but they are functionally very similar, and flies of both species can function with either one of the two forms. It might seem that the two alleles are equivalent, their different frequencies being due to random drift. However, the experiment shown in Figure 5.8 indicates that genetic coadaptation is responsible for the different allelic frequencies in the two species. Laboratory populations of each species were set up, with the

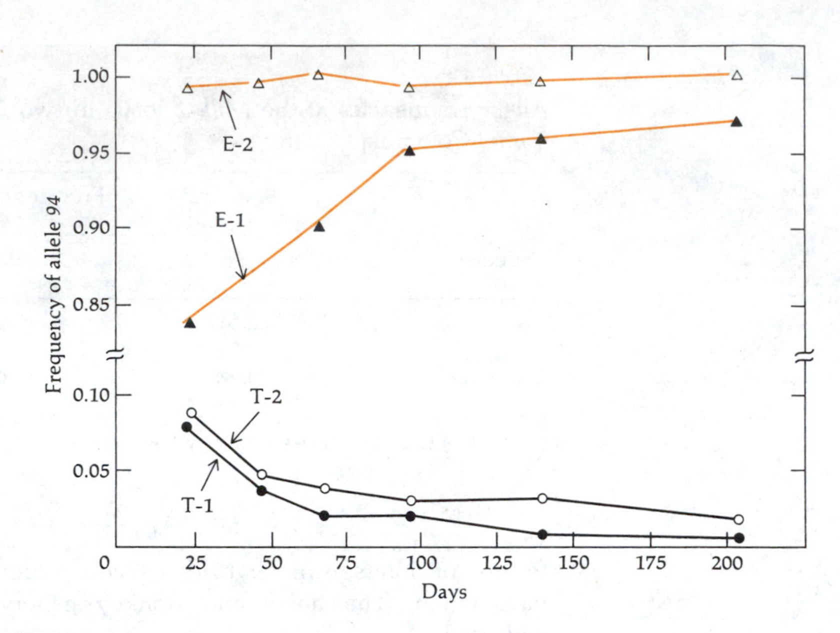

Figure 5.8
Natural selection in laboratory populations of *Drosophila equinoxialis* (E-1 and E-2) and *D. tropicalis* (T-1 and T-2). Two alleles, *86* and *94*, of the *Mdh-2* locus are present in all four populations. Natural selection tends to restore the frequencies occurring in nature: allele *94* increases in frequency in population E-1, where its initial frequency was lower than that in nature. In populations T-1 and T-2, however, where its initial frequency was greater than that in nature, allele *94* decreases in frequency.

frequency of the rare allele in a given species being artificially increased. Natural selection acts to restore the natural frequencies, although the experimental environment is the same for all populations. In *D. tropicalis*, allele *94*, which is rare in nature, decreases in frequency, while the same allele is favored in *D. equinoxialis*, a species in which allele *94* is common in nature. Alleles coding for polypeptides with a given electric charge are favored in one species but selected against in the other. The different directions of the process of natural selection in the two species must be due to the different genetic backgrounds in which the alleles are present (i.e., the different distributions of alleles present at the other loci), since the experimental environment is the same for all four laboratory populations.

Genetic coadaptation is a property of the gene pool of a species as a whole, but also of local populations. Alleles may be favored in one locality but not in another, because they interact well with other alleles in the first population but not in the second. An illustrative example is provided by the African swallowtail butterfly, *Papilio dardanus* (Figure 5.9). This species has several female phenotypes that mimic different butterfly species noxious to bird predators. The males, however, have a constant,

Figure 5.9
Mimicry and coadaptation in *Papilio dardanus* butterflies. (1) *P. dardanus* male. (2) *P. dardanus* female, nonmimetic. (3) and (4) *Amauris albimaculata* and its mimetic, *P. dardanus*, the "cenea" phenotype. (5) and (6) *Amauris niavius dominicanus* and its mimetic, *P. dardanus*, the "hippocoonides" phenotype. (7) and (8) F_1 offspring of a cross between cenea and hippocoonides from different African regions. When the butterflies come from the same region, the allele responsible for the cenea phenotype is dominant over that for hippocoonides, so that F_1 and F_2 progenies yield only one or the other phenotype. However, when cenea and hippocoonides from different regions are crossed, the progenies include intermediate, nonmimetic phenotypes, as exemplified by (7) and (8). [From C. A. Clarke and P. M. Sheppard, *J. Genet.* 56:236 (1959); *Heredity* 14:73, 163 (1960).]

1

2

3

4

5

6

7

8

nonmimetic phenotype. Birds find *P. dardanus* butterflies palatable, but avoid mimetic forms that they confuse with the noxious species. Several mimetic forms exist in some localities, while only one is found in other localities, depending on which noxious species happen to be present. Crosses can be made between two mimetic strains, which we may call A and B. What is interesting is that crosses between A and B give different results depending on whether or not the two strains come from the same locality. If both parents come from the same locality, only perfect female mimics are produced in the F_1, F_2, and backcross generations. When the two mimicking strains come from different regions, however, the F_1 female progenies are intermediate in appearance between the two female phenotypes of the parental strains, and the F_2 and backcross progenies also show intermediate phenotypes.

The mimetic patterns are determined for the most part by two major gene loci. At one locus there are two alleles, one determining the presence, the other the absence, of the "tails" that are typical of swallowtail butterflies. The other locus consists of several alleles, each determining the main color pattern of one mimetic form. There are, moreover, a number of modifier-gene loci that affect the expression of the major genes. At these modifier loci, alleles have been selected that maximize the mimetic characteristics of the butterflies. This, however, is accomplished by different sets of alleles in different local populations. Because in nature *P. dardanus* butterflies from different regions do not intercross, natural selection has not coadapted the sets of modifier alleles from separate regions. When mimetic forms from different regions are intercrossed, alleles that are not mutually coadapted are joined together, and imperfectly mimicking forms arise.

Linkage Disequilibrium

Genetic coadaptation may exist between some alleles but not others within the same population. In the case of polymorphic loci, certain alleles at one locus may be coadapted with some alleles at different loci, but not with others.

Assume that there are two loci, A and B, and that at each locus there are two alleles, A_1 A_2, and B_1 B_2. Assume further that alleles A_1 and B_1 interact well with each other so that they produce well-adapted phenotypes, and that the same is true for A_2 and B_2, but that the combinations A_1B_2 and A_2B_1 yield poorly adapted phenotypes. The adaptation of the population would be increased if the alleles would always (or most often) be transmitted in the combinations A_1B_1 and A_2B_2, and never (or rarely) in the combinations A_1B_2 and A_2B_1.

When alleles at different loci are not associated at random, they are said to be in *linkage disequilibrium*. When alleles at different loci *are* associated at random (i.e., in proportion to their frequencies), the loci are in linkage equilibrium.

Assume that the allelic frequencies at two loci are

First locus: $A_1 = p$ $A_2 = q$

Second locus: $B_1 = r$ $B_2 = s$

so that $p + q = 1$ and $r + s = 1$. If the alleles at the two loci are associated at random, we expect the four possible gametic classes to have frequencies that are the product of the frequencies of the alleles involved, that is,

$$A_1B_1 = pr$$

$$A_2B_2 = qs$$

$$A_1B_2 = ps$$

$$A_2B_1 = qr$$

Because these are the only possible kinds of gametes, the sum of their frequencies must be 1. Indeed,

$$pr + qs + ps + qr = p(r + s) + q(r + s) = p + q = 1$$

If the alleles are associated at random, the product of the frequencies of the two *coupling* gametes ($pr \times qs = pqrs$) is the same as the product of the frequencies of the two *repulsion* gametes ($ps \times qr = pqrs$). However, if the alleles are *not* randomly associated, the two products will be different; the extent of linkage disequilibrium, d, is measured by the difference between the two products:

$$d = (\text{freq. of } A_1B_1)(\text{freq. of } A_2B_2) - (\text{freq. of } A_1B_2)(\text{freq. of } A_2B_1)$$

The condition for linkage equilibrium is, therefore, $d = 0$.

Linkage disequilibrium is complete only when two gametic combinations exist—either only the two coupling gametes (A_1B_1 and A_2B_2) or only the two repulsion gametes (A_1B_2 and A_2B_1). The maximum absolute value that d can have is 0.25, namely, when linkage disequilibrium is complete and the allelic frequencies are 0.5 at both loci (Table 5.5). Note that, if the allelic frequencies at the two loci are different, complete linkage is not possible. For example, if the frequency of A_1 is 0.5 but the frequency of B_1 is 0.6, then not all B_1 alleles can be associated with either A_1 or A_2, but some must be associated with each of these two alleles.

According to the Hardy-Weinberg law, the equilibrium genetic frequencies at any one autosomal locus are reached in one single generation of random mating (or in two, if the allelic frequencies are different in the two sexes). This is not so when two loci are considered simultaneously (Box 5.1). However, linkage disequilibrium decreases with every generation of random mating—unless there is some process opposing the

Table 5.5
Maximum possible values of linkage disequilibrium, d, for three different cases. The frequency designated for each case is that of the most common allele and is the same at both loci.

Case	Frequency of Gametic Combination:				
	A_1B_1	A_2B_2	A_1B_2	A_2B_1	d
1. Frequency 0.5					
Coupling	0.5	0.5	0	0	$(0.5)(0.5) - (0)(0) = 0.25$
Repulsion	0	0	0.5	0.5	$(0)(0) - (0.5)(0.5) = -0.25$
2. Frequency 0.6					
Coupling	0.6	0.4	0	0	$(0.6)(0.4) - (0)(0) = 0.24$
Repulsion	0	0	0.6	0.4	$(0)(0) - (0.6)(0.4) = -0.24$
3. Frequency 0.9					
Coupling	0.9	0.1	0	0	$(0.9)(0.1) - (0)(0) = 0.09$
Repulsion	0	0	0.9	0.1	$(0)(0) - (0.9)(0.1) = -0.09$

approach to linkage equilibrium. Permanent linkage disequilibrium may result from natural selection if some gametic combinations result in higher fitness than other combinations. Assume, for example, that the two coupling combinations produce, in either homozygous or heterozygous condition, viable zygotes, while the two repulsion combinations are lethal even in heterozygous combination; complete linkage disequilibrium would follow, even if the two loci were unlinked. Situations as extreme as this are not likely, however. Because the approach to linkage equilibrium is facilitated by the extent of recombination, the less closely linked two loci are, the greater the strength of natural selection required to maintain linkage disequilibrium. Consequently, in natural populations, linkage disequilibrium is more common between closely linked loci.

Supergenes

Linkage disequilibrium is decreased by recombination. The possibility of maintaining favorable allelic combinations in linkage disequilibrium is,

Box 5.1 Random Mating with Two Loci

It was shown in Chapter 3 that the genotypic equilibrium frequencies at any one autosomal locus are reached in one generation of random mating. This is not so when two loci are considered simultaneously. If the frequency of recombination between two loci is c, then the value of the linkage disequilibrium, d, decreases by the fraction cd in one generation of random mating (assuming that there is no selection). That is, if the value of d in one generation is d_0, its value in the following generation will be

$$d_1 = (1 - c)d_0$$

When two loci are unlinked, then $c = 0.5$, and the linkage disequilibrium value will be halved with every generation of random mating. If $c < 0.5$, then the approach to equilibrium will be slower. For example, if the frequency of recombination is 0.1, the value of d in one generation will be 90% of its value in the previous generation. The approach to equilibrium for various values of c, assuming random mating, is shown in Figure 5.10.

---

therefore, enhanced by reduction of the frequency of recombination between the loci involved. This may be accomplished through translocations and inversions. Assume that two loci, *A* and *B*, are located in different chromosomes; a translocation might bring them together in the same chromosome. Assume, now, that the two loci are separated within the same chromosome by a number of loci that we represent as *FG* · · · *MN*, so that the gene sequence along the chromosome is

$$\cdots AFG \cdots MNB \cdots$$

An inversion comprising the segment *FG* · · · *MNB* might bring together *A* and *B*; the new gene sequence would be

$$\cdots ABNM \cdots GF \cdots$$

Whenever linkage disequilibrium is favored by natural selection, chromosomal rearrangements increasing linkage between the loci will also be favored by natural selection. The term *supergene* is used to refer to several closely linked gene loci that affect a single trait or a series of interrelated traits.

A supergene is responsible for the expression of two flower phenotypes, known as "pin" and "thrum," found in the primrose and other species of the genus *Primula* (Figure 5.11). Pin-thrum polymorphisms were made famous by Darwin, who gave a detailed account of them in 1877. The pin phenotype is characterized by a long style above the ovary, which places the stigma at the same level as the mouth of the corolla; the pollen-bearing anthers are halfway down the corolla tube. The thrum phenotype has a short style, so that the stigma is halfway down the corolla tube, while the stamens are long, placing the anthers at the mouth of the

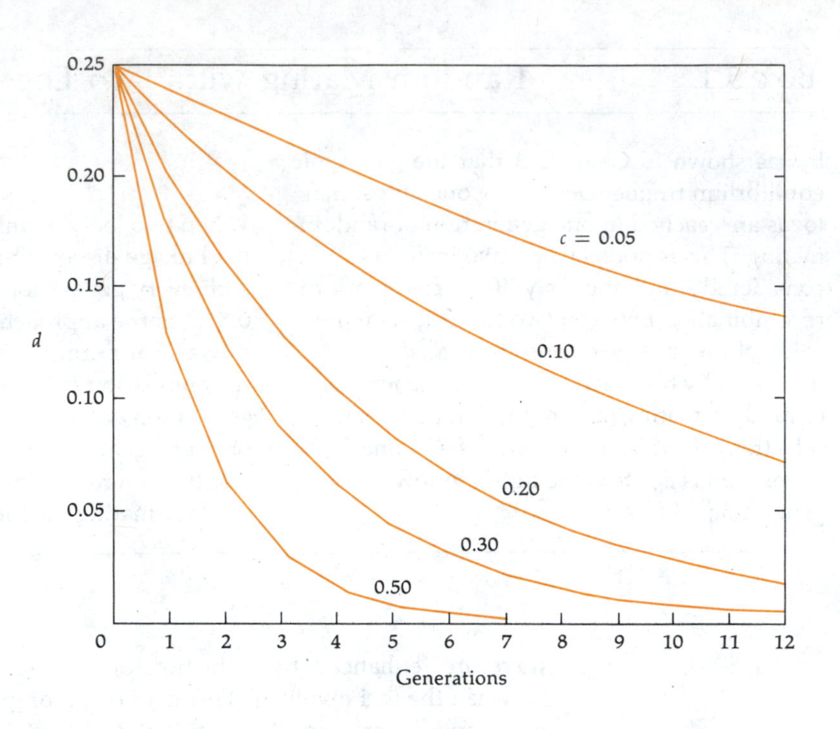

Figure 5.10
Decrease in linkage disequilibrium, *d*, over the generations for various levels of linkage (recombination frequency, *c*, from 0.05 to 0.50). The value of *d* after *t* generations of random mating is given by $d_t = (1 - c)^t d_0$.

corolla. Pin and thrum are phenotypically different in other ways also, such as the configuration of the stigma and the size of the pollen grains. Moreover, they differ physiologically: thrum pollen is more successful in fertilization when deposited on pin stigmas than on thrum stigmas; conversely, pin pollen fertilizes thrum flowers more successfully than it does pin flowers.

The pin-thrum phenomenon is known as *heterostyly* (meaning "different styles"). Heterostyly promotes cross-pollination. An insect that visits both pin and thrum flowers will receive pollen from one type of flower on parts of its body that get close to the stigma of the other type. The physiological differences reinforce the chances of cross-fertilization.

Pin and thrum phenotypes behave, as a rule, as if they were controlled by a single gene locus, with two alleles: *S* (for thrum) is dominant over *s* (for pin). Thrum plants, however, are generally heterozygous (*Ss*); when they are selfed or intercrossed, they produce pin and thrum plants in a typical 3:1 Mendelian ratio. Pin plants are homozygous (*ss*), and produce only pin types when selfed or intercrossed. In nature, most crosses are between thrum (*Ss*) and pin (*ss*) plants, producing thrum and pin progenies in about a 1:1 ratio. Pin and thrum flowers are found in approximately similar frequencies in natural populations.

The set of traits characteristic of the thrum or pin phenotypes are not determined by a single gene locus, however, but by several closely linked loci making up a supergene. The existence of multiple gene loci could be

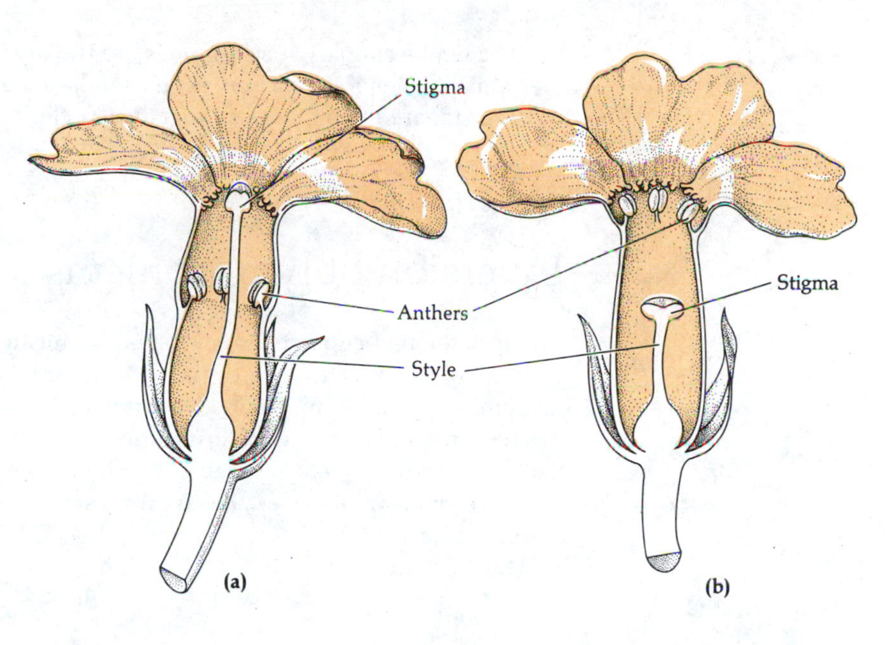

Figure 5.11
Two phenotypes in the primrose, *Primula officinalis*. **(a)** The pin phenotype has a highly placed stigma but low-placed anthers. **(b)** The thrum phenotype is just the opposite. This complementary arrangement facilitates cross-fertilization between the two phenotypes.

suspected, because the phenotypic and physiological differences between the thrum and pin types are multiple, and it has been confirmed by examination of large progenies from experimental thrum × pin crosses. Examples of mixtures between components of the two complex phenotypes are occasionally found, owing to recombination within the supergene. In nature, mixed phenotypes are occasionally found as well, but these remain rare, owing to their low fitness relative to that of the thrum and pin phenotypes. The supergene has become established in *Primula* because it makes possible the joint transmission of sets of alleles that produce adaptive phenotypes. The supergene control saves *Primula* populations from a high proportion of ill-adapted phenotypes.

A well-known example of a supergene is the set of gene loci controlling the color and the presence or absence of bands on the shell of the snail *Cepaea nemoralis*, studied by Arthur J. Cain and Philip M. Sheppard. The gradual process of formation of supergenes by means of successive translocations and inversions is evidenced by several species of grouse locusts. Robert K. Nabours has shown that the color patterns are determined by alleles at some 25 gene loci. In one species, *Acridium arenosum*, 13 of the genes are spread throughout one single chromosome and recombine fairly freely; in another species, *Apotettix eurycephalus*, the corresponding genes are combined into two groups (supergenes) of closely linked genes, the recombination between the groups being only 7%; and in a third species, *Paratettix texanus*, 24 of the 25 genes are all tightly linked, forming a single supergene. The formation of supergenes has advanced most in this last species.

Bacterial examples of supergenes are the operons. Genes involved in the same biochemical function, such as those controlling the synthesis of tryptophan, are often clustered together in close linkage within the genome.

Inversion Polymorphisms

The formation of supergenes is a way of reducing crossing over, thereby facilitating the maintenance of linkage disequilibrium. Linkage disequilibrium can also be maintained by inversion polymorphisms, another mechanism for reducing genetic recombination. Assume, as we did above, that A_1B_1 and A_2B_2 are favorable allelic combinations, while A_1B_2 and A_2B_1 are unfavorable ones. Let us represent the gene sequence in the chromosome as

$$\cdots DEAF \cdots NBOP \cdots$$

Assume that an inversion of the segment from E to O takes place and that the alleles A_1 and B_1 are included in the segment. We would then have the following sequence:

$$\cdots DOB_1N \cdots FA_1EP \cdots$$

(Subscripts at loci other than A and B are not added, because we are not concerned with alleles at these other loci.)

Assume now that an individual heterozygous for the inversion and the original chromosome sequence carries alleles A_2 and B_2 in the original chromosome sequence, i.e., that this individual has the following genetic constitution:

$$\cdots DOB_1N \cdots FA_1EP \cdots / \cdots DEA_2F \cdots NB_2OP \cdots$$

Recombination is suppressed in the progenies of inversion heterozygotes, because recombinant gametes either lack some genes or have them duplicated. Thus, the individual described above will produce only two kinds of functional gametes, one containing the alleles A_1 and B_1, and the other containing A_2 and B_2. Natural selection may, then, favor original chromosome sequences that have alleles A_2 and B_2, and inverted chromosome sequences that have alleles A_1 and B_1. The population can then consist of only three types of individuals: (1) homozygotes for the chromosomal inversion, and therefore for alleles A_1 and B_1; (2) homozygotes for the original chromosome sequence, and therefore for alleles A_2 and B_2; and (3) heterozygotes for the inverted sequence and the original sequence. Only the two gametic combinations A_1B_1 and A_2B_2

Table 5.6
Relative frequencies of third chromosomes with different chromosomal arrangements in populations of *Drosophila pseudoobscura* in various localities.

Locality	Frequency (%) of Chromosomal Arrangement:								
	ST	*AR*	*CH*	*PP*	*TL*	*SC*	*OL*	*EP*	*CU*
Methow, Washington	70.4	27.3	0.3	–	2.0	–	–	–	–
Mather, California	35.4	35.5	11.3	5.7	10.7	0.9	0.5	0.1	–
San Jacinto, California	41.5	25.6	29.2	–	3.4	0.3	–	–	–
Fort Collins, Colorado	4.3	39.9	0.2	32.9	12.3	–	2.1	7.2	–
Mesa Verde, Colorado	0.8	97.6	–	0.5	–	–	–	0.2	–
Chiricahua, Arizona	0.7	87.6	7.8	3.1	0.6	–	–	–	–
Central Texas	0.1	19.3	–	70.7	7.7	–	2.4	–	–
Chihuahua, Mexico	–	4.6	68.5	20.4	1.0	3.1	0.7	–	–
Durango, Mexico	–	–	74.0	9.2	3.1	13.1	–	–	–
Hidalgo, Mexico	–	–	–	0.9	31.4	1.7	13.5	1.7	48.3
Tehuacán, Mexico	–	–	–	–	20.2	1.1	–	3.2	74.5
Oaxaca, Mexico	–	–	10.3	–	7.9	–	0.9	1.6	71.4

After J. R. Powell, H. Levene, and Th. Dobzhansky, *Evolution* 26:553 (1973).

exist in all three kinds of individuals.

Inversion polymorphisms have been studied in many species of *Drosophila*. Some species have polymorphisms in all chromosomes—e.g., the European species *D. subobscura* and the American tropical species *D. willistoni*—whereas others have inverted segments concentrated mostly in one chromosome—e.g., the North American species *D. pseudoobscura*, which exhibits extensive polymorphism in only one of the five chromosomes, the third (Table 5.6).

As shown in Table 5.6, the frequencies of the various chromosomal arrangements in *Drosophila pseudoobscura* vary from one locality to another. Moreover, the frequencies may change from month to month throughout

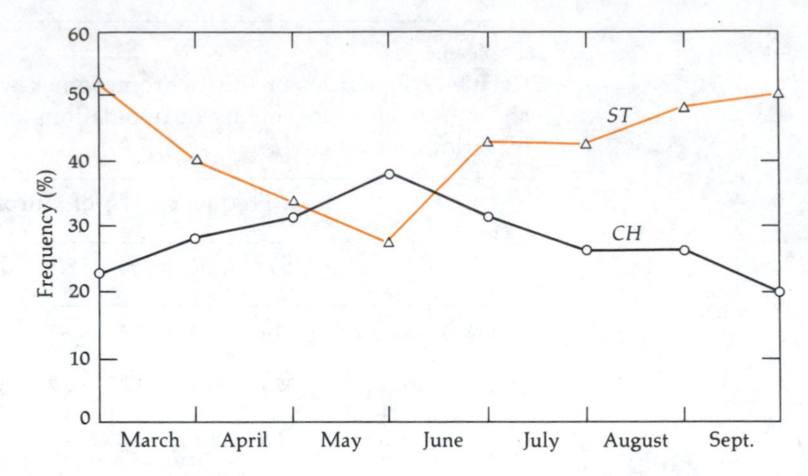

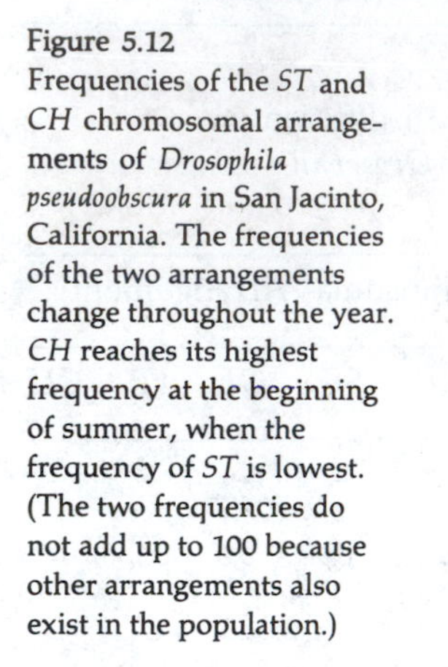
Figure 5.12
Frequencies of the *ST* and *CH* chromosomal arrangements of *Drosophila pseudoobscura* in San Jacinto, California. The frequencies of the two arrangements change throughout the year. *CH* reaches its highest frequency at the beginning of summer, when the frequency of *ST* is lowest. (The two frequencies do not add up to 100 because other arrangements also exist in the population.)

the year (Figure 5.12). These changes are seasonal, and are thus repeated in successive years. This suggests that the chromosomal arrangements differ in the sets of alleles they carry, and that these differences are adaptive: one arrangement is adaptively superior to the other during some period of the year but inferior during some other period. This hypothesis was tested with laboratory populations that were started with known frequencies of the chromosomal arrangements and allowed to breed freely within the laboratory cage. Typical results are shown in Figure 5.13. The frequencies of the inversions change rapidly in the early generations and more slowly later on, eventually reaching an equilibrium with both chromosomal arrangements present. From the rate of change and the equilibrium frequencies, the fitnesses of the three genotypes can be estimated; these are 1 for the heterozygote (*ST/CH*), 0.89 for one homozygote (*ST/ST*), and 0.41 for the other homozygote (*CH/CH*).

These results show that heterozygote superiority contributes to the maintenance of chromosomal polymorphism. Other laboratory experiments have shown that the fitnesses of the two homozygous genotypes depend on the temperature and on the density of the population, which may account for the seasonal oscillations observed in nature.

Direct evidence that chromosomal inversions differ in their allelic content has been obtained in *D. pseudoobscura* by Satya Prakash and Richard C. Lewontin. Two gene loci, *Pt-10* and *α-Amy*, coding for two proteins, were examined by gel electrophoresis. It was discovered that the allelic frequencies were quite different in different chromosomal arrangements (Table 5.7).

Inversion polymorphisms have been observed in natural populations of mosquitoes, black flies, midges, and other dipterans. It is uncertain how

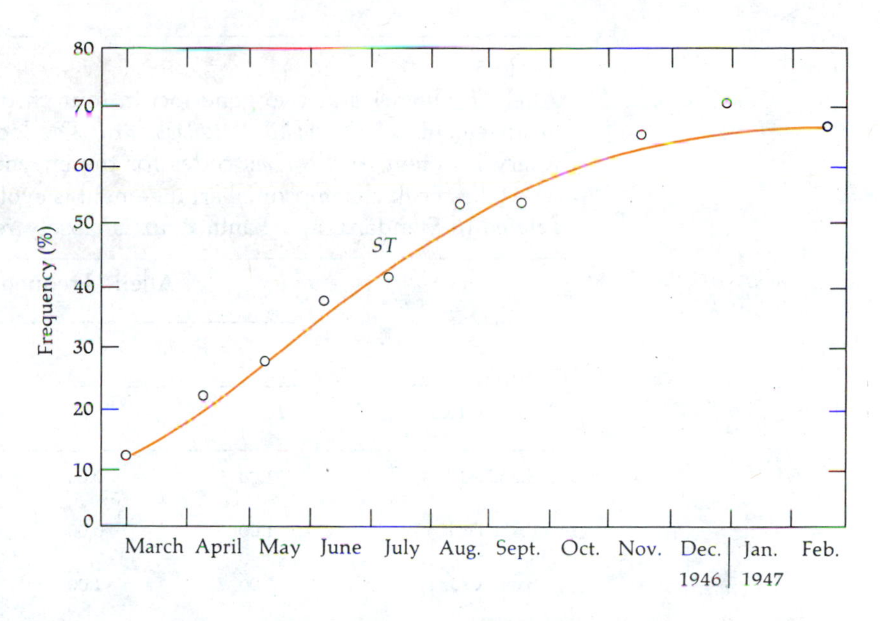

Figure 5.13
Change in the frequency of the *ST* chromosomal arrangement of *Drosophila pseudoobscura* in a laboratory population. Two chromosomal arrangments, *ST* and CH, are present in the population. The frequency of *ST* gradually increases from its initial frequency of 12% to an equilibrium frequency of about 70%. Correspondingly, *CH* decreases from its initial frequency of about 88% to an equilibrium frequency of about 30%.

widespread such polymorphisms are in other organisms less favorable for chromosomal studies, although inversion heterozygotes are known in many animals, such as grasshoppers, and in some plants.

Geographic Differentiation

Environmental conditions are infinitely variable. The weather and many physical and chemical conditions, as well as the foods, competitors, parasites, and predators may all vary to a greater or lesser extent. Natural selection promotes adaptation to local conditions, and this results in genetic differentiation between geographically separated populations.

The adaptive character of local differentiation is apparent in the experiments with the cinquefoil, *Potentilla glandulosa* (Figure 1.3). Plants collected at different altitudes are genetically different, as demonstrated by their different growth habits in identical environments. Moreover, the genetic differences are adaptive—plants grow best in the environment most similar to their natural habitat.

Table 5.8 shows the result of an experiment with two populations of *Drosophila serrata,* one collected in a temperate habitat (near Sydney, Australia, at latitude 34°S) and the other in a tropical habitat (Popondetta, New Guinea, at latitude 9°S). The populations were kept for about one year (18 generations) at two temperatures. At 19°C, a temperature considerably lower than those experienced in the tropics, the Sydney flies

Table 5.7
Allelic frequencies at two gene loci in four chromosomal arrangements of *Drosophila pseudoobscura.* The locus *Pt-10* codes for a larval protein, and *α-Amy* codes for the enzyme α-amylase. The Pikes Peak chromosomal arrangement is evolutionarily closely related to Standard, and Santa Cruz is closely related to Tree Line.

Chromosomal Arrangement	Allelic Frequency at Locus: *Pt-10*		*α-Amy*	
	104	*106*	*84*	*100*
Standard	1.00	0.00	0.15	0.85
Pikes Peak	1.00	0.00	0.00	1.00
Santa Cruz	0.00	1.00	1.00	0.00
Tree Line	<0.01	>0.99	>0.90	0.05

After S. Prakash and R. C. Lewontin, *Proc. Natl. Acad. Sci. USA* 59:398 (1968).

performed better, maintaining a greater population size than the flies from Popondetta. At 25°C, a temperature more common in the tropics, the differences disappeared. (See also Table 2.1).

The variation in skin pigmentation of different human populations is probably the result of past adaptation to local conditions. Humans, like other mammals, require vitamin D for calcium fixation and bone growth; an insufficient supply of vitamin D causes the bone disease called rickets. Vitamin D is produced in the deep layers of the skin under the stimulus of ultraviolet radiation from the sun. If the skin is heavily pigmented, the amount of UV radiation at high latitudes may not produce enough vitamin D, because the radiation is absorbed by the skin pigment. Consequently, natural selection has favored light skin pigmentations at higher latitudes. In tropical regions, too much vitamin D may be produced (as well as some skin damage) in lightly pigmented skins; hence, natural selection has favored deeper pigmentation. Figure 5.14 gives one additional example of adaptive geographic differentiation in human populations.

As a result of genetic differences among geographically separate populations of the same species, groupings may arise that include populations genetically more similar to each other than they are to populations placed in different groups. The genetic differences on which the geographic groupings are based may or may not be expressed in the

Table 5.8
Number of flies of *Drosophila serrata* populations from two different localities, maintained at two temperatures in population cages of the same size and with the same amount of food. The values given are the means and standard errors for 18 generations.

Locality	Population Size at Temperature: 19°C	25°C
Sydney, Australia	1803 ± 87	1782 ± 76
Popondetta, New Guinea	1580 ± 52	1828 ± 90

After F. J. Ayala, *Genetics* 51:527 (1965).

visible phenotype. The differences in human skin pigmentation or in beetle color and pattern (Figure 2.5) are examples of conspicuously expressed genetic differences. Differences in the frequencies of human blood groups or of chromosomal arrangements in *Drosophila* (Table 5.6) are examples of "hidden" genetic differences.

The Concept of Race

Geographically separate groups of populations are sometimes called *races*, which may be defined as *genetically distinct populations of the same species*. The concept of race, particularly when applied to human populations, has been much misunderstood—even abused—and deserves clarification.

Racial classification may be useful in order to recognize that geographic populations are genetically differentiated to some extent (as a consequence of drift and of adaptation to local conditions). Sometimes races are identified by a single trait (e.g., wing pattern in butterflies, and skin pigmentation or blood group in humans), but races are populations that have somewhat differentiated *gene pools*. The differences among races must involve the gene pool as a whole, and therefore allelic frequencies at many loci. Differences in one locus or trait may serve as indicators of overall genetic differentiation, but alone they do not constitute a sufficient fundament for distinguishing among races. Indeed, parents and their offspring may differ at a locus whenever this is polymorphic; for example, two A blood-group parents (I^Ai) may have O blood-group children (*ii*).

Races are populations of the same species and are thus not reproductively isolated from one another. The formation of new species

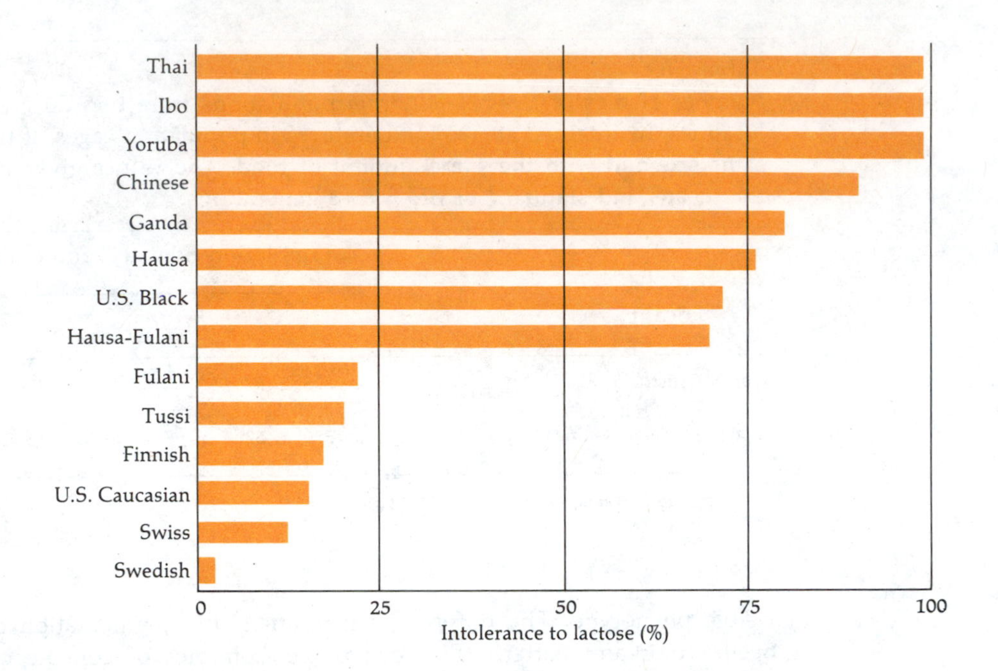

Figure 5.14
Intolerance to lactose in various human populations. The proportion of adult individuals tolerant to this milk sugar is highest in peoples that have used milk and dairy products in their diets for millennia. Populations that traditionally have not used dairy products in their diets are nearly 100% intolerant to lactose as adults. (After N. Kretchmer, *Scientific American,* October, 1972, p. 70.)

often involves transitional stages of racial differentiation. But races are not necessarily incipient species, because the process of racial differentiation is reversible. Racial differences may, and often do, decrease with time or even become obliterated. In humans, for example, racial differentiation has decreased in the last few centuries through migration and intermixture.

The formation and preservation of races requires that gene flow be limited; otherwise, races will fuse into a single gene pool. Usually, gene flow is curtailed by geographic separation. (Exceptions to this rule are possible through human choice. The differentiation of human races may be retained even where they are sympatric, because people choose their mates predominantly within their own race. As another example, people keep dog breeds separate, even when the breeds live in the same locality, by preventing interbreeding.) Sometimes, abrupt geographic barriers exist that facilitate the formation and identification of races, e.g., terrestrial organisms living on separate islands or aquatic ones living in separate lakes. Nevertheless, different degrees of gene flow and of genetic differentiation exist around various boundaries, some of which may include further subdivisions. The data in Table 5.6 serve to illustrate this point.

The geographic locations of the populations in Table 5.6 range from the North (Washington) toward the South (California), then to the East (Colorado, Arizona, and Texas), and then to the South again (Mexico). There is considerable differentiation in the frequency of chromosomal arrangements throughout the range. The frequency of *ST* is high in Washington, intermediate in California, and low or zero in the other localities. The frequency of *AR* is intermediate in Washington, California, and Fort Collins, high in Mesa Verde and Chiricahua, and becomes low and eventually zero as we move farther south; and so on for the other chromosomal arrangements. Some transitions in the frequency of chromosomal arrangement would be smoother if data for intermediate populations were added to the table.

The genetic differences reflected in the chromosomal frequencies can be the basis for racial differentiation in *D. pseudoobscura.* But how many races are there? One possible classification would distinguish four races: (1) a northern and central race, from Methow through Fort Collins, characterized by the presence of *AR* in intermediate frequencies; (2) a second race, from Mesa Verde through Chiricahua, with *AR* in high frequency; (3) a third race, from Central Texas through Durango, characterized by the presence of either *CH* or *PP* in high frequencies; and (4) a fourth race, from Hidalgo through Oaxaca, identified by the occurrence of the *CU* arrangement. But we might choose to divide the third race into two, distinguished by the high frequency of *CH* in one and of *PP* in the other. Or we might choose to separate the first two races not between Fort Collins and Mesa Verde, but between San Jacinto and Fort Collins; we would then have a northwestern race, characterized by relatively high frequencies of *ST*, and a central race, characterized by high frequencies of *AR*. This exercise illustrates a very important point: *the amount of genetic differentiation* required to distinguish among races and, consequently, the *number of races* and the *position of the boundaries* are largely arbitrary matters. Racial classification recognizes the existence of genetic differentiation within a species, but, rather than sharp differences, there is often a gradual transition (a *cline*) from less to more genetic differentiation along geographic lines.

Human Races

Once the concept of race is understood, it comes as no surprise to learn that there are multiple classifications of human races. Some classifications recognize only three races, others, more than fifty.

The ethnic diversity of mankind was recognized by Carolus Linnaeus, who identified four human varieties: African, American, Asiatic, and European. The familiar five "color" races were established by Johann Friederich Blumenbach in 1775: white, or Caucasian; yellow, or Mongolian; black, or Ethiopian; red, or American; and brown, or Malayan. Although the identifying characteristic was skin color, it is clear that ethnic groups differ in many other characteristics, such as facial features, hair, body build, and so on. The correspondence between the various

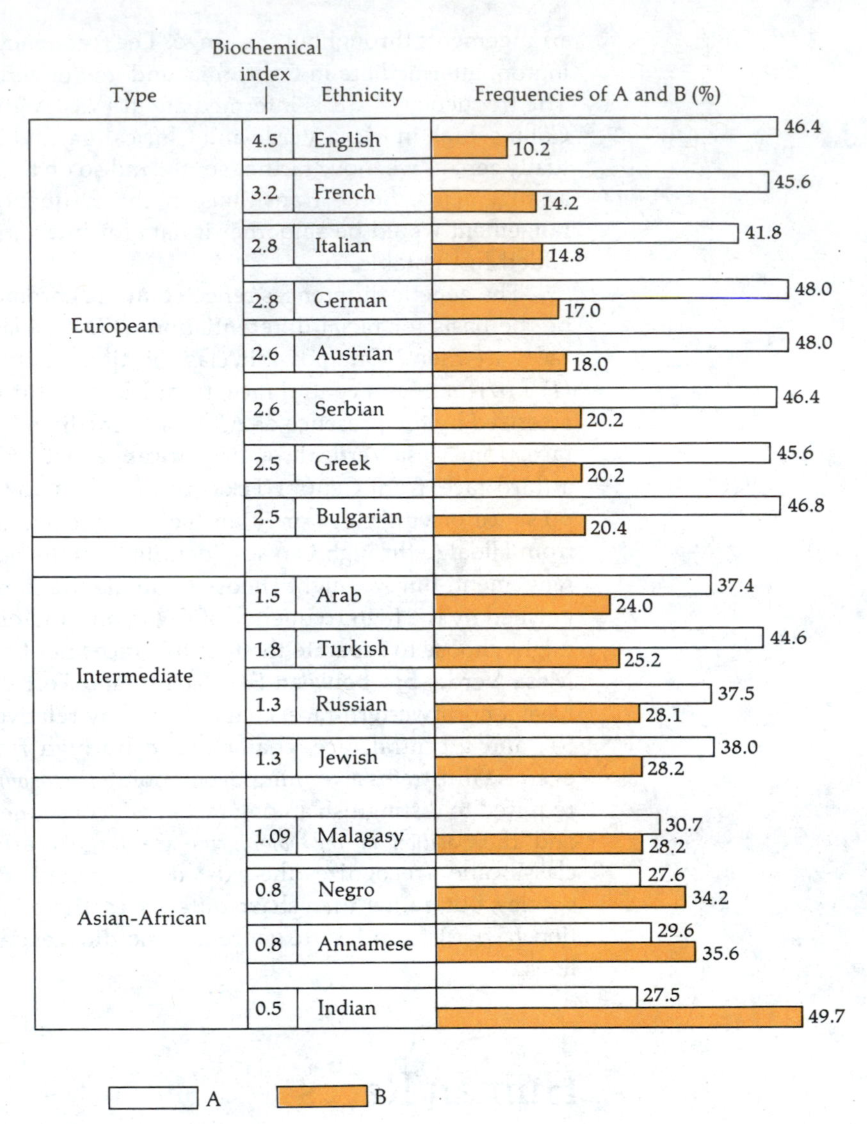

Figure 5.15
Ethnic differences manifested in the ABO blood groups. This graph represents the first use of genetic markers to identify racial differences. The numbers given are the frequencies of two blood groups, A and B. The biochemical index is the ratio of A to B. [After L. Hirszfeld and H. Hirszfeld, *Anthropologie* 29:505 (1919).]

characteristics is far from precise. In parts of India, for example, Caucasian facial features are found in people with black skin.

L. Hirszfeld and H. Hirszfeld suggested in 1918 that the ABO blood groups could be used for analysis of ethnic origins (Figure 5.15). The data then available for Old World peoples showed that the B blood group (genotypes I^BI^B and I^Bi) increases in frequency from about 10% in England to 50% in India; the A blood group (I^AI^A and I^Ai) has about the same frequency throughout Europe, a somewhat lower frequency in Russia and the Middle East, and a still lower frequency in Africa and India. A

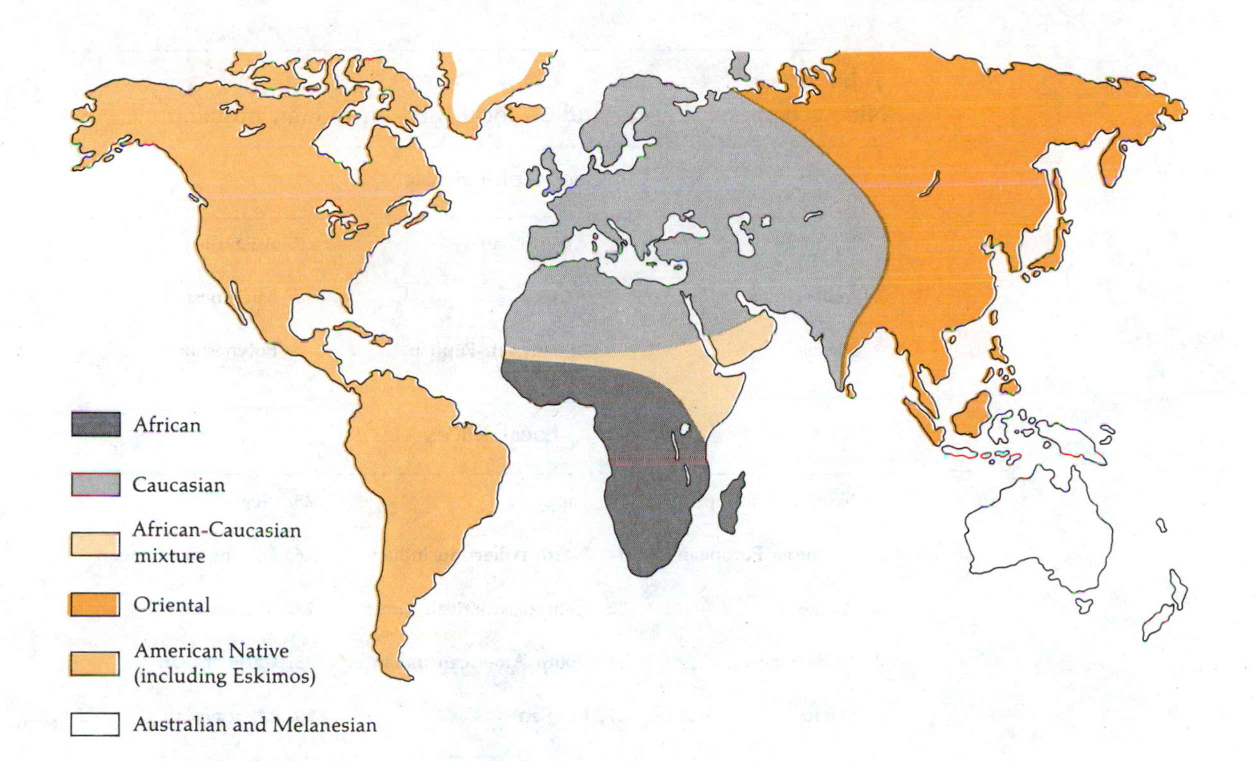

Figure 5.16
Geographic distribution of the main human ethnic groups. Five major groups can be distinguished: African, Caucasian, Oriental, American Native, and Australian and Melanesian. Admixtures between ethnic groups occur in various places, notably in Africa between Africans and Caucasians. (After W. F. Bodmer and L. L. Cavalli-Sforza, *Genetics, Evolution, and Man*, W. H. Freeman, San Francisco, 1976.)

"biochemical index" (the ratio of the frequencies of blood groups A and B) was used to distinguish three racial groups: European, Intermediate, and Asian-African.

The frequency of the I^B gene throughout the world is shown in Figure 3.10. A racial classification based on blood-group gene frequencies does not, of course, imply that people with different blood groups belong to different races, but rather that allele frequency differences at the blood-group locus reflect overall differentiation in the gene pool. It has turned out, however, that the variations in ABO frequencies are not as great as variations in other blood groups, such as the Rhesus (*R*), Duffy (*Fy*), and Diego (*Di*), which are therefore more informative about ethnic groups.

Geographic boundaries help to identify three major racial groups: Africans, Caucasians, and a highly heterogeneous group called Easterners. The Easterners can be subdivided into three groups: American Natives, Orientals, and Australians and Melanesians. The resulting five groups largely overlap the five "color" races of Blumenbach. Caucasians are a fairly homogeneous group ranging from western Europe to western Russia and through the Middle East to India, where there is a gradual transition to Easterners. In northern and central Africa, however, populations show various degrees of admixture between Caucasians and Africans (Figure 5.16).

Racial classifications recognize genetic heterogeneity among populations. The question is how much diversity to recognize. If only a few

Table 5.9
Nine geographical races and 34 local races in human populations.

Geographical Races		
1. European	4. Amerindian	7. Australian
2. Indian	5. African	8. Micronesian
3. Asiatic	6. Melanesian-Papuan	9. Polynesian

Local Races		
1. Northwest European	13. Lapp	25. Negrito
2. Northeast European	14. North American Indian	26. Melanesian-Papuan
3. Alpine	15. Central American Indian	27. Murrayian
4. Mediterranean	16. South American Indian	28. Carpentarian
5. Hindu	17. Fuegian	29. Micronesian
6. Turkic	18. East African	30. Polynesian
7. Tibetan	19. Sudanese	31. Neo-Hawaiian
8. North Chinese	20. Forest Negro	32. Ladino
9. Classic Mongoloid	21. Bantu	33. North American Black
10. Eskimo	22. Bushman and Hottentot	34. South African Black
11. Southeast Asiatic	23. African Pygmy	
12. Ainu	24. Dravidian	

After S. M. Garn, *Human Races*, C. C. Thomas, Springfield, Ill., 1961.

subdivisions are recognized, as in Figure 5.16, then there will be too much heterogeneity within some of them. On the other hand, with ever finer splitting, the differences become less sharp and the boundaries more blurred. An eclectic solution proposed by Stanley M. Garn distinguishes 9 "geographical races" and 34 "local races" in human populations (Table 5.9).

How genetically different are human races? In a variety of human populations, some 25 gene loci have been studied that are polymorphic in at least one racial group. The average heterozygosity of an individual gives a measure of the amount of genetic variation in a population, because it estimates the probability that two genes picked up at random at a given locus are different from each other. For any given human group, the average heterozygosity per individual at the 25 polymorphic loci ranges from 28 to 30%. The probability that two genes taken at random, each from a different racial group, are different (which is also the probability of heterozygosity in the progeny of an interracial cross) is about 35–40%. This is a small increase (from about 29% to about 37%) over the heterozygosity within groups. That is, the additional genetic differentiation *between* human races is relatively small in comparison to the genetic differentiation *within* groups. As was stated in the previous section, knowing the racial group of an individual provides little information about its genetic makeup. Each person has a unique genotype—it is different from that of every other person, whether or not they belong to the same race.

Problems

1. A cattle rancher breeds a bull (A) with its own daughter (C). What is the inbreeding coefficient of the offspring (D), assuming that A and B are unrelated?

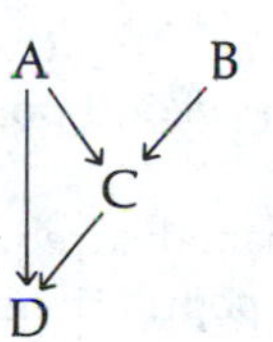

2. Calculate the inbreeding coefficient of the offspring (G) of an uncle-niece mating:

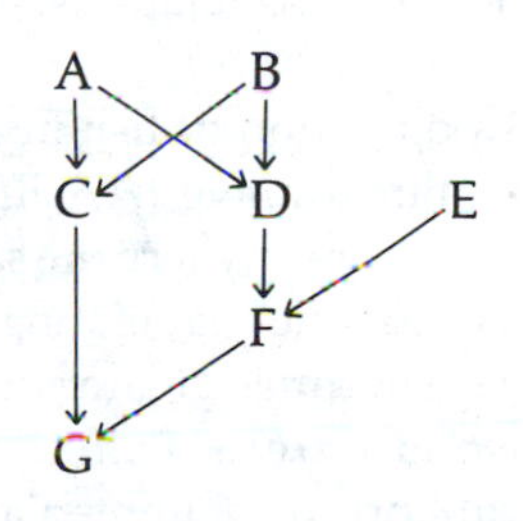

3. Calculate the inbreeding coefficient of the offspring (K) of double first cousins:

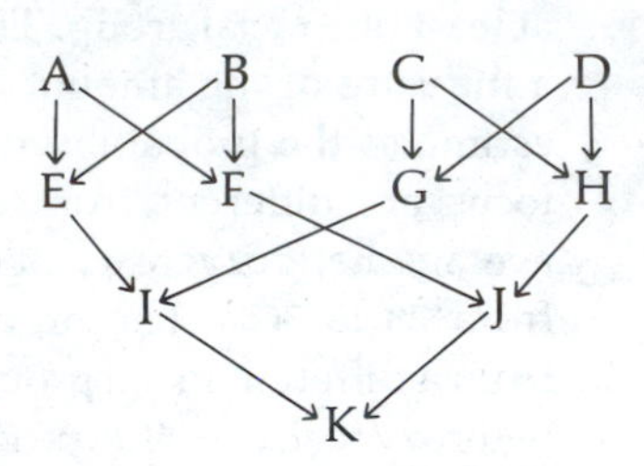

4. The frequencies of two autosomal alleles, A and a, in each of three plant populations are 0.80 and 0.20, respectively. The inbreeding coefficients in the three populations are 0, 0.40, and 0.80. What is the frequency of heterozygotes in each population?

5. In a small population, the numbers of individuals of the three possible genotypes at a locus are 28 AA, 24 Aa, and 48 aa. Calculate the inbreeding coefficient, assuming that inbreeding alone is responsible for any deviations of the genotypic frequencies from Hardy-Weinberg expectations.

6. Assume that the mutation rate to the recessive lethal allele responsible for cystic fibrosis is $u = 4 \times 10^{-4}$. Assume also that a population is at equilibrium with respect to this allele. What is the expected frequency of the disease in the offspring of first-cousin marriages? Assume now that in another population the mutation rate is twice as high ($u = 8 \times 10^{-4}$), owing to long-term exposure to background radiation, and that this population is also at equilibrium with respect to this allele. What is the expected frequency of the disease in the offspring of first-cousin marriages in this second population?

7. In Japan, the frequency of $L^M L^N$ heterozygotes for the M-N blood groups in large populations, in which random mating may be assumed, is 0.4928. However, in a town where matings between relatives are common, the frequency of MN individuals is 0.4435. Calculate the inbreeding coefficient in this town, assuming that the allele frequencies are the same as in the large populations.

8. A population of fish in a lake is fixed at each of two unlinked loci for the dominant allele (AA BB). A canal is built that connects the lake with a smaller one, where the same fish species is fixed for the recessive alleles (aa bb). Assume that mating is random thereafter, that the original population was ten times greater in the large lake than in the small lake, and that the two loci are not affected by natural selection. What is the linkage disequilibrium, d, immediately after the populations of the two lakes become mixed, and after five generations of random mating?

9. In the previous problem, assume that the two loci are in the same chromosome and have a recombination frequency $c = 0.10$. What will be the value of d after five generations of random mating? How many generations will be necessary to reduce d to the value achieved for unlinked loci in five generations of random mating?

10. Two linked esterase loci were examined in a population of *Drosophila montana*. At each locus there were two alleles, one exhibiting activity, the other being inactive ("null"). The observed numbers of each two-allele combination in a sample of 474 gametes were as follows:

	Locus 1	
Locus 2	Active	Null
Active	31	273
Null	97	73

Calculate the linkage disequilibrium value, d.

11. An experimental population of barley (*Hordeum vulgare*) was established by intercrossing 30 barley varieties from various parts of the world. The population was thereafter maintained by spontaneous self-fertilization. Two loci (A and B) coding for esterases were examined at various times; two alleles (*1* and *2*) were present at each locus. The gametic frequencies in three generations were as follows (several thousand gametes were examined in each generation):

Generation	A_1B_1	A_2B_2	A_1B_2	A_2B_1
4	0.453	0.019	0.076	0.452
14	0.407	0.004	0.098	0.491
26	0.354	0.003	0.256	0.387

Calculate the value of d for these three generations. What process(es) is (are) likely to be responsible for changing the linkage disequilibrium as the generations proceed?

6

Quantitative Characters

Continuous Variation

Mendel's discovery of the basic laws of heredity was made possible by choosing contrasting characters that were easily distinguishable from one another. The peas were yellow or green, round or wrinkled; the flowers were axial or terminal; the plants were dwarf or tall, and so on. There were two alternative forms for each trait, determined by different alleles of the same gene. However, not all traits appear in such clearly distinguishable alternative forms. Humans, or pine trees, do not come in only two height classes, tall and short; rather, they vary continuously over a large range of heights. Height, weight, fertility, and longevity are a few examples of many traits that exhibit more or less *continuous variation*. The occurrence of continuous variation is due to (1) interactions between different genes, and (2) interactions between genes and the environment.

The interaction between the genes and the environment was already pointed out in Chapter 1, where we introduced the distinction between genotype and phenotype. The genotype of an organism is the genetic information it has inherited; the phenotype is its appearance, which we can observe. The ultimate effect of a gene on the phenotype depends on the environmental conditions but also on the actions of other genes. In this chapter we deal with characters exhibiting continuous variation, in an

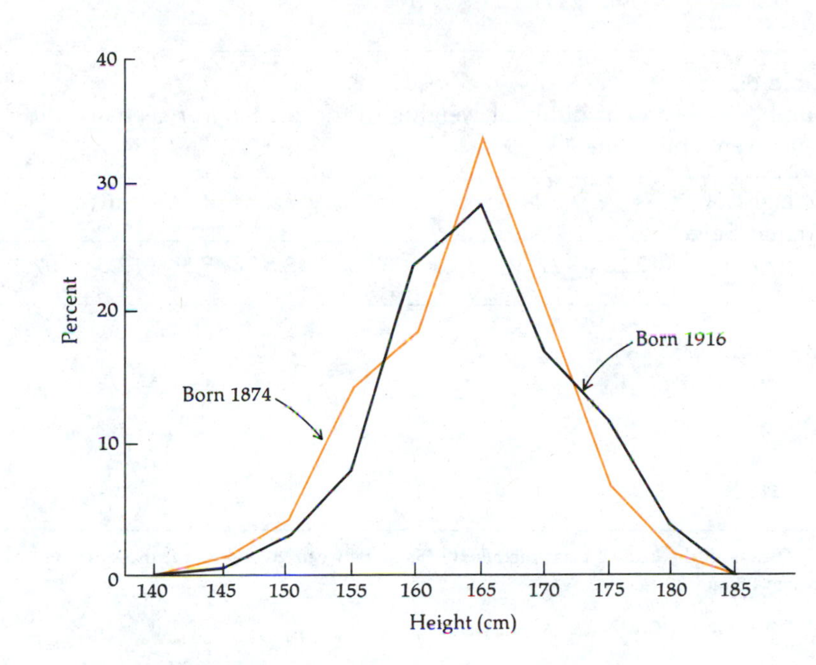

Figure 6.1
Distribution of height in 20-year-old Italian males. The men have been divided into classes differing by 5 cm. The mean height is 163.7 cm for men born in 1874, and 166.1 cm for men born in 1916. More than 200,000 men were included in each sample. A few individuals were excluded because they fell outside the height range shown in the figure.

effort to disentangle the environmental and genetic factors that account for such variation.

There are two kinds of variation among organisms of the same species. With respect to some traits, variation is discrete, or *discontinuous*: organisms fall into one or another of a few clearly distinguishable classes. The traits studied by Mendel are of this kind. With respect to other traits, variation is *continuous*: organisms vary more or less continuously over a range. Most people are between 145 and 185 cm tall; although they can be divided into height classes differing, for example, by 5 cm, there are people with all intermediate heights. *Drosophila* females may lay from a few to several hundred eggs; corn ears may have from a few score to several hundred seeds. Traits exhibiting continuous variation are sometimes called *quantitative*, or *metric*, characters because differences between individuals are quantitative, or small (requiring precise measurement), rather than qualitative, or large (requiring only simple observation, without precise measurement). One characteristic of quantitative traits is that their numerical values are usually distributed on a bell-shaped curve called a *normal distribution* (Figure 6.1; see Appendix A.IV).

In the early 1900s, geneticists raised the questions of whether quantitative variation is hereditary and, if so, whether it is inherited according to Mendelian laws. The questions were soon settled—quantitative variation is due in part to environmental influences and in part to genetic differences that are inherited like other genes. The Danish geneticist

Table 6.1
Number of beans of different weights in the progeny of seeds of different weights from Johannsen's pure line 13.

Weight of Parental Seed (cg)	Number of Progeny Beans of Weight (cg):*									Mean Weight of Progeny (cg)
	22.5	27.5	32.5	37.5	42.5	47.5	52.5	57.5	62.5	
27.5		1	5	6	11	4	8	5		44.5
37.5	1	2	6	27	43	45	27	11	2	45.3
47.5		5	9	18	28	19	21	3		43.4
57.5		1	7	17	16	26	17	8	3	45.8

*Beans in the 22.5-cg class include all beans between 20 and 25 cg; those in the 27.5-cg class include all beans between 25 and 30 cg, etc.

Wilhelm Johannsen showed in 1903 that continuous variation was partly environmental and partly genetic (he subsequently formulated the distinction between genotype and phenotype). In 1906 the mathematician George Udny Yule suggested that several gene loci each having a small effect could account for quantitative variation. This was experimentally confirmed in the following years by two geneticists, the Swede Herman Nilsson-Ehle and the American Edward M. East.

Johannsen found that the weight of beans (*Phaseolus vulgaris*) in a commercial seed lot ranged from "light" (15 centigrams) to "heavy" (90 cg). By allowing self-pollination of the beans over several generations, he established several lines, each highly homozygous. He planted seeds of different weights but *all from a single line* and weighed the seeds produced by each plant. Although the original seeds differed in size, the beans produced by all plants were of the same mean weight (Table 6.1). For example, the mean weight of beans produced by 57.5-cg seeds was 45.8 cg, not significantly greater than 44.5 cg, the mean weight of beans produced by seeds weighing only 27.5 cg. Thus he demonstrated that weight variation among beans of a single plant was environmental—as it had to be, since each line was homozygous and thus lacked genetic variation that could segregate in the progenies.

Johannsen demonstrated, moreover, that genetic differences contribute to bean weight by showing that beans from different lines had different average sizes. Table 6.2 shows the mean weight of seeds from different lines. From each line he used seeds of different weights: for example, he used seeds weighing 20 cg, 40 cg, and 60 cg in line 7. All

Table 6.2
Mean weight of beans from different pure lines of Johannsen's.

Weight of Parental Seed (cg)	Mean Weight (cg) in Line Number:					
	19	18	13	7	2	1
20		41.0		45.9		
30	35.8	40.7	47.5			
40	34.8	40.8	45.0	49.5	57.2	
50			45.1		54.9	
60			45.8	48.2	56.5	63.1
70					55.5	64.9
Mean weight of all progeny of a single line:	35.1	40.8	45.4	49.2	55.8	64.2

progeny of a given line had the same mean weight, regardless of how heavy the parental seeds were, thus confirming the results shown in Table 6.1. But the mean weights of beans from *different* lines (last row in Table 6.2) were consistently different; for example, the mean weights of all beans from lines 19 and 1 were 35.1 cg and 64.2 cg, respectively.

Seed Color in Wheat

The experiments of Johannsen demonstrated that both the environment and the heredity contribute to continuous variation. However, the experiments did *not* show that the hereditary differences between pure lines were due to Mendelian genes. It was conceivable that quantitative characters were inherited in a different fashion than discrete characters, as some biologists were still arguing. In 1909 Nilsson-Ehle showed that continuous variation occurs when a trait is determined by several genes, each having a small effect, that behave according to Mendelian laws.

Nilsson-Ehle had several true-breeding lines of wheat with kernels ranging from white through various shades of red to dark red. Crosses between white and light red varieties produced F_1 offspring that were

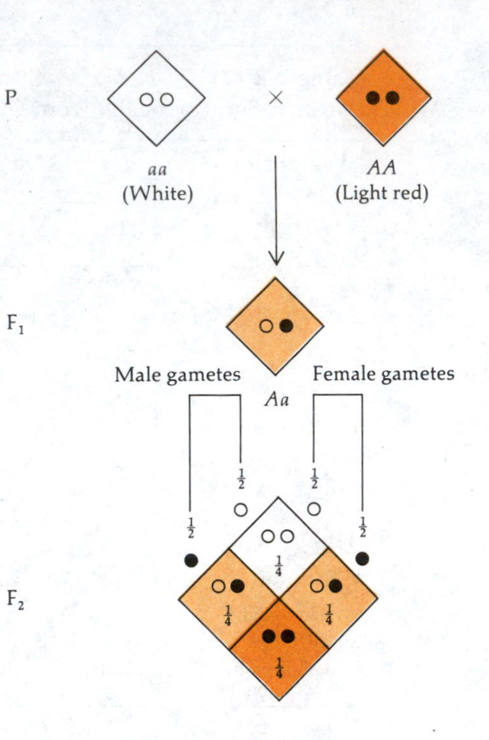

Figure 6.2
Kernel color in the F_2 of a cross between a white and a light red variety of wheat. Capital letters represent alleles for redness, so that two *A* alleles produce light-red kernels, but one *A* and one *a* produce kernels intermediate between white and light red.

intermediate in color between the parents; the F_2 consisted of white, intermediate, and light red kernels in the approximate proportions 1:2:1. When white and red varieties were crossed, the F_1 was light red, and in the F_2 there were five kinds of kernels, ranging from white to red; the proportion of white kernels was approximately 1/16 in each of several crosses. Crosses between white and dark red varieties produced F_1 offspring intermediate between the parents; the F_2 consisted of seven classes, ranging from white to dark red, with kernels of intermediate color being the most common, and white kernels representing about 1/64 of the total number of kernels. Nilsson-Ehle correctly explained his results as being due to three pairs of genes with two alleles each, one for white and one for red, in such a way that the alleles for redness each contributed a small amount of color to the phenotype.

Let us denote the alleles at the three loci as *A a*, *B b*, and *C c*, with the capital letters representing alleles for redness and the lower-case letters representing alleles for whiteness. The genotypes of the varieties can be represented as *aa bb cc* for white, *AA bb cc* for light red, *AA BB cc* for red,

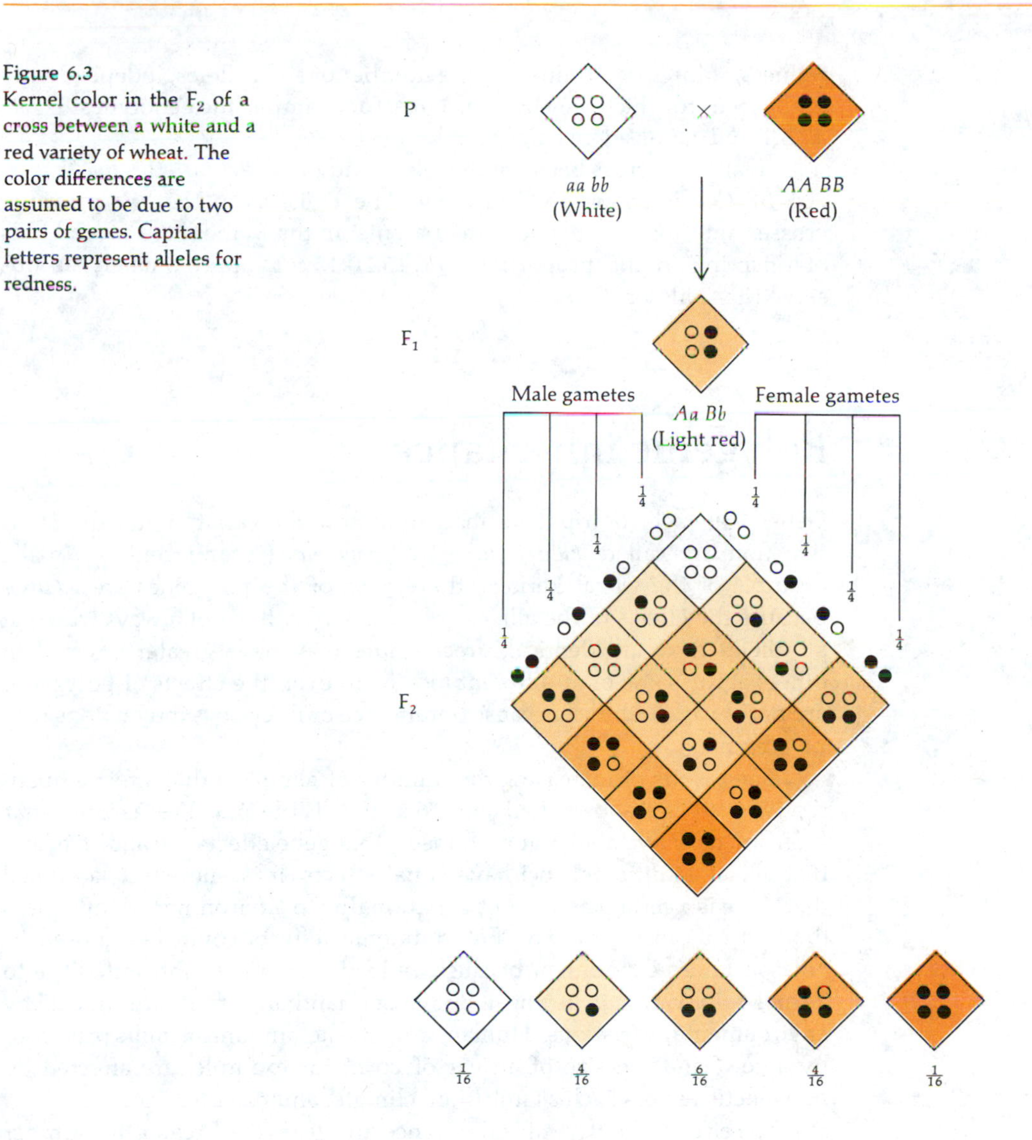

Figure 6.3
Kernel color in the F_2 of a cross between a white and a red variety of wheat. The color differences are assumed to be due to two pairs of genes. Capital letters represent alleles for redness.

and *AA BB CC* for dark red. The cross between the white and light red varieties is *aa bb cc* × *AA bb cc*. Since the two varieties are identical at the *B* and *C* loci, we need be concerned only with the *A* locus. The results of the cross are shown in Figure 6.2. The F_1 is intermediate in color between the two parents because it has one gene for red and one for white. In the F_2 the familiar ratio 1:2:1 appears.

The cross between the white and red varieties (*aa bb cc* × *AA BB cc*) is shown in Figure 6.3, where the *C* locus has been ignored because the two varieties are identical with respect to it. If, as assumed, each capital-letter allele (dark circles in the figure) contributes an equal amount of redness, color is determined by the number of such alleles, independently of the locus to which they belong. Thus, for example, the genotypes *AA bb*, *aa BB*, and *Aa Bb* all give light red kernels.

Finally, the cross between the white and dark red varieties (*aa bb cc* × *AA BB CC*) is shown in Figure 6.4. The F_1 is, as in the two previous crosses, intermediate between the parents. In the F_2 there are seven kinds of offspring in the proportions 1:6:15:20:15:6:1, approximately as observed by Nilsson-Ehle.

Polygenic Inheritance

Genes that each contribute a small amount to the variation in a quantitative trait are called *multiple factors* or *polygenes* ("many genes"). In the example of the wheat kernels, the effects of the polygenes are *additive* because the effects of the alleles are cumulative. It is not always true that all alleles have an identical effect—some may have greater effect than others at the same or at different loci. Moreover, the effects of polygenes are not always additive, because dominance or interlocus interactions may be present.

The results of increasing the number of gene loci that affect a quantitative trait are shown in Figure 6.5 and Table 6.3. We assume that there are only two alleles at each locus, that gene effects are additive, and that alleles at different loci have equal effects. It is, moreover, assumed that there is a certain amount of variation due to environmental influences (bottom row in Figure 6.5). Environmental effects could be ignored in Nilsson-Ehle's experiment because, indeed, they may contribute little to kernel-color variation in wheat. But most quantitative traits are affected by environmental variations. Human weight, the amount of milk produced by a cow, and the size of an ear of corn, for example, are affected by nongenetic factors such as nutrition, climate, and disease.

The effects of polygenic inheritance are clear. The greater the number of gene loci that cumulatively affect a trait, the more nearly continuous the variation in the trait will be. Figure 6.5 shows that the number of genetic classes in the F_2 is the number of alleles plus 1: three classes with one locus (two alleles), five with two loci, seven with three loci, thirteen with six loci. Environmental factors contribute to make the distribution more nearly continuous.

Quantitative traits tend to be normally distributed because the proportion of genotypes is greater in the intermediate classes than in the extreme classes. This tendency becomes stronger as the number of loci affecting the trait increases. Consequently the *variance* (which measures the dispersion of the distribution—see Appendix A.III) decreases as the number of genes increases. It can be seen in Figure 6.5 (and in Table 6.3) that the proportion of individuals falling in each extreme class is $(\frac{1}{2})^2 = \frac{1}{4}$ for one locus, $(\frac{1}{2})^4 = \frac{1}{16}$ for two loci, $(\frac{1}{2})^6 = \frac{1}{64}$ for three loci, etc.

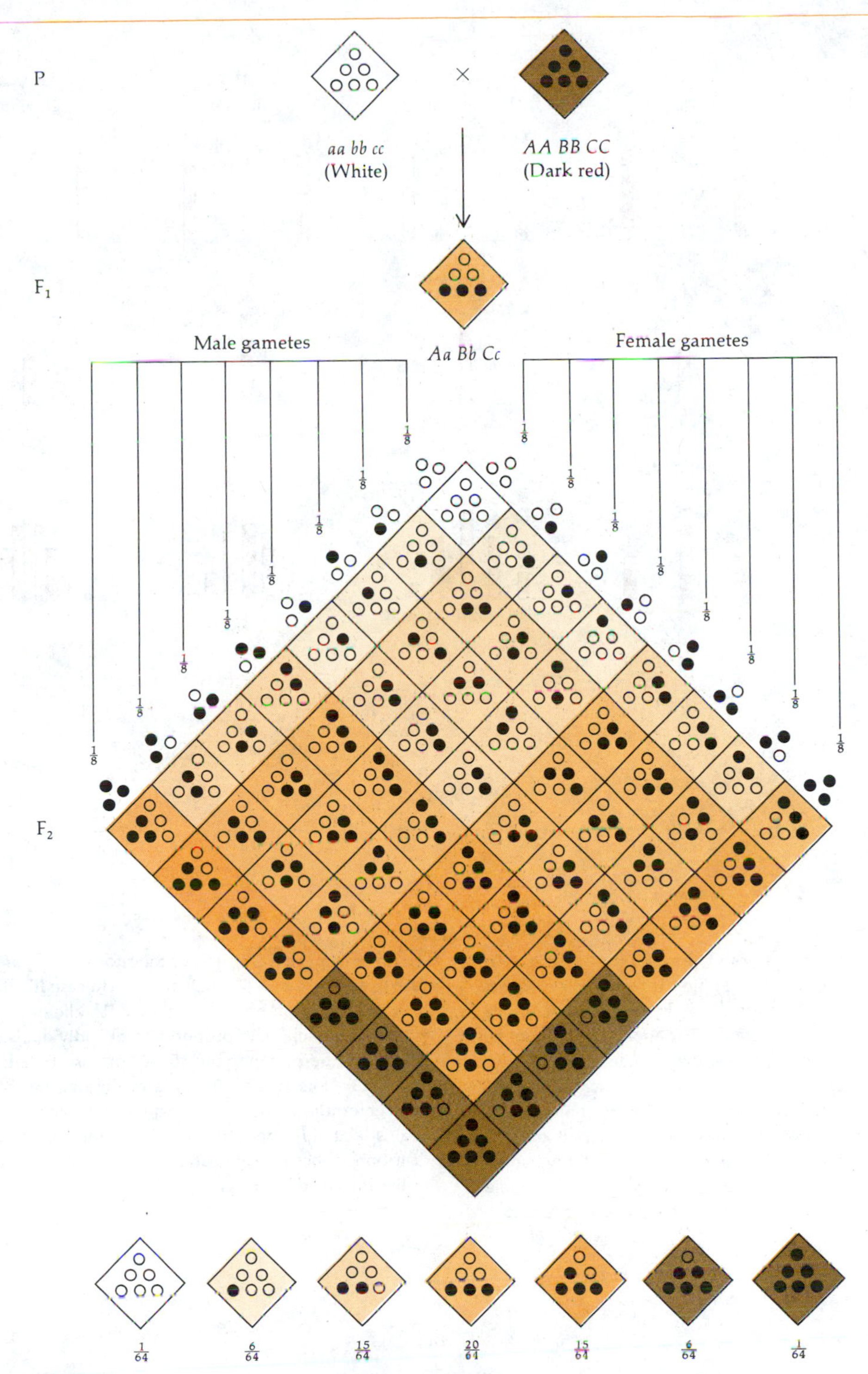

Figure 6.4
Kernel color in the F_2 of a cross between a white and a dark red variety of wheat. The color differences are assumed to be due to three pairs of genes. Seven different genotypes appear in the F_2 in the proportions indicated at the bottom of the figure. Capital letters represent alleles for redness.

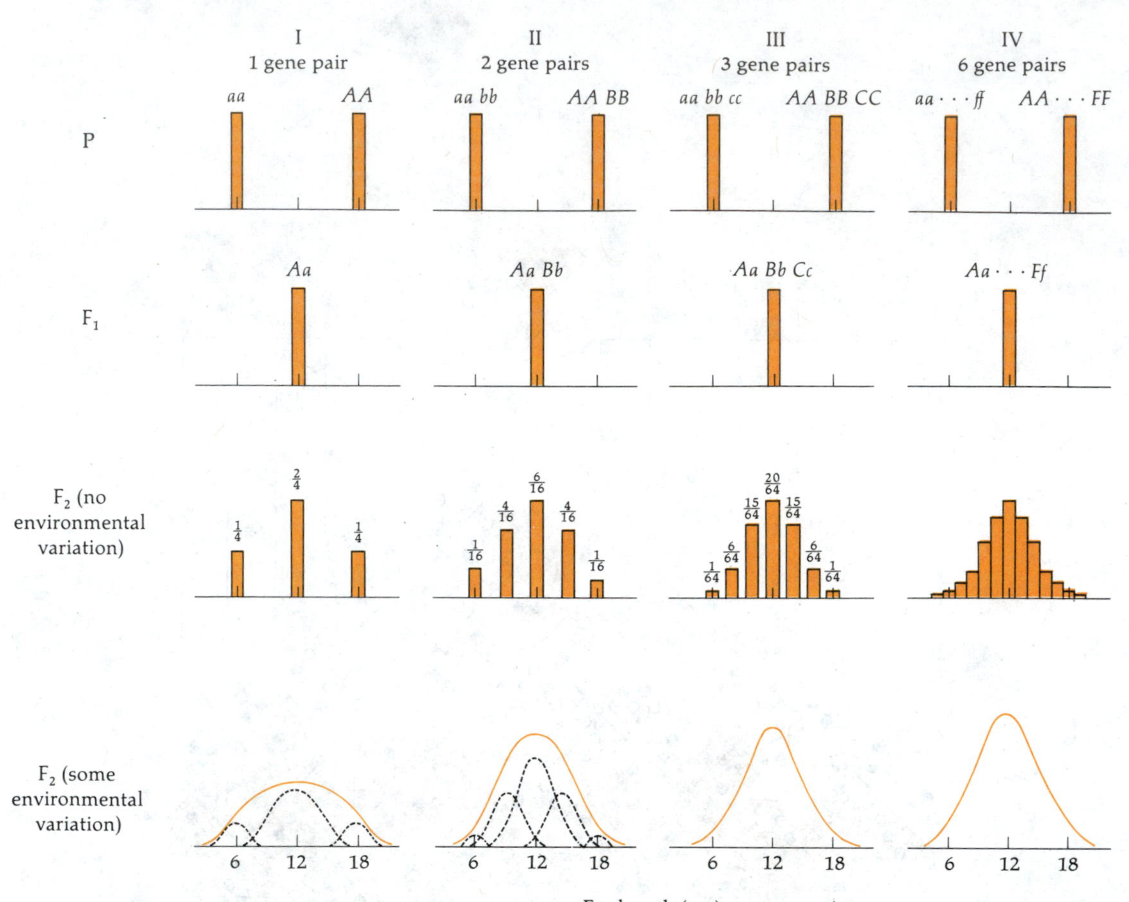

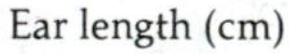

Figure 6.5
Crosses between two lines of plants differing at one, two, three, and six gene loci. The first three rows show the expected distributions for the P, F_1, and F_2 generations in the absence of environmental variation. The bottom row shows the F_2 distributions as they would appear when the environment contributes to the phenotypic variation. The parents in this hypothetical example are corn plants with ears that are 6 cm long in the short-eared parent and 18 cm long in the long-eared parent. In case I, the difference is assumed to be due to a single pair of genes, so that allele *A* causes an increase of 6 cm each time it occurs over the length of 6 cm characteristic of the short-eared parent; in case II, alleles *A* and *B* add 3 cm each; in case III, alleles *A*, *B*, and *C* add 2 cm each; and in case IV, alleles *A*, *B*, . . . , *F* add 1 cm each. The proportions of individuals falling in each of the F_2 genotypic classes of case IV are given in Table 6.3. Note that an increase in the number of gene pairs determining the same difference between two parental strains leads to a decrease in the variance of the F_2 distribution, because a greater proportion of individuals fall in the intermediate classes.

Table 6.3
Frequency distribution expected in the F_2 from hypothetical crosses between two lines of corn differing in ear length by 12 cm, when this difference is determined by one, two, three, and six pairs of genes with equal additive effects.

Gene Pairs	Frequency of Ear of Length (cm):* 6	7	8	9	10	11	12	13	14	15	16	17	18	Total
1	$\frac{1}{4}$						$\frac{2}{4}$						$\frac{1}{4}$	1
2	$\frac{1}{16}$			$\frac{4}{16}$			$\frac{6}{16}$			$\frac{4}{16}$			$\frac{1}{16}$	1
3	$\frac{1}{64}$		$\frac{6}{64}$		$\frac{15}{64}$		$\frac{20}{64}$		$\frac{15}{64}$		$\frac{6}{64}$		$\frac{1}{64}$	1
6	$\frac{1}{4096}$	$\frac{12}{4096}$	$\frac{66}{4096}$	$\frac{220}{4096}$	$\frac{495}{4096}$	$\frac{792}{4096}$	$\frac{924}{4096}$	$\frac{792}{4096}$	$\frac{495}{4096}$	$\frac{220}{4096}$	$\frac{66}{4096}$	$\frac{12}{4096}$	$\frac{1}{4096}$	1

*The fractions are the terms in the binomial expansion of $(\frac{1}{2} + \frac{1}{2})^n$, where n is the number of alleles (twice the number of loci, or gene pairs). The denominator is simply 2^n.

Owing to environmental variation, quantitative traits do not usually fall into classes that precisely reflect the genotype. This often makes it impossible to ascertain the number of genes affecting a quantitative trait by the kind of analysis performed by Nilsson-Ehle. The number of genes involved can, however, be estimated from the proportion of F_2 individuals falling in the parental classes.

East used the polygenic model of inheritance to explain variation in the length of flowers in *Nicotiana longiflora*. He crossed two varieties whose flowers had average lengths of 40.5 mm and 93.3 mm. The F_1 was intermediate in length, as expected. The F_2, consisting of 444 plants, had a broader distribution than the F_1, also as expected, but none of the F_2 flowers was either as short or as long as the average lengths of the parental flowers. With four pairs of genes, the expected proportion of F_2 individuals falling in each parental class is $(1/2)^8 = 1/256$. Because none of the 444 F_2 plants fell in the parental classes, it can be concluded that more than four gene pairs are involved in the flower-length differences between the two parental varieties (Figure 6.6).

Skin pigmentation in humans varies from light to dark. The most extreme differences occur between Caucasians and African Blacks. Although differences in skin color exist within each group, they are small relative to the differences between the groups. Matings between Blacks and Caucasians produce children of intermediate skin color; matings between F_1 individuals and their backcrosses to Blacks or Caucasians produce children with a range of skin pigmentations. The distribution of these F_2 and backcross individuals is approximately as expected if the difference in skin pigmentation between Blacks and Caucasians were due to three or four pairs of genes, each contributing equally to skin color (Figure 6.7).

Genetic and Environmental Variation

Francis Galton (1822–1911) used the terms *nature* and *nurture* to refer to the roles played by heredity and environment in determining quantitative traits. Both types of influence—genetic and environmental—are usually present. A well-fed mouse will be larger than a starved one, and a well-schooled person will have a higher IQ than one raised in a deprived environment. But variation among individuals can also be due to genetic differences. It is interesting—and practical, for example, in plant and animal breeding—to separate the genetic from the environmental effects on quantitative traits.

We might ask the question: Is it possible to ascertain to what extent a given trait is determined by the genotype and to what extent by the environment? Reflecting on this question makes us realize that it is not well formulated. Any trait depends completely for its development on both heredity *and* environment. For an individual to develop, it must have a genotype—the genetic constitution of the zygote—but development can

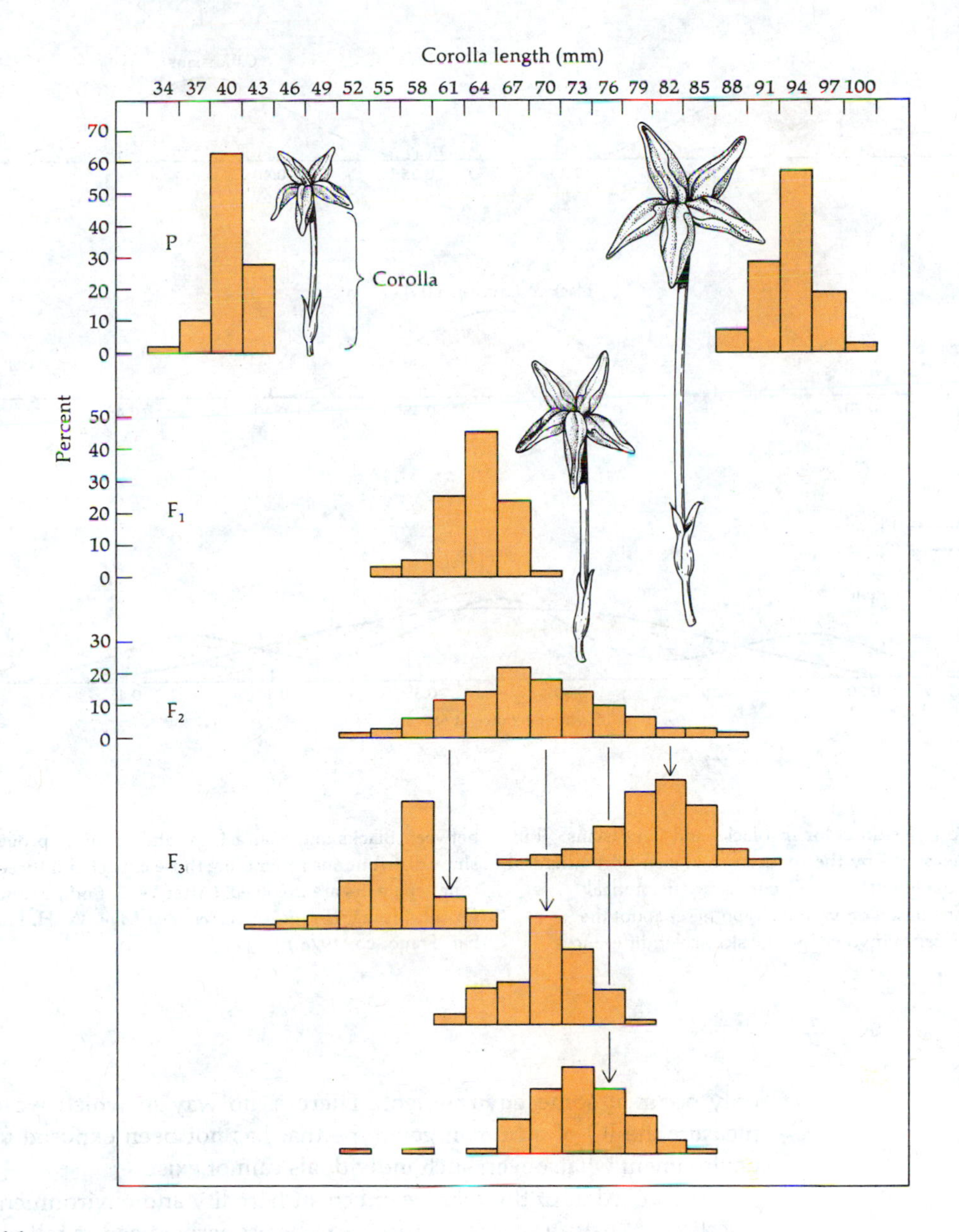

Figure 6.6
Corolla length in *Nicotiana longiflora*. The distribution of the proportion of individuals falling in each class is shown for the P, F_1, F_2, and F_3 generations. Corolla length varies continuously, but the flowers are grouped in classes each covering a range of 3 mm. The distributions of the F_1 and F_2 have means intermediate between the two parental strains, but the F_2 has a greater variance. The F_3 progenies demonstrate that part of the variance observed in the F_2 is genetic, because F_2 plants with different corolla lengths give rise to different distributions whose mean corolla lengths correspond approximately to the lengths of the F_2 plants from which they are produced.

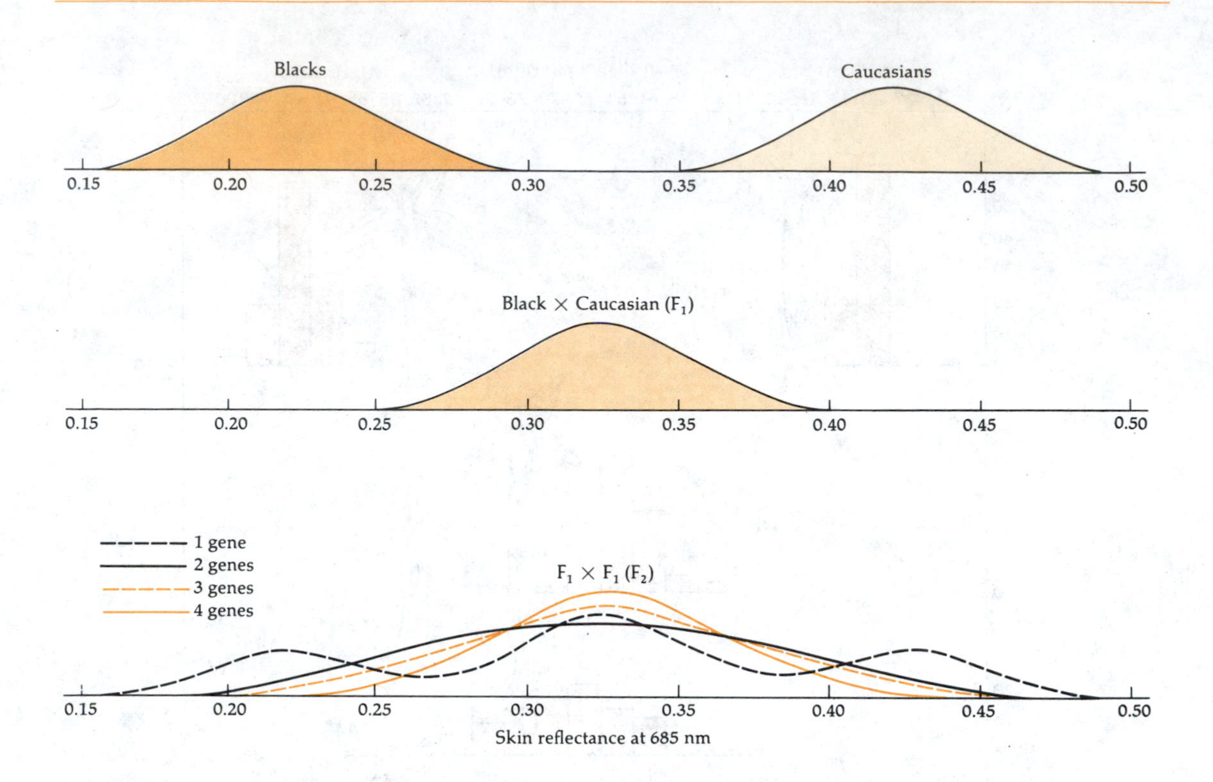

Figure 6.7
Distribution of skin color in Blacks and Caucasians. Skin color is measured by the reflectance of skin to red light of 685 nm wavelength. The F_2 curves are theoretical expectations based on various hypotheses about the number of genes involved in the skin-color differences between Blacks and Caucasians. Studies of F_2 progenies show distributions resembling those expected if three or four gene pairs are involved. (After W. F. Bodmer and L. L. Cavalli-Sforza, *Genetics, Evolution, and Man*, W. H. Freeman, San Francisco, 1976.)

only occur in some environment. There is no way in which we could measure the IQ of a human genotype that has not been exposed to any environment whatsoever; such individuals cannot exist.

The question of the relative effects of heredity and environment, the question of "nature versus nurture," can be properly raised as follows: To what extent is the *variation* among individuals with respect to a trait due to *genetic* variation (i.e., to genetic differences among the individuals), and to what extent is it due to *environmental* variation (i.e., to environmental differences)? It will soon be apparent why it is important to realize that this question, rather than the one formulated above, is the one being answered when geneticists investigate the relative effects of heredity and environment.

The fraction of the phenotypic variation in a trait that is due to genetic differences can be measured by the *heritability* of the trait, a concept advanced by the American geneticist Jay L. Lush. Let us use the following symbols:

H = heritability
V_T = the total phenotypic variance observed in a trait
V_G = the fraction of the phenotypic variance that is due to genetic differences among individuals
V_E = the fraction of the phenotypic variance that is due to differences in the environmental conditions to which the individuals were exposed

We thus have $V_T = V_G + V_E$, and, by definition,

$$\text{Heritability} = \frac{\text{Genetic variance}}{\text{Phenotypic variance}} \qquad H = \frac{V_G}{V_T} = \frac{V_G}{V_G + V_E}$$

Measuring the total phenotypic variance of a trait in a group of individuals is usually not difficult. (First the mean value of the trait is calculated; then the differences between each value and the mean are obtained and squared; the average of these squared differences is the variance—see Appendix A.III.) However, partitioning the total variance into the environmental and genetic components is not a simple matter. Geneticists use a variety of methods that will not be reviewed here. The rationales of the *twin method* and the *mass selection method* are presented in Boxes 6.1 and 6.2. Here we shall use East's data on flower length to illustrate the concept (Figure 6.6).

The varieties of *Nicotiana longiflora* crossed by East were homozygous. Hence the variance within each parental group is all environmental. The variance among the F_1 offspring is all environmental as well; all F_1 individuals are genetically identical to each other (although not homozygous) because all gametes produced by each parental strain are identical. The average variance within each of the two parental varieties and the F_1 offspring is 8.76, which is, therefore, an estimate of the environmental variance (V_E) in the environment where the experiments were conducted. Thus we have $V_E = 8.76$.

The genes inherited from the two parental strains segregate in the F_2. Consequently, the phenotypic variance of the F_2 consists of both genetic and environmental variance. The total phenotypic variance (V_T) of the F_2 offspring is 40.96; thus $V_T = 40.96$. Because $V_T = V_G + V_E$, it follows that

$$V_G = V_T - V_E = 40.96 - 8.76 = 32.20$$

The heritability of flower length in East's experiment is therefore

$$H = \frac{V_G}{V_T} = \frac{32.20}{40.96} = 0.79$$

Box 6.1 Heritability by the Twin Method

One way of estimating heritability is to measure the phenotypic variance in groups of relatives with known degrees of relatedness, such as twins or sibs or first cousins. There are two kinds of twins: identical and fraternal. *Identical*, or monozygotic, twins arise from a single zygote that splits in two early in embryogenesis and develops into two genetically identical individuals. *Fraternal*, or dizygotic, twins arise from two independent zygotes, i.e., two eggs fertilized by two sperm. The genetic relationship between fraternal twins is the same as that between ordinary full-sibs, because in both instances zygotes form from different eggs from the same mother and different sperm from the same father. Fraternal twins, like full-sibs, share on the average half their genes (for each gene received by one sib from the mother, the probability that the other sib will receive the same gene is one-half, and the same for the paternal genes).

Identical twins are always of the same sex; fraternal twins may be both female, one female and one male, or both male, in the proportions 1:2:1. In order to estimate heritability, sets of identical twins of one sex are compared with sets of fraternal twins of the same sex as the identical twins. Let V_i = phenotypic variance between identical twins and V_f = phenotypic variance between fraternal twins. Because the identical twins are genetically identical, their variance is all environmental: $V_i = V_E$

The variance between fraternal twins is part genetic and part environmental. However, because they have half of their genes in common, their genetic variance will be half that of unrelated individuals:

$$V_f = \tfrac{1}{2}V_G + V_E$$

Therefore the difference between the two phenotypic variances estimates half the genetic variance:

$$V_f - V_i = \tfrac{1}{2}V_G + V_E - V_E = \tfrac{1}{2}V_G$$

Twice that difference divided by the total phenotypic variance is an estimate of heritability:

$$H = \frac{2(V_f - V_i)}{V_T} = \frac{2(\frac{1}{2}V_G)}{V_T} = \frac{V_G}{V_T}$$

There are, nevertheless, some problems. The environmental variance has been assumed to be the same for identical and for fraternal twins of the same sex, because in both cases the twins are born at the same time, are of the same sex, live together, go to the same schools, etc. But it is possible that identical twins are treated more similarly than fraternal twins, precisely because identical twins are genetically more alike and thus tend to react more similarly. If the V_E of fraternal twins is greater than the V_E of identical twins, the difference between the two phenotypic variances contains not only genetic variance but also some residual environmental variance. The formula given above will then overestimate heritability.

Another difficulty is that twins are treated more similarly than unrelated individuals because the twins are raised in the same family, cultural milieu, etc. This and the problem raised in the previous paragraph can be largely circumvented by studying twins who have been adopted by different families. Although the proportion of such cases is limited, a number of them have been studied by geneticists.

Yet other problems remain that may also affect estimates of heritability. These problems need not concern us here, however, because the purpose of this discussion is only to understand the rationale and basic methodology in estimating heritability using twin studies.

Box 6.2 Heritability by the Mass Selection Method

Plant and animal breeders improve domestic stocks by *mass selection*, one form of artificial selection. The procedure is simply to breed the next generation from those individuals that are best with respect to the trait being improved. Dairies breed the cows producing more milk, racers breed the swiftest horses, farmers use for seed wheat plants with the greatest yield, and so on. Artificial selection has been practiced for more than 10,000 years. If phenotypic variation is due in part to genetic differences among individuals, mass selection will improve the genetic characteristics of the stock.

Assume that the trait under consideration is normally distributed, as is often true for quantitative traits. For purposes of illustration, we might think of flower length in *Nicotiana longiflora* (Figure 6.6). We need to be concerned with two measures: the *selection differential*, which is the difference between the mean of the selected parents and the mean of the population, and the *selection gain*, which is the difference between the mean of the progeny of the selected parents and the mean of the parental generation (Figure 6.8).

If the variation is all due to environmental causes, the distribution of the selected parents' offspring will be the same as that of the parental populations (Figure 6.8, left). This was the situation observed by Johannsen when he bred beans of different sizes from the same inbred line (Table 6.1). If the phenotypic variation is all genetic, on the other hand, the mean of the offspring will be the same as that of the selected parents (Figure 6.8, center). This would be the case with respect to kernel color in wheat (Figure 6.4). The most common situation is that the variation is partly genetic and partly environmental. In such cases there will be a selection gain, but the mean of the offspring will not be as high as that of the selected parents. Heritability will then be estimated by the ratio between the selection gain and the selection differential. If we let G represent the selection gain and D represent the selection differential, we have

$$H = \frac{G}{D}$$

For example, the number of certain abdominal bristles in a strain of *Drosophila melanogaster* is 38. Flies with an average of 42.8 bristles were used to breed the next generation; therefore $D = 42.8 - 38 = 4.8$. The mean number of bristles in the offspring of the selected parents was 40.6; therefore $G = 40.6 - 38 = 2.6$. The heritability is then estimated as

$$H = \frac{G}{D} = \frac{2.6}{4.8} = 0.54$$

This estimate agrees very well with an estimate of $H = 0.52$ obtained in a separate experiment by comparing half-sibs (which share one-quarter of their genes) with full-sibs (which share one-half of their genes).

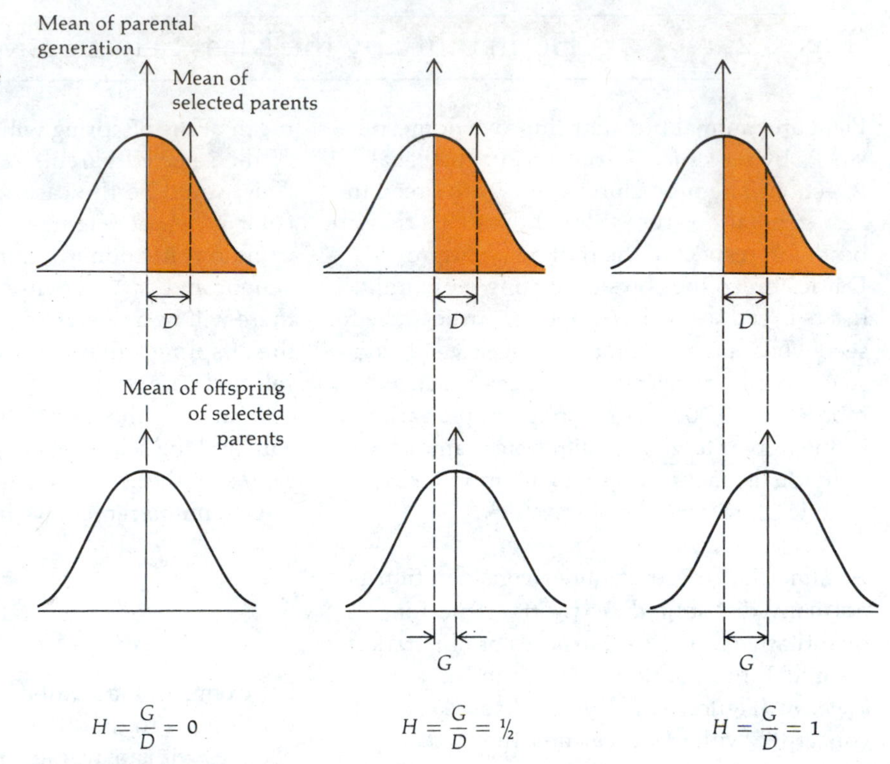

Figure 6.8
Selection differential (D) and selection gain (G). The selection gain divided by the selection differential estimates the heritability (H) in artificial selection experiments.

It would be incorrect, however, to say that flower length in *N. longiflora* is 79% determined by genes and 21% determined by the environment. What was said earlier must now be emphasized: Heritability measures *not the degree to which a trait is determined by genes,* but rather *the proportion of the phenotypic variation among individuals that is due to genetic variation.*

It must be pointed out that interactions between genes (dominance between alleles at the same locus as well as epistasis between alleles at different loci) affect heritability estimates, because such interactions are not additive. How to deal with these genetic interactions is a matter for texts more advanced than this one; awareness of their existence is sufficient here.

There is, however, another class of interactions affecting heritability estimates that is most important for understanding the significance of these estimates, namely, the interactions between genes and environments. Heritability estimates are valid only for the particular environment in which they are obtained. In other environments, they may be quite different. Consider, for example, the experiment illustrated in Figure 1.4. In the normal environment used for selecting bright and dull rats, the genetic differences between the two strains result in a considerable difference in phenotype: bright rats perform much better than dull rats. However, the difference between the two strains disappears when the rats are raised in a deprived environment. This means that the genetic differences between the strains are not expressed as phenotypic differences in the restricted environment; therefore, the heritability for brightness is greater in the normal environment than in the restrictive environment. Additional examples showing that heritability estimates are valid only for the environment in which they were obtained are given in the following section.

Despite their restricted significance, heritability estimates are quite useful in plant and animal breeding because they indicate the amount of response that can be expected in artificial selection of desirable traits. Heritability estimates also give us an idea of the role played by genetic differences in determining phenotypic variation among organisms living in the same environment. Some heritability estimates for various traits in animals and plants are given in Table 6.4. These values might, of course, be different if they were estimated under environmental conditions different from those under which they *were* obtained.

Heritability in Different Populations

Heritability is a *population-specific* measurement. It does not measure any invariant property of organisms, but only the relative contributions of genetic differences and environmental differences to phenotypic variation. If the genetic variation or the environmental variation is changed, heritability estimates will also change. Thus, measuring heritability for a group of organisms in two different environments, or for two different populations in the same environment, is likely to yield different results. These qualifications have already been stated, but they will be further developed here because of (1) their fundamental importance in understanding the concept of heritability and (2) their implications with respect to issues such as the much-debated question of whether IQ differences between human racial groups are genetic.

Consider the following "thought experiments," which we shall assume to have been done with *Potentilla glandulosa* (Figure 1.3). This plant can be reproduced by cuttings that make it possible to obtain a group of individuals genetically identical to each other when they are all derived from parts of a single plant.

EXPERIMENT 1. We cut one plant into many equal parts and place these parts in a very heterogeneous hillside where plants may be exposed to considerable environmental differences in the quality of the soil, the

Table 6.4
Examples of heritability for traits in various organisms (estimated by various methods).

Trait	Heritability
Amount of white spotting in Frisian cattle	0.95
Slaughter weight in cattle	0.85
Plant height in corn	0.70
Root length in radishes	0.65
Egg weight in poultry	0.60
Thickness of back fat in pigs	0.55
Fleece weight in sheep	0.40
Ovarian response to gonadotropic hormone in rats	0.35
Milk production in cattle	0.30
Yield in corn	0.25
Egg production in poultry	0.20
Egg production in *Drosophila*	0.20
Ear length in corn	0.17
Litter size in mice	0.15
Conception rate in cattle	0.05

amounts of sunlight and moisture, etc. We measure phenotypic variation in a certain trait, say, the total weight (biomass) of each plant. Because the plants are genetically identical, the variation is all environmental. The heritability of the trait as measured in these plants is therefore zero.

EXPERIMENT 2. We collect plants in different localities, so that they are genetically heterogeneous. We plant a small cutting from each plant in an experimental garden. We provide the plants with optimal conditions of soil, fertilizer, moisture, light, etc. We take great care that all plants receive uniform treatment. We estimate heritability in these plants for the same trait as in Experiment 1; we might obtain a very high value, say, 0.95, as we would expect, because the plants are genetically very heterogeneous and the environment quite uniform.

EXPERIMENT 3. We use cuttings from the same set of plants used for Experiment 2 and proceed similarly, except that we provide the plants with poor soil, no fertilizer, and marginal levels of moisture and light. Nevertheless, we make sure that all plants are treated uniformly. At the end of the growing season, the plants are very small owing to the poor environmental conditions. We estimate heritability and obtain a high value, as in Experiment 2, and for the same reasons: the plants are genetically heterogeneous but the environment is uniform.

EXPERIMENT 4. We use cuttings from the same set of plants as in Experiments 2 and 3 and plant them on a hillside where the plants grow naturally. We estimate heritability at the end of the growing season and obtain an intermediate value, say, 0.60. This experiment corresponds better than any of the other three to the usual conditions found for animals

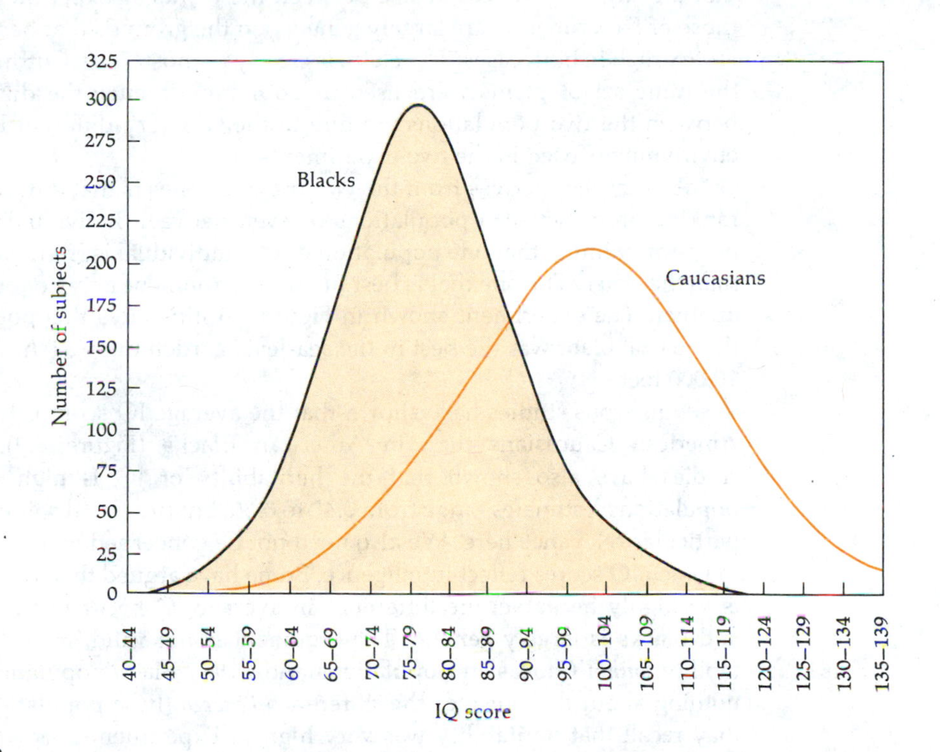

Figure 6.9
Distribution of IQ in Blacks and Caucasians. A sample of 1800 Black children from schools in Alabama, Florida, Georgia, Tennessee, and South Carolina is compared with a representative sample of Caucasians. The mean IQs in these distributions are 80.7 and 101.8. This difference of 21.1 points is larger than that observed in most other studies. The difference in mean IQ between U.S. Blacks and Caucasians is often around 15 points.

and plants in nature, where there is variation in both the genotype and the environment.

These thought experiments illustrate, first, a point already made: heritability estimates do *not* measure the degree to which a trait is determined by genes, but rather the proportion of the *phenotypic variation* among individuals that is due to *genetic variation.* Estimates of heritability are population-specific, valid only for a given population in a given environment. In all four experiments we have estimated the heritability of the same trait, the total biomass of a plant. Yet the heritability of the trait is 0 in Experiment 1, 0.95 in Experiments 2 and 3, and 0.60 in Experiment 4. The heritability of the trait is different when estimated in different groups of individuals (as in Experiments 1 versus 2) or in different environments (as in Experiments 3 versus 4).

The experiments also illustrate a point of practical importance. From the fact that the heritability of a trait is high in each of two populations, *it does not follow* that average differences between the two populations are largely due to genetic differences. Assume that somebody would claim that the large differences in size between the plants of Experiment 2 and those of Experiment 3 are largely genetic, on the grounds that heritability is very high in both cases. The claim is clearly preposterous. Cuttings from the same set of plants were used in both experiments; the differences between the two populations are due to the two very different kinds of environments used in the two experiments.

A point that derives from the two previous ones is that differences in ranking order between populations or even between individuals do not necessarily imply that one population or one individual is *genetically* better than the other. The one that is best in one environment may not be best in another. The experiment shown in Figure 1.3 illustrates this point well: the coastal plant was the best in the sea-level garden but was the worst at 10,000 feet.

Numerous studies have shown that the average IQ score is higher in American Caucasians than in American Blacks (Figure 6.9). Other studies have also shown that the heritability of IQ is high in both populations; estimates range from 0.40 to 0.80, but the actual value is of no particular relevance here. We also need not be concerned with the extent to which IQ scores reflect intelligence. Some have argued that, because IQ is so highly heritable, the difference in average IQ between Caucasians and Blacks is largely genetic. The argument is not valid. High IQ heritability *within* Caucasian populations and *within* Black populations tells nothing about the cause of the difference *between* these populations. We may recall that heritability was very high in Experiment 2 as well as in Experiment 3, and that the plants of Experiment 2 performed much better than those of Experiment 3. Nevertheless, the difference was not genetic at all, but rather the two sets of plants were genetically identical and the difference between the two sets was entirely environmental.

Human populations are genetically different from each other. There is also considerable genetic heterogeneity within any one human population. These differences affect IQ as well as many other characters. But heritability studies do not allow us to conclude that one population is genetically "better" than another with respect to IQ. Moreover, IQ rankings might be quite different if cultural environments were significantly changed.

Problems

1. In maize, the development of scutellum color requires the presence of any two of the three genes S_2, S_3, and S_4. Give the ratios expected with respect to scutellum color in the F_2 of the following crosses:

   (a) $S_2S_2s_3s_3s_4s_4$ (colorless) $\times$ $s_2s_2S_3S_3s_4s_4$ (colorless)
   (b) $S_2S_2s_3s_3S_4S_4$ (colored) $\times$ $s_2s_2S_3S_3S_4S_4$ (colored)

2. Artificial selection often ceases to be effective after it is practiced for a number of generations. When two independently selected stocks in which selection is no longer effective are intercrossed, a selection response may be obtained in the following generations when selection is practiced in progenies from the crosses. How can this result be explained?

3. When pure parental strains differing with respect to a size character are intercrossed, the F_1 is usually no more variable than the parents, whereas the F_2 is considerably more variable. Why?

4. Assume that two highly inbred strains of oats consistently yield about 4 g and 10 g per plant, respectively. The two strains are intercrossed, and the F_1 is selfed. About 1/64 of the F_2 plants yield about 10 g per plant. How many genes are likely to be responsible for the difference between the two original inbred strains?

5. When the F_2 plants from the previous problem are selfed, it is observed that F_3 families differ markedly in their variability. Some have as little variation as the original parents, some have somewhat more variation, and some have as much variation as the F_2 itself. How do you explain this? Would you expect any F_3 family to exhibit more variability than the F_2? Why?

6. Assume that the difference between a corn plant with 6-cm-long ears and one with 18-cm-long ears is due to: (1) two pairs of genes; (2) three pairs of genes; (3) four pairs of genes. Assume also that in each case the genes have

equal and cumulative effects on ear length and are inherited independently. The two plants are crossed, and the F_1 is backcrossed to the long-eared parent. What proportion of the progeny is expected to produce 18-cm-long ears in each of the three cases?

7. Nilsson-Ehle crossed two types of oats, one with white seeds, the other with black seeds. The F_1 between them had black seeds. The F_2 consisted of 560 plants as follows: 418 black, 106 gray, and 36 white. How can the inheritance of seed color be explained in this case?

8. The following heritability estimates were obtained for a strain of cultivated strawberry: (1) yield, 0.48; (2) firmness of fruit, 0.46; (3) size of fruit, 0.20. The mean values and standard deviations in the parental generation were: (1) 380 ± 91 g; (2) 4.4 ± 0.6 (the method and units of measurement for firmness need not concern us here); (3) 11.3 ± 3.0 g. Assume that we produce a new generation, using as parents plants that are two standard deviations above the mean. What will be the expected gain for each of the three traits after one generation of selection?

9. The mean weight of 6-week-old mice in a laboratory population is 21.5 grams. Two sets of parents are used to produce the following generation: (1) heavy mice, with a mean weight of 27.5 g, and (2) light mice, with a mean weight of 15.5 g. The mean weights of the progenies when 6 weeks old are: (1) 22.7 g and (2) 18.1 g. Calculate the heritability of weight in each of the two sets of progenies.

10. The method for estimating heritability by the twin method, given in Box 6.1, requires knowing the total phenotypic variance for the trait in the population. When only the variance between twins is known, heritability can be estimated by the following formula:

$$H = \frac{V_f - V_i}{V_f}$$

where V_f and V_i are the phenotypic variances between fraternal twins and between identical twins, respectively. The values of V_f and V_i are calculated by obtaining the means of the squares of the differences between the two members of twin pairs.

The differences in IQ between the two twins of 10 identical pairs and 10 fraternal pairs are as follows (all 20 pairs of twins are males, and the two twins were raised together in every case):

Pair	1	2	3	4	5	6	7	8	9	10
Identical twins	4	7	5	3	6	1	9	7	3	7
Fraternal twins	12	4	9	7	7	11	13	10	9	9

Estimate the heritability of IQ based on this set of data.

7

Speciation and Macroevolution

Anagenesis and Cladogenesis

The process of evolution has two dimensions: (1) *anagenesis,* or evolution within a lineage, and (2) *cladogenesis,* or diversification. Changes occurring in a lineage as time passes are anagenetic evolution. They are often due to natural selection, which promotes adaptation to physical or biotic changes in the environment. Cladogenetic evolution occurs when a lineage splits into two or more lineages. The great diversity of the living world is the result of cladogenetic evolution, which results in adaptation to a greater variety of niches, or ways of life. The most fundamental cladogenetic process is *speciation,* the process by which a species splits into two or more species.

The previous chapters have been concerned with evolution as it occurs within a species; this is sometimes called *microevolution* (small-scale evolution). Correspondingly, evolution above the species level is called *macroevolution* (large-scale evolution). The genetic study of macroevolution has been made possible by the advances in molecular biology. The classical methods of Mendelian genetics ascertain the presence of genes by observing segregation in the progenies of crosses between individuals differing in some trait. But interspecific crosses are usually impossible and, when they do take place, the hybrid offspring are usually inviable or sterile. Genetic comparisons between different species can now be made by direct examination of their DNA or of the proteins encoded by the DNA.

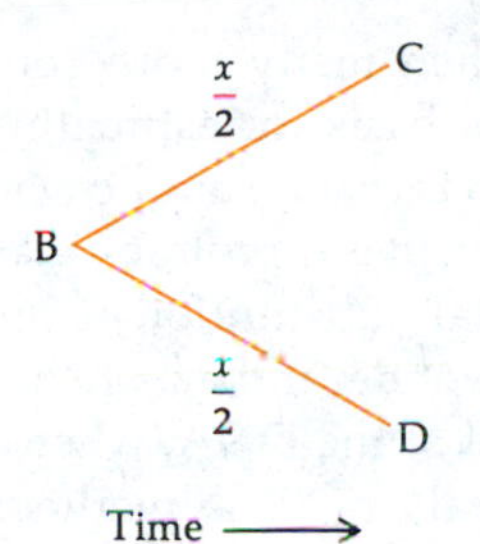

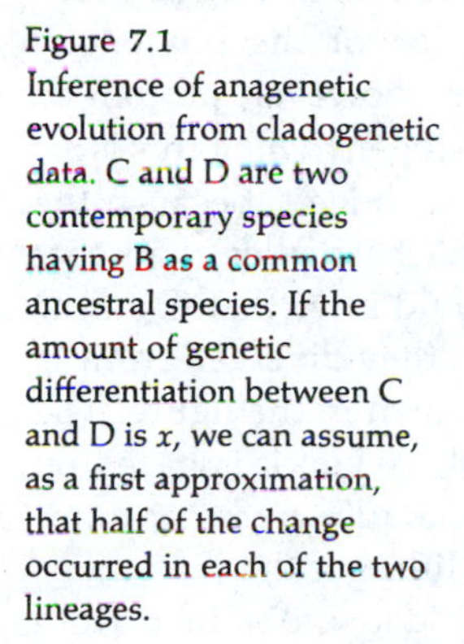

Figure 7.1
Inference of anagenetic evolution from cladogenetic data. C and D are two contemporary species having B as a common ancestral species. If the amount of genetic differentiation between C and D is x, we can assume, as a first approximation, that half of the change occurred in each of the two lineages.

It might at first seem that the genetic study of anagenetic evolution is impossible, because it would require the study of organisms that lived in the past. Extinct organisms are sometimes preserved as fossils, but their DNA and proteins are usually disintegrated. Nevertheless, the study of cladogenesis provides information about anagenesis. Consider two contemporary species, C and D, evolved from a common ancestral species, B. Assume that we find that C and D differ by x amino acid substitutions in a certain protein, say, myoglobin. It is reasonable to assume, as a first approximation, that $x/2$ substitutions have taken place in each of the two evolutionary lineages, i.e., from B to C and from B to D (Figure 7.1).

The assumption that equal amounts of change have occurred in the two lineages can be dropped. Suppose that a third contemporary species, E, is compared with C and D and that the numbers of amino acid differences between the myoglobin molecules of the three species are as follows:

C and D: 4

C and E: 11

D and E: 9

If the *phylogeny* (evolutionary history) of the three species is as shown in Figure 7.2, we can estimate the number of substitutions that occurred in each of its branches. Let us use x and y to denote the number of amino acid differences between B and C and between B and D, respectively, and z to denote the number of differences between A and B *plus* those between A and E. We have, then, the following three equations:

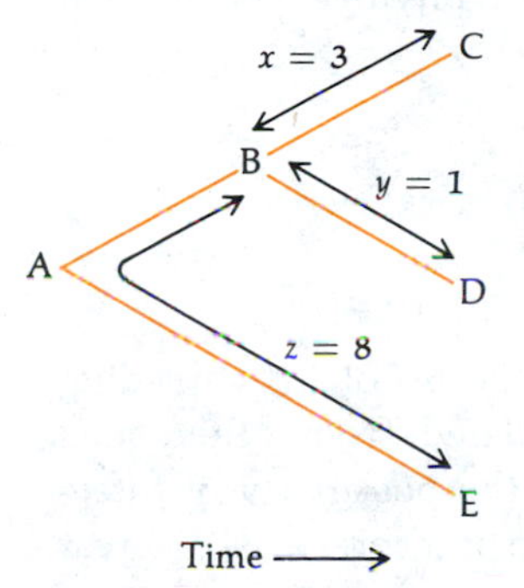

Figure 7.2
Estimated amounts of anagenetic change in the phylogeny of three contemporary species.

$$x + y = 4$$

$$x + z = 11$$

$$y + z = 9$$

Subtracting the third equation from the second, we get

$$x - y = 2$$

Adding this equation to the first, we get

$$2x = 6 \quad \text{or} \quad x = 3$$

Therefore

$$y = 4 - x = 1$$

$$z = 11 - x = 8$$

The procedure becomes more complicated when many more contemporary species are involved, but the conceptual basis for estimating anagenetic change is the same. An inherent problem is that an amino acid substitution that occurred in the past—say, from leucine to proline—may subsequently have been canceled by the reciprocal substitution—from proline to leucine—at the same position in the polypeptide, and may hence go undetected. The same problem exists at the level of the DNA, when a nucleotide substitution is canceled by the reciprocal one. The methods used to correct for such hidden substitutions need not be discussed here.

We have assumed above that the configuration of the phylogeny shown in Figure 7.2 was previously known. In fact, however, protein or DNA information may serve to reconstruct the phylogeny when this is *not* known from other sources or when it is doubtful. Indeed, because the number of substitutions between C and D is much smaller than that between either one of these species and E, we would infer that C and D have diverged from each other more recently than they diverged from E. We would thus arrive at the same phylogeny as shown in the figure. The reconstruction of phylogenetic history may be risky when it is based on the analysis of a single protein or DNA molecule, because a greater number of substitutions may have taken place in some lineages than in others, or at different times. However, the cumulative evidence obtained from many proteins studied in many species tends to converge upon phylogenies that are usually in good agreement with phylogenies deduced from morphological and paleontological evidence.

The Concept of Species

In sexually reproducing organisms, a *species* is a group of interbreeding natural populations that are reproductively isolated from other such groups. Species are natural systems, defined by the possibility of interbreeding between their members. The ability to interbreed is of great evolutionary import, because it establishes species as discrete and independent *evolutionary units*. Consider an adaptive mutation or some other genetic change originating in a single individual. Over the generations, this may spread by natural selection to all members of the species, but not to individuals of other species. This can be stated differently: individuals of a species share in a common gene pool, which is not, however, shared in by individuals of other species. Owing to reproductive isolation, different species have independently evolving gene pools.

Reproductive isolation is the criterion of speciation in sexual organisms. An ancestral species becomes transformed into two descendant species when an array of populations able to interbreed becomes segregated into two reproductively isolated arrays. It should not be surprising that reproductive isolation is used as the fundamental criterion to

Table 7.1
Classification of reproductive isolating mechanisms (RIMs).

1. *Prezygotic RIMs,* which prevent the formation of hybrid zygotes.
   a. *Ecological isolation:* populations occupy the same territory but live in different habitats, and thus do not meet.
   b. *Temporal isolation:* mating or flowering occur at different times, whether in different seasons or at different times of day.
   c. *Behavioral isolation* (also called *ethological isolation,* from the Greek *ethos,* "behavior," or *sexual isolation*): sexual attraction between females and males is weak or absent.
   d. *Mechanical isolation:* copulation or pollen transfer is forestalled by the different size or shape of genitalia, or the different structure of flowers.
   e. *Gametic isolation:* female and male gametes fail to attract each other, or the spermatozoa or pollen are inviable in the sexual ducts of animals or in the stigmas of flowers.

2. *Postzygotic RIMs,* which reduce the viability or fertility of hybrids.
   a. *Hybrid inviability:* hybrid zygotes fail to develop or at least to reach sexual maturity.
   b. *Hybrid sterility:* hybrids fail to produce functional gametes.
   c. *Hybrid breakdown:* the progenies of hybrids (F_2 or backcross generations) have reduced viability or fertility.

define species—reproductive isolation allows gene pools to evolve independently.

The biological properties of organisms that prevent interbreeding are called *reproductive isolating mechanisms* (RIMs). A classification of RIMs is shown in Table 7.1. Reproductive isolating mechanisms may be classified as prezygotic and postzygotic. *Prezygotic* RIMs impede hybridization between members of different populations, and thus prevent the formation of hybrid zygotes. *Postzygotic* RIMs reduce the viability or fertility of hybrids. Prezygotic and postzygotic RIMs all serve the same purpose: they forestall gene exchange between populations. But there is an important difference between them: the waste of reproductive effort is greater for postzygotic RIMs than for prezygotic RIMs. If a hybrid zygote is produced but is inviable (*hybrid inviability*—see Table 7.1), two gametes that could have been used in nonhybrid reproduction have been wasted. If the hybrid is viable but sterile (*hybrid sterility*), the waste includes not only the gametes but also the resources used by the hybrid in its development. The waste is even greater in *hybrid breakdown,* because it involves the resources used not only by the hybrids but also by their progenies. One prezygotic RIM, *gametic isolation,* may also involve reproductive waste when gametes fail to form viable zygotes. The other prezygotic RIMs avoid gametic waste, but some energy may be wasted in unsuccessful courtship (*behavioral isolation*) or in unsuccessful attempts to copulate (*mechanical isolation*). Natural selection promotes the development of prezygotic RIMs between populations already isolated by postzygotic RIMs, whenever the populations coexist in the same territory and there is opportunity for the formation of hybrid zygotes. This occurs precisely because reproductive waste is reduced or altogether eliminated when prezygotic RIMs exist.

The reproductive isolating mechanisms listed in Table 7.1 do not all occur between any two species, but it is usually true that two or more mechanisms, not just a single one, are involved in the reproductive isolation between species. Some RIMs are more common in plants (e.g., temporal isolation), others, in animals (e.g., behavioral isolation); but even among closely related species, different sets of RIMs are often involved when different pairs of species are compared. This is an example of the opportunism of natural selection: the evolutionary function of RIMs is to prevent interbreeding, but how this is accomplished depends on environmental circumstances as well as on the genetic variability available.

The Process of Speciation

Species are reproductively isolated groups of populations. The question of how species come about is, therefore, equivalent to the question of how reproductive isolation arises between groups of populations. In general, reproductive isolation usually starts as an incidental byproduct of genetic divergence, but is completed when it becomes directly promoted by natural selection. Speciation may occur in a variety of ways, but two main stages can be recognized in the process (Figure 7.3).

STAGE I. The onset of the speciation process requires, first, that gene flow between two populations of the same species be somehow interrupted—completely, or nearly so. Absence of gene flow enables the two populations to become genetically differentiated as a consequence of their adaptation to different local conditions or to different ways of life (and also as a consequence of genetic drift, which may play a lesser or greater role, depending on the circumstances). The interruption of gene flow is necessary, because otherwise the two populations would in fact share in a common gene pool and fail to become genetically different. As populations become ever more genetically different, reproductive isolating mechanisms appear, because different gene pools are not mutually coadapted; hybrid individuals will have disharmonious genetic constitutions and will have reduced fitness in the form of reduced viability or fertility.

Thus, the two characteristics of the first stage of speciation are that: (1) reproductive isolation appears primarily in the form of postzygotic RIMs, and (2) these RIMs are a byproduct of genetic differentiation—reproductive isolation is not directly promoted by natural selection at this stage.

Genetic differentiation, and the consequent appearance of postzygotic RIMs, is usually a gradual process. As such, it makes the decision as to whether or not the process of speciation has already begun between two

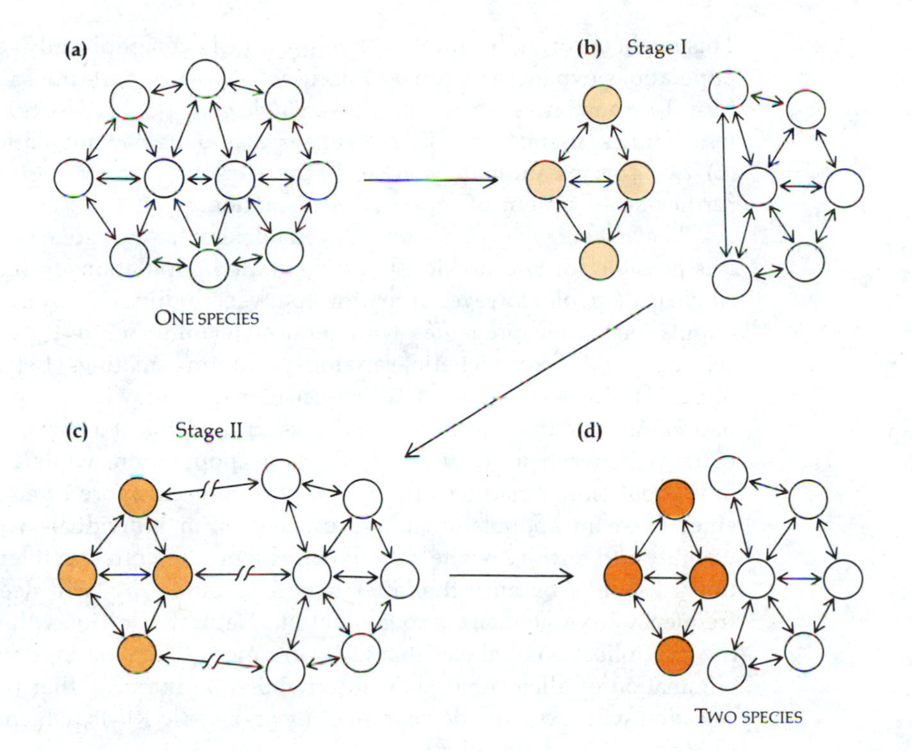

Figure 7.3
Generalized model of the process of speciation. **(a)** Local populations of a single species are represented by circles; the arrows indicate that gene flow occurs between populations. **(b)** The populations have become separated into two groups between which there is no gene flow. These groups gradually become genetically different, as indicated by the shading of the populations on the left. As a consequence of this genetic differentiation, reproductive isolating mechanisms arise between the two groups. This is the first stage of speciation. **(c)** Individuals from different population groups are able to intermate. However, owing to the pre-existing reproductive isolating mechanisms, little if any gene flow takes place, as indicated by the broken arrows. Natural selection favors the development of additional reproductive isolating mechanisms, particularly prezygotic ones, which prevent matings between individuals from different population groups. This is the second stage of speciation. **(d)** Speciation has been completed because the two groups of populations are fully reproductively isolated. There are now two species that can coexist without gene exchange.

populations a somewhat arbitrary matter. Populations may be considered to be in the first stage of speciation if RIMs have appeared between them. Local populations of a given species are often genetically somewhat different, but are not thought to be in the first stage of speciation if their differentiation is small and does not result in the appearance of RIMs.

STAGE II. This stage encompasses the completion of reproductive isolation. Assume that the external conditions that interfered with gene flow between two populations in the first stage of speciation disappear.

This might occur, for example, if two previously geographically separated populations expand and come to occupy, at least in part, the same territory. Two outcomes are possible: (1) a single gene pool arises, because the loss of fitness in the hybrids is not very great and the two populations fuse; (2) two species ultimately arise, because natural selection favors the further development of reproductive isolation.

The first stage of speciation is reversible: if it has not gone far enough, it is possible for two previously differentiated populations to fuse into a single gene pool. However, if matings between individuals from different populations leave progenies with reduced viability or fertility, natural selection will favor genetic variants promoting matings between individuals of the same population. Consider the following simplified situation. Assume that there are two alleles, A_1 and A_2, at a locus; A_1 favors matings between individuals of the same population, while A_2 favors interpopulational matings. Then A_1 will be present more often in progenies from intrapopulational crosses, that is, in individuals with good viability and fertility, while A_2 will be present more often in interpopulational hybrids; because the latter have low fitness, A_2 will decrease in frequency from generation to generation. Natural selection will result in the multiplication of alleles that favor intrapopulation matings and in the elimination of alleles that favor interpopulation matings; that is, natural selection will favor the development of prezygotic RIMs, which prevent the formation of hybrid zygotes.

The two characteristics of the second stage of speciation, then, are that: (1) reproductive isolation develops mostly in the form of prezygotic RIMs, and (2) the development of prezygotic RIMs is directly promoted by natural selection. These two characteristics of Stage II stand in contrast to the two characteristics of Stage I, pointed out above.

Nevertheless, speciation may take place without the occurrence of Stage II. In the absence of gene exchange, populations may develop complete reproductive isolation if the process of genetic differentiation continues long enough—for example, when the populations remain separated indefinitely on two islands. However, the speciation process is accelerated by Stage II because natural selection directly promotes the development of reproductive isolation.

Geographic Speciation

The general model of speciation just outlined may be realized in different ways, or modes, which can be classified as geographic speciation or quantum speciation. In *geographic speciation,* Stage I begins owing to geographic separation between populations. Terrestrial organisms may be separated by water (such as rivers, lakes, and oceans), mountains, deserts, or any kind of territory uninhabitable by the populations; freshwater organisms may be kept separate if they live in different river systems or unconnected lakes; marine organisms may be separated by land, by water of greater or lesser depth than the organisms can tolerate, or by water of different salinity.

As a result of natural selection, geographically separate populations become adapted to local conditions and thus become genetically differentiated. Random genetic drift may also contribute to genetic differentiation, particularly when populations are small or are derived from only a few individuals. If geographic separation continues for some time, incipient reproductive isolation may appear, particularly in the form of postzygotic RIMs; the populations will then be in the first stage of speciation.

The second stage of speciation begins when previously separated populations come into contact, at least over part of their geographic distributions. This may happen, for example, by topographic changes on the earth's surface, by ecological changes in the intervening territory that make it habitable by the populations, or by migration of members of one population into the territory of the other. Matings between individuals from different populations may then take place. Depending on the strength of the pre-existing RIMs and on the extent of hybridization, the two populations may fuse into a single gene pool or may develop additional (prezygotic) RIMs and become separate species.

The two stages of the process of geographic speciation can be illustrated with a group of closely related species of *Drosophila* that live in the American tropics (Figure 7.4). This group, collectively called the *Drosophila willistoni* group, consists of 15 species, 6 of which are *sibling species* (i.e., they are morphologically virtually indistinguishable). One of these siblings is *D. willistoni* itself, which consists of two subspecies: *D. w. quechua* lives in continental South America west of the Andes, and *D. w. willistoni* lives east of the Andes. There is incipient reproductive isolation between them, particularly in the form of hybrid sterility: when males and females from the two subspecies are intercrossed in the laboratory, the results depend on the direction of the mating:

♀ *D. w. willistoni* × ♂ *D. w. quechua* → fertile female and male progeny

♀ *D. w. quechua* × ♂ *D. w. willistoni* → fertile females but sterile males

If these two subspecies were to come into contact in nature and intercross, natural selection would favor the development of prezygotic RIMs, because the male progenies of all crosses between *quechua* females and *willistoni* males are sterile. In other words, these subspecies are two groups of populations in the first stage of geographic speciation.

The first stage of speciation is also found in another species of the group. *Drosophila equinoxialis* consists of two geographically separated subspecies: *D. e. equinoxialis* inhabits continental South America, and *D. e.*

Figure 7.4
Geographic distribution of six closely related species of the *Drosophila willistoni* group. *D. willistoni* and *D. equinoxialis* each consist of two subspecies. These subspecies represent populations in the first stage of geographic speciation.

caribbensis inhabits Central America and the Caribbean islands. Laboratory crosses between these two subspecies always produce fertile females but sterile males, regardless of the direction of the cross. Thus, there is somewhat greater reproductive isolation between the two subspecies of *D. equinoxialis* than between the two subspecies of *D. willistoni*. Natural selection in favor of prezygotic RIMs would be stronger in *D. equinoxialis*, because all hybrid males are sterile.

It is worth noting that prezygotic RIMs do *not* exist between the subspecies of *D. willistoni* or of *D. equinoxialis*. Reproductive isolation

Figure 7.5
Geographic distribution of the six semispecies of *Drosophila paulistorum*. These semispecies represent populations in the second stage of geographic speciation. Speciation has been virtually completed in places where two or three semispecies coexist without interbreeding.

between the subspecies is, therefore, far from complete, and they are not considered different species.

Stage II of the speciation process can be found within yet another species of the *D. willistoni* group. *Drosophila paulistorum* is a species consisting of six *semispecies*, or incipient species, two or three of which are sympatric in many localities (Figure 7.5). The semispecies exhibit hybrid sterility similar to that found in *D. equinoxialis*: crosses between males and females of two different semispecies yield fertile females but sterile males. But two or three semispecies have come into geographic contact in many places, and there the second stage of speciation has advanced to the point that ethological isolation is complete, or nearly so. When females and males from two different semispecies are placed together in the laboratory, the results depend on the geographic origin of the flies. When both semispecies are from the same locality, only *homogamic* matings (matings between members of the same semispecies) occur; when they are from different localities, however, *heterogamic* matings (matings between members of different semispecies) as well as homogamic matings occur, indicating that ethological isolation is not yet complete. The semispecies of *D. paulistorum* thus provide a remarkable example of the action of natural selection during the second stage of speciation: reproductive isolation has been completed where the semispecies are sympatric, but not elsewhere, because the genes involved have not yet spread fully throughout each semispecies.

Quantum Speciation

In geographic speciation, Stage I entails the gradual genetic divergence of geographically separated populations. The development of postzygotic RIMs as byproducts of genetic divergence usually requires a long period of time: thousands, perhaps millions, of generations. However, there are other modes of speciation in which the first stage, and the appearance of postzygotic RIMs, may require only relatively short periods of time. *Quantum speciation* (also called *rapid speciation* and *saltational speciation*) refers to these accelerated modes of speciation, particularly in the first stage.

One form of quantum speciation is polyploidy, which involves the multiplication of entire chromosome complements. Polyploid individuals may arise in just one or a few generations. Polyploid populations are reproductively isolated from their ancestral species and are thus new species. In polyploidy, the suppression of gene flow that is required for the onset of the first stage of speciation is due not to geographic separation but to cytological irregularities. Reproductive isolation in the form of hybrid sterility does not require many generations, but follows immediately, owing to chromosomal imbalance. If diploids and their fertile polyploid derivatives exist near each other and hybridization occurs, natural selection will favor the development of prezygotic isolating mechanisms (Stage II), which prevent interfertilization and the waste of gametes.

Modes of quantum speciation other than polyploidy are known to occur in plants. One instance of quantum speciation, studied by Harlan Lewis, involves the two diploid species *Clarkia biloba* and *C. lingulata*. Both species are native to California, but *C. lingulata* has a narrow distribution, being known from only two sites in the central Sierra Nevada, at the southern periphery of the distribution of *C. biloba*. The two species are outcrossers, although capable of self-fertilization, and are similar in external morphology, although there are differences in petal shape. How-

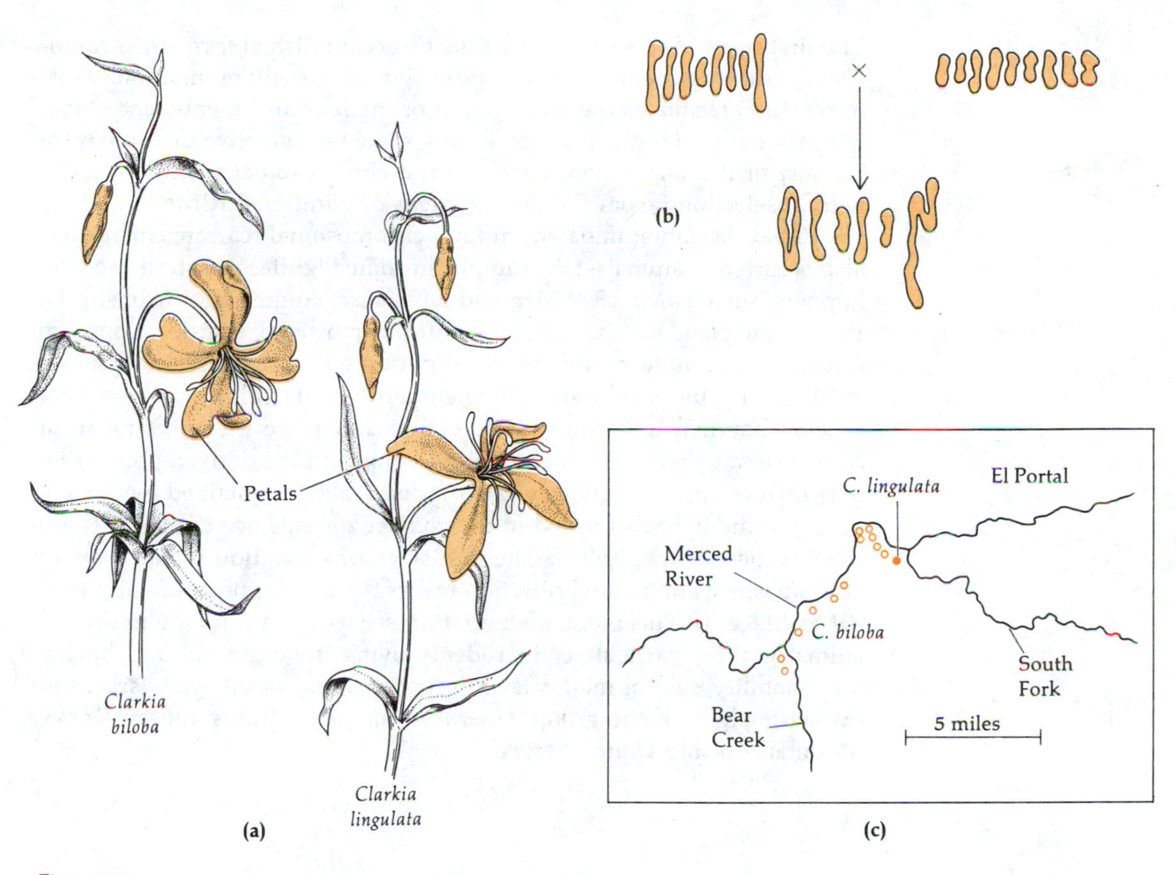

Figure 7.6
Two species of annual plants: *Clarkia lingulata* has arisen from *C. biloba* by quantum speciation. **(a)** Flowering branches of the two species, showing a difference in petal shape, which is bilobed in *C. biloba* but not in *C. lingulata*. **(b)** Paired chromosomes, at meiotic metaphase, of *C. biloba* (eight pairs, left), and of *C. lingulata* (nine pairs, right), and of the F_1 hybrid between the two. The chromosomes differ by at least two reciprocal translocations and a fission (or fusion). **(c)** A small portion of the Merced River canyon in the Sierra Nevada of California, west of Yosemite Valley, showing the southernmost populations of *C. biloba* (circles) and the two known populations of *C. lingulata* (dot).

ever, their chromosomal configurations differ by a translocation, several inversions, and there is an extra chromosome in *C. lingulata*, homologous to parts of two chromosomes of *C. biloba* (Figure 7.6). The narrowly distributed species, *C. lingulata*, has arisen from *C. biloba* by a rapid series of events involving extensive chromosomal reorganization. Chromosomal rearrangements, such as translocations, fusions, and fissions, reduce the fertility of individuals heterozygous for the new arrangements.

The first stage of speciation may thus be accomplished through chromosomal rearrangements without extensive allelic differentiation. Self-fertilization facilitates the propagation of the rearrangements; once there is a population of individuals exhibiting some reproductive isolation from the rest of the population, owing to the chromosomal rearrangements, natural selection favors the development of additional RIMs.

Rapid speciation initiated through chromosomal rearrangements has also occurred in animals, for example, in some flightless Australian grasshoppers, such as *Moraba scurra* and *M. viatica,* studied by Michael J. D. White. Incipient species differing by chromosomal translocations are found in adjacent territories. A translocation establishes itself at first in a small colony by genetic drift. If members of this colony possess high fitness, they may subsequently spread and displace the ancestral form from a certain area. The ancestral population and the derived population may then coexist contiguously, their individuality maintained by the low fitness of the hybrids formed in the contact zones, since the hybrids are translocation heterozygotes. The first stage of speciation is thus rapidly accomplished, and natural selection favors the development of additional RIMs (Stage II). This mode of speciation seems to be common in several animal groups, particularly in rodents living underground and having little mobility, such as mole rats of the group *Spalax ehrenbergi* in Israel and pocket gophers of the group *Thomomys talpoides* in the southern Rocky Mountains of the United States.

Genetic Differentiation During Speciation

The discovery that genes code for proteins and the development of the techniques of gel electrophoresis have made it possible to estimate the amount of genetic change during the speciation process. Before these techniques became available, there was already evidence suggesting that a fair amount of allelic substitution might be involved in speciation, since it was known that even closely related species are genetically quite different. For example, Erwin Baur had crossed two species of snapdragons, *Antirrhinum majus* and *A. molle,* which produce fertile hybrids. Considerable phenotypic variability appeared in the F_2. Most individuals showed various combinations of the parental traits, but some had characteristics present in neither parent, although they were found in other species of *Antirrhinum* or related genera. Baur estimated that more than 100 genetic differences exist between *A. majus* and *A. molle*. But it was not possible to estimate what proportion of the genes are different—invariant genes cannot be detected by Mendelian methods.

Box 7.1 Genetic Identity and Genetic Distance

Electrophoretic studies provide data in the form of genotypic frequencies that can be readily converted to allelic frequencies. Assume that A and B are two different populations and K is a given gene locus, and that i different alleles are observed in the two populations. Let us represent the frequencies of the alleles in population A as a_1, a_2, a_3, etc., and in population B as b_1, b_2, b_3, etc. The genetic similarity between the two populations at this locus can be measured by I_K, defined as follows:

$$I_K = \frac{\sum a_i b_i}{\sqrt{\sum a_i^2 \sum b_i^2}}$$

where the symbol Σ means "summation of"; $a_i b_i$ represents the products $a_1 b_1$, $a_2 b_2$, $a_3 b_3$, etc.; a_i^2 means a_1^2, a_2^2, a_3^2, etc.; and b_i^2 means b_1^2, b_2^2, b_3^2, etc. The formula for I_K calculates the (normalized) probability that two alleles, one taken from each population, are identical.

Let us consider some simple examples. First assume that only one allele is observed, with frequency 1 in both populations. Then $a_1 = 1$, $b_1 = 1$, and therefore

$$I_K = \frac{1 \times 1}{\sqrt{1^2 \times 1^2}} = \frac{1}{1} = 1$$

Not surprisingly, the value of I_K is 1, indicating that the two populations are identical at this locus.

Assume now that we observe two different alleles, the first having a frequency of 1 in population A, and the second, a frequency of 1 in population B. Then $a_1 = 1$, $b_1 = 0$, $a_2 = 0$, $b_2 = 1$, and therefore

$$I_K = \frac{(1 \times 0) + (0 \times 1)}{\sqrt{(1^2 + 0^2)(0^2 + 1^2)}} = \frac{0 + 0}{\sqrt{1 \times 1}} = \frac{0}{1} = 0$$

The value of I_K is 0, indicating that the two populations are genetically completely different at this locus.

Now consider the case where two alleles exist in both populations, with frequencies $a_1 = 0.2$, $a_2 = 0.8$ ($a_1 + a_2 = 1$), and $b_1 = 0.7$, $b_2 = 0.3$ ($b_1 + b_2 = 1$). We then obtain

$$I_K = \frac{(0.2 \times 0.7) + (0.8 \times 0.3)}{\sqrt{(0.2^2 + 0.8^2)(0.7^2 + 0.3^2)}}$$

$$= \frac{0.14 + 0.24}{\sqrt{0.68 \times 0.58}}$$

$$= 0.605$$

The value of I_K lies between 1 and 0, as we would expect, since the two populations share common alleles, although not in identical frequencies.

In order to estimate the genetic differentiation between two populations, several loci need to be studied. Let I_{ab}, I_a, and I_b be the arithmetic means (the averages), over all loci, of $\Sigma a_i b_i$, Σa_i^2, and Σb_i^2, respectively. Then the *genetic identity*, I, between the two populations can be measured, as proposed by M. Nei, by

$$I = \frac{I_{ab}}{\sqrt{I_a I_b}}$$

and the *genetic distance*, D, between the populations can be measured by

$$D = -\ln I$$

Assume that the three examples of differentiation at single loci given above actually correspond to three different loci studied in two populations. Then we have

$$I_{ab} = \frac{1 + 0 + 0.38}{3} = 0.460$$

$$I_a = \frac{1 + 1 + 0.68}{3} = 0.893$$

$$I_b = \frac{1 + 1 + 0.58}{3} = 0.860$$

Therefore

$$I = \frac{0.460}{\sqrt{0.893 \times 0.860}} = 0.525$$

and

$$D = -\ln 0.525 = 0.644$$

That is, it is estimated that 0.644 allelic substitutions per gene locus (or 64.4 allelic substitutions per 100 loci) have occurred in the separate evolution of the two populations. More than three gene loci need to be studied in order to obtain an acceptable estimate of genetic differentiation between any two populations, but the three loci are sufficient to show how genetic identity and genetic distance are calculated.

Estimates of genetic differentiation between two populations can be obtained by studying, in both populations, a sample of proteins, chosen without knowing whether or not they are different in the populations. The genes coding for the proteins then represent a random sample of all the structural genes with respect to the differentiation between the populations. The results obtained from the study of a moderate number of gene loci can therefore be extrapolated to the whole genome.

An efficient technique for studying protein variation in natural populations is gel electrophoresis (Box 2.1), which provides estimates of genotypic and allelic frequencies in populations. A useful method, proposed by Masatoshi Nei, of estimating genetic differentiation between populations, using electrophoretic data, is shown in Box 7.1. Two parameters are used: (1) *genetic identity*, I, which estimates the proportion of genes that are identical in structure in two populations, and (2) *genetic distance*, D, which estimates the number of allelic substitutions per locus that have occurred in the separate evolution of two populations. There is an allelic substitution when one allele is replaced by a different allele, or when a set of alleles is replaced by a different set. The method takes into account the fact that not all observed allelic substitutions are complete: an allele may have been partly replaced by a different one, but the original allele may still exist in greater or lesser frequency.

Genetic identity, I, may range in value from zero (no alleles in common) to one (the same alleles and in the same frequencies are found in both populations). Genetic distance, D, may range in value from zero (no allelic change at all) to infinity; D can be greater than one because each locus may experience complete allelic substitution more than once as evolution goes on for long periods of time.

The statistics I and D can be used to measure genetic differentiation during the speciation process. We shall consider geographic speciation

Table 7.2
Genetic differentiation between populations of the *Drosophila willistoni* group at various levels of evolutionary divergence. Levels 2 and 3 represent Stage I and Stage II, respectively, of the process of geographic speciation. *I* estimates the degree of genetic similarity, *D*, the degree of genetic differentiation. The values given are the means and standard errors for several comparisons.

Level of Comparison	*I*	*D*
1. Local populations	0.970 ± 0.006	0.031 ± 0.007
2. Subspecies	0.795 ± 0.013	0.230 ± 0.016
3. Incipient species	0.798 ± 0.026	0.226 ± 0.033
4. Sibling species	0.563 ± 0.023	0.581 ± 0.039
5. Morphologically different species	0.352 ± 0.023	1.056 ± 0.068

From F. J. Ayala, *Evol. Biol.* 8:1 (1975).

first. The *Drosophila willistoni* group of species was used as a model of geographic speciation, because both stages of the process can be identified therein. This group of species has been extensively studied using electrophoretic techniques. The results are summarized in Table 7.2, in which five levels of evolutionary divergence are represented. The first level involves comparisons between populations living in different localities but without any reproductive isolation between them; the genetic identity is 0.970, indicating a very high degree of genetic similarity.

The second level involves comparisons between different subspecies (such as *D. w. willistoni* and *D. w. quechua*, and *D. e. equinoxialis* and *D. e. caribbensis*). These populations are in the first stage of speciation and exhibit postzygotic RIMs in the form of hybrid sterility. They also exhibit a fair amount of genetic differentiation, $I = 0.795$ and $D = 0.230$; complete allelic substitutions have occurred, on the average, in 23 of every 100 gene loci.

The third level of evolutionary divergence in Table 7.2 involves comparisons between the incipient species of the *D. paulistorum* complex. These are populations in the second stage of speciation, exhibiting some prezygotic as well as postzygotic RIMs. Apparently these populations are not genetically more differentiated than those in the first stage of speciation. This means that the second stage of speciation has not required much genetic change, which is perhaps not surprising. During the first stage of speciation, reproductive isolation comes about as a byproduct of genetic change: a fair amount of genetic change needs to take place over the whole genome before postzygotic RIMs develop. However, during the second stage of speciation, natural selection directly favors the development of prezygotic RIMs; only a few genes—those affecting courtship and mating behavior, for example—need to be changed to accomplish this.

The fourth level in Table 7.2 involves comparisons between sibling species (such as *D. willistoni* and *D. equinoxialis*). Despite their morphological similarity, these species are genetically quite different; about 58 allelic substitutions have occurred, on the average, per 100 loci. Species are independently evolving groups of populations. Once the process of speciation is completed, species will continue to diverge genetically. The results of this gradual process of divergence are also apparent in the comparisons between morphologically different species of the *D. willistoni* group (fifth level in Table 7.2). On the average, somewhat more than one allelic substitution per gene locus has occurred in the evolution of these nonsibling species.

Using the techniques of gel electrophoresis, comparisons between populations at various levels of evolutionary divergence have been carried out in many kinds of organisms during the past few years. Evolution is a complex process determined by the environmental conditions as well as by the nature of the organisms, and thus the amount of genetic change corresponding to a given level of evolutionary divergence is likely to vary from organism to organism, from place to place, and from time to time. The results of electrophoretic studies confirm this variation, but also show some general patterns (Table 7.3). With few exceptions, the genetic distance between populations in either the first or the second stage of speciation is about 0.20 (most comparisons fall in the range 0.16 to 0.30), for organisms as diverse as insects, fishes, amphibians, reptiles, and mammals. These results are consistent with the conclusions derived from the study of the *Drosophila willistoni* group: the first stage of the geographic speciation process requires a fair amount of genetic change (of the order of 20 allelic substitutions per 100 gene loci), whereas little additional genetic change is required during the second stage.

How much genetic change takes place in the quantum mode of speciation? It is clear that when a new species arises by polyploidy, no genetic changes other than the chromosome duplications are required; the new species has the alleles present in the parental species, and no others. However, because most polyploid species start from only one individual of each parental species, they possess at the beginning less genetic variation than the parental species (the founder effect; see Chapter 3, page 78).

Other modes of quantum speciation start with chromosomal rearrangements that cause either partial or total hybrid sterility. As in polyploidy, such rearrangements do not necessarily involve changes in allelic constitution, although there is often a reduction of genetic variation because the derivative population starts from only one or a few in-

Table 7.3
Genetic differentiation at various stages of evolutionary divergence in several groups of organisms. The average genetic identity is given first, followed by the average genetic distance (in parentheses).

	I (D)			
Organisms	Local populations	Subspecies	Incipient species	Species and closely related genera
Drosophila	0.987 (0.013)	0.851 (0.163)	0.788 (0.239)	0.381 (1.066)
Other invertebrates	0.985 (0.016)	–	–	0.465 (0.878)
Fishes	0.980 (0.020)	0.850 (0.163)	–	0.531 (0.760)
Salamanders	0.984 (0.017)	0.836 (0.181)	–	0.520 (0.742)
Reptiles	0.949 (0.053)	0.738 (0.306)		0.437 (0.988)
Mammals	0.944 (0.058)	0.793 (0.232)	0.769 (0.263)	0.620 (0.559)
Plants	0.966 (0.035)	–	–	0.510 (0.808)

Calculated from data in F. J. Ayala, *Evol. Biol.* 8:1 (1975).

dividuals. The first stage of speciation is therefore accomplished with little or no genetic change at the level of the individual genes.

What about genetic change in the second stage of quantum speciation? The second stage of speciation is similar in both geographic and quantum speciation. In both cases, the populations already exhibit postzygotic RIMs and are developing prezygotic isolation by natural selection. If, in geographic speciation, the second stage requires genetic changes in only a small fraction of the genes, the same should be true in quantum speciation. Experimental results confirm this prediction (Table 7.4). The first comparison is between the two annual plant species, *Clarkia biloba* and *C. lingulata,* discussed earlier as examples of quantum speciation. These species remain genetically quite similar: $I = 0.880$ and $D = 0.128$, indicating that only about 13 allelic substitutions per 100 gene loci have occurred in their separate evolution.

The second comparison is also between two annual plants, *Stephanomeria exigua* and *S. malheurensis;* the latter was derived from the former only very recently. Leslie Gottlieb has shown that the original and derivative populations differ by one chromosomal translocation and by their mode of reproduction: the original species reproduces by outcrossing, while the derivative species reproduces by selfing. As expected, the two species are genetically very similar (about 6 allelic substitutions per 100 loci).

Table 7.4
Genetic differentiation in quantum speciation. Little differentiation is observed between species or incipient species arisen by quantum speciation.

Populations Compared	*I*	*D*
Plants		
Clarkia biloba vs. *C. lingulata**	0.880	0.128
Stephanomeria exigua vs. *S. malheurensis*†	0.945	0.057
Rodents		
Spalax ehrenbergi†	0.978	0.022
Thomomys talpoides†	0.925	0.078

*Comparison between two recently arisen species.
†Comparison between incipient species, i.e., populations completing the second stage of speciation.

After F. J. Ayala, *Evol. Biol.* 8:1 (1975).

The third and fourth comparisons in Table 7.4 involve rodents. *Spalax ehrenbergi* is a species of mole rat consisting of four groups of populations differing in their number of chromosomes (52, 54, 58, and 60). The populations are largely allopatric, although they are in contact with one another in narrow zones at the edges of their distributions, and some hybridization takes place there. The differences in chromosome number due to chromosomal fusions or fissions provide effective postzygotic RIMs; moreover, some ethological isolation has developed: laboratory tests show greater preference for matings between individuals of the same chromosomal type, although they appear morphologically indistinguishable. These four populations in the second stage of quantum speciation are, on the average, genetically very similar: only about 2 allelic substitutions per 100 gene loci have taken place in their separate evolution.

Thomomys talpoides is a species of pocket gopher consisting of more than eight populations differing in their chromosomal arrangements; they live in the north central and northwest United States and neighboring areas of southern Canada. As in the case of *Spalax,* the populations of *Thomomys* are mostly allopatric, but are in contact at the edges of their distributions. The chromosomal rearrangements keep the populations from interbreeding in the zones of contact. Nevertheless, the average genetic distance between these populations is quite small (about 8 allelic substitutions per 100 loci).

In conclusion, quantum speciation can occur with little change at the level of the genes; that is, neither Stage I nor Stage II requires substantial allelic evolution in this mode of speciation. This result, in turn, confirms the conclusion reached with respect to geographic speciation, namely, that Stage II—when natural selection directly promotes prezygotic RIMs—does not require major genetic changes.

Genetic Change and Phylogeny: DNA Hybridization

Species are independent evolutionary units because they are reproductively isolated. Species evolve independently of each other and are likely to become genetically more and more different as time proceeds. It was pointed out at the beginning of this chapter that the amount of genetic change that has occurred in the branches of a phylogeny can be inferred by measuring the amount of genetic differentiation between living species. But the configuration of a phylogeny, if it is not known, can also be inferred from the amounts of genetic differentiation between species. This is because evolution is a gradual process; hence, species that are genetically quite similar to each other are more likely to share a recent common ancestor with each other than with species from which they are genetically more different.

The degree of genetic differentiation between species can be measured directly by examining the DNA sequence of their genes or, indirectly, by examining the proteins encoded by structural genes. Simple techniques have recently been developed for determining the nucleotide sequence of DNA molecules. These techniques have not yet been extensively applied to phylogenetic studies, because specific genes must first be obtained from the organisms to be compared. The isolation of specific genes is still a laborious enterprise.

A technique that estimates the overall similarity between the DNA of various organisms is *DNA hybridization.* Radioactively labeled DNA that has been "melted" (i.e., dissociated) and fractionated can be reacted with DNA from a different species. Homologous sequences will hybridize to form duplexes; the extent of the reaction gives an estimate of the proportion of DNA sequences that are homologous. The repetitive DNA sequences are usually first removed so that only single-copy DNA is employed in the tests.

Sequences forming duplexes need not be complementary for every nucleotide. The proportion of noncomplementary nucleotides in interspecific DNA duplexes can be estimated by the rate at which the DNA strands separate at increasing temperatures. The critical parameter, called *thermal stability* (T_s), is the temperature at which 50% of the duplex DNA has dissociated (Figure 7.7). A difference (ΔT_s) between the T_s values of hybrid and control DNA molecules of 1°C corresponds approximately

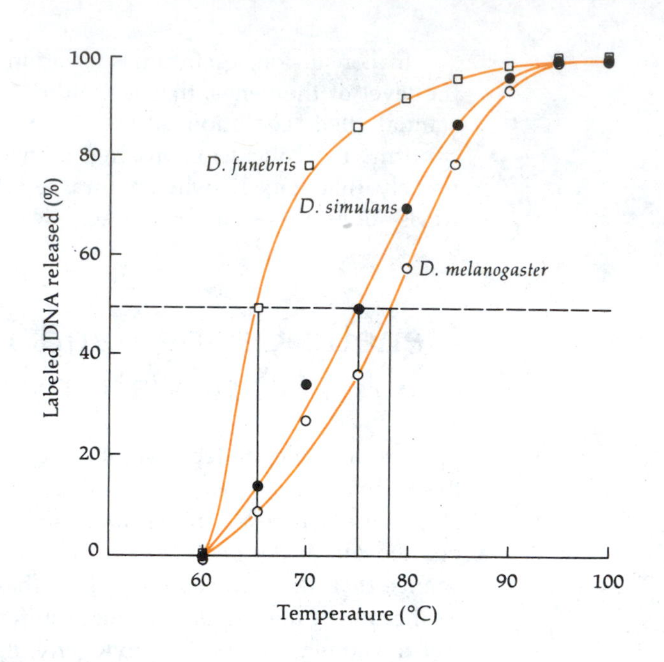

Figure 7.7
Thermal stability profiles of DNA duplexes having one strand from *Drosophila melanogaster* and the other from the species indicated. The thermal stability (T_s) value for the nonhybrid duplex DNA is 78°C; for the *D. melanogaster*/*D. simulans* duplex it is 75°C; and for the *D. melanogaster*/*D. funebris* duplex it is 65°C. Because $\Delta T_s = 1°C$ corresponds approximately to 1% mismatched nucleotides, the proportion of nucleotide pairs different from *D. melanogaster* is estimated as 3% for *D. simulans* and 13% for *D. funebris*. [After C. D. Laird and B. J. McCarthy, *Genetics* 60:303 (1968).]

to 1% mismatched nucleotides. The results of comparing the DNA of various primates first with human DNA and then with green-monkey DNA (Table 7.5) serve to estimate the percentage of nucleotide-pair substitutions that have occurred during primate evolution (Figure 7.8).

Protein-Sequence Phylogenies

Cytochrome *c* is a protein involved in cellular respiration; it is found in the mitochondria of animals and plants. The amino acid sequences of cytochrome *c* in humans, rhesus monkeys, and horses are shown in Figure 7.9. The amino acid at position 66 is isoleucine in humans, but threonine in rhesus monkeys and horses. Humans and rhesus monkeys have identical amino acids at the other 103 positions, but differ from horses in 11 additional amino acids (Table 7.6). It is known that the evolutionary divergence between the human and rhesus monkey lineages occurred well after these diverged from the horse lineage. The number of amino acid substitutions that have occurred in the various branches of the phylogeny are shown in Figure 7.10.

The genetic code (Table 1.3) makes it possible to calculate the minimum number of nucleotide differences required to change from a codon for one amino acid to a codon for another. At position 19 of cytochrome *c*, humans and rhesus monkeys have isoleucine, but horses have valine. Isoleucine may be encoded by any one of the three codons AUU, AUC, and AUA, and valine, by any one of the four codons GUU,

Table 7.5
Percent nucleotide differences between the DNA of various primates and the DNA of humans and green monkeys.

	Tester DNA from:	
Species Tested	Human	Green monkey
Human	0	9.6
Chimpanzee	2.4	9.6
Gibbon	5.3	9.6
Green monkey	9.5	0
Rhesus monkey	–	3.5
Capuchin	15.8	16.5
Galago	42.0	42.0

After D. E. Kohne, J. A. Chiscon, and B. H. Hoyer, *J. Human Evol.* 1:627 (1972).

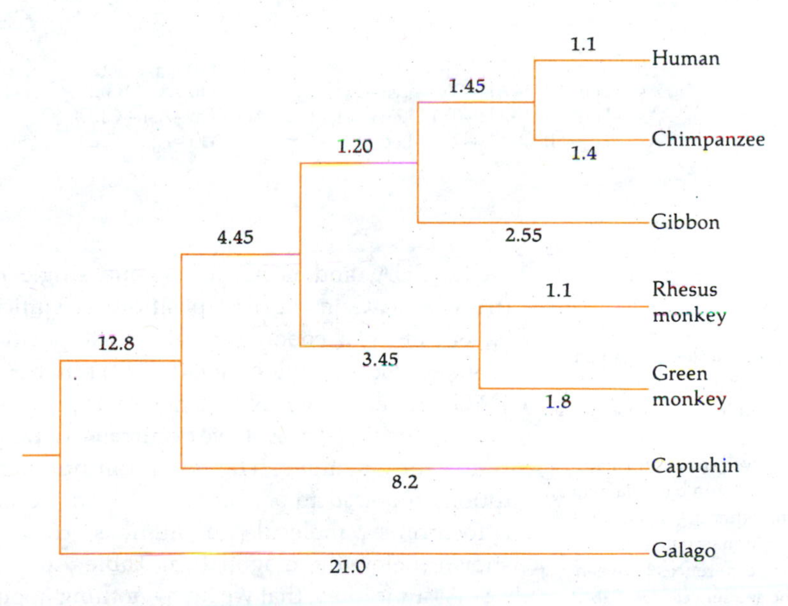

Figure 7.8
Phylogeny of primate species, based on thermal stability profiles of DNA hybrid duplexes. The numbers on the branches are estimated percent nucleotide-pair substitutions. [After D. E. Kohne, J. A. Chiscon, and B. H. Hoyer, *J. Human Evol.* 1:627 (1972).]

	1–8	9 10 20
Human	——	Gly–Asp–Val–Glu–Lys–Gly–Lys–Lys–Ile–Phe–Ile–Met–
Rhesus monkey	——	Gly–Asp–Val–Glu–Lys–Gly–Lys–Lys–Ile–Phe–Ile–Met–
Horse	——	Gly–Asp–Val–Glu–Lys–Gly–Lys–Lys–Ile–Phe–Val–Gln–

21 30 40
Lys–Cys–Ser–Gln–Cys–His–Thr–Val–Glu–Lys–Gly–Gly–Lys–His–Lys–Thr–Gly–Pro–Asn–Leu–
Lys–Cys–Ser–Gln–Cys–His–Thr–Val–Glu–Lys–Gly–Gly–Lys–His–Lys–Thr–Gly–Pro–Asn–Leu–
Lys–Cys–Ala–Gln–Cys–His–Thr–Val–Glu–Lys–Gly–Gly–Lys–His–Lys–Thr–Gly–Pro–Asn–Leu–

41 50 60
His–Gly–Leu–Phe–Gly–Arg–Lys–Thr–Gly–Gln–Ala–Pro–Gly–Tyr–Ser–Tyr–Thr–Ala–Ala–Asn–
His–Gly–Leu–Phe–Gly–Arg–Lys–Thr–Gly–Gln–Ala–Pro–Gly–Tyr–Ser–Tyr–Thr–Ala–Ala–Asn–
His–Gly–Leu–Phe–Gly–Arg–Lys–Thr–Gly–Gln–Ala–Pro–Gly–Phe–Thr–Tyr–Thr–Asp–Ala–Asn–

61 70 80
Lys–Asn–Lys–Gly–Ile–Ile–Trp–Gly–Glu–Asp–Thr–Leu–Met–Glu–Tyr–Leu–Glu–Asn–Pro–Lys–
Lys–Asn–Lys–Gly–Ile–Thr–Trp–Gly–Glu–Asp–Thr–Leu–Met–Glu–Tyr–Leu–Glu–Asn–Pro–Lys–
Lys–Asn–Lys–Gly–Ile–Thr–Trp–Lys–Glu–Glu–Thr–Leu–Met–Glu–Tyr–Leu–Glu–Asn–Pro–Lys–

81 90 100
Lys–Tyr–Ile–Pro–Gly–Thr–Lys–Met–Ile–Phe–Val–Gly–Ile–Lys–Lys–Lys–Glu–Glu–Arg–Ala–
Lys–Tyr–Ile–Pro–Gly–Thr–Lys–Met–Ile–Phe–Val–Gly–Ile–Lys–Lys–Lys–Glu–Glu–Arg–Ala–
Lys–Tyr–Ile–Pro–Gly–Thr–Lys–Met–Ile–Phe–Ala–Gly–Ile–Lys–Lys–Lys–Thr–Glu–Arg–Glu–

101 110 112
Asp–Leu–Ile–Ala–Tyr–Leu–Lys–Lys–Ala–Thr–Asn–Glu
Asp–Leu–Ile–Ala–Tyr–Leu–Lys–Lys–Ala–Thr–Asn–Glu
Asp–Leu–Ile–Ala–Tyr–Leu–Lys–Lys–Ala–Thr–Asn–Glu

Figure 7.9
The primary structures of cytochrome *c* in humans, rhesus monkeys, and horses. In these organisms, cytochrome *c* consists of 104 amino acids (positions 9–112; positions 1–8 are reserved for amino acids that exist in bacteria, wheat, and other organisms, but not in mammals). Amino acid differences between the sequences (see Table 7.6) are highlighted with color.

GUC, GUA, and GUG. Thus, one single nucleotide-pair substitution (from A to G in the first position) is sufficient to change a codon for isoleucine to a codon for valine. At position 20, humans and rhesus monkeys have methionine (AUG), while horses have glutamine (CAA or CAG); therefore, at least two nucleotide-pair substitutions (in the first and second positions) must have occurred to change the methionine codon to a codon for glutamine. The minimum numbers of nucleotide-pair substitutions required to account for the amino acid differences between the cytochrome *c* molecules of humans, rhesus monkeys, and horses are shown (below the diagonal) in Table 7.6.

Assume, now, that we knew nothing about the phylogeny of humans, rhesus monkeys, and horses. The data in Table 7.6 suggests that the configuration shown in Figure 7.10 is the most likely. Evolution is, on the whole, a gradual process of change. Thus, species that are genetically more similar are likely to have diverged from each other more recently than they

Table 7.6
Number of amino acid differences (above the diagonal) and minimum number of nucleotide differences (below the diagonal) between the cytochrome *c* molecules of humans, rhesus monkeys, and horses. The cytochrome *c* in these organisms has 104 amino acids (see Figure 7.9).

	Human	Rhesus Monkey	Horse
Human	–	1	12
Rhesus monkey	1	–	11
Horse	15	14	–

did from species that are genetically less similar. Figure 7.11 shows two possible phylogenies of humans, rhesus monkeys, and horses, as well as the nucleotide substitutions required in each branch. The configurations of these phylogenies would seem very unlikely even if no information other than the amino acid sequences of the cytochrome *c* molecules were available.

The minimum numbers of nucleotide differences necessary to account for the amino acid differences in the cytochrome *c* molecules of 20 organisms are given in Table 7.7. A phylogeny based on that data matrix, as well as the minimum numbers of nucleotide changes required in each

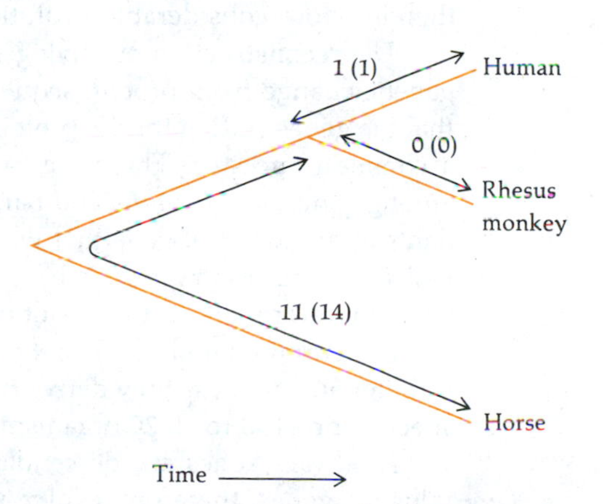

Figure 7.10
Anagenetic change in the evolution of cytochrome *c* from humans, rhesus monkeys, and horses. The numbers indicate the amino acid substitutions (and, in parentheses, the minimum number of nucleotide substitutions) that have taken place in each branch of the phylogeny.

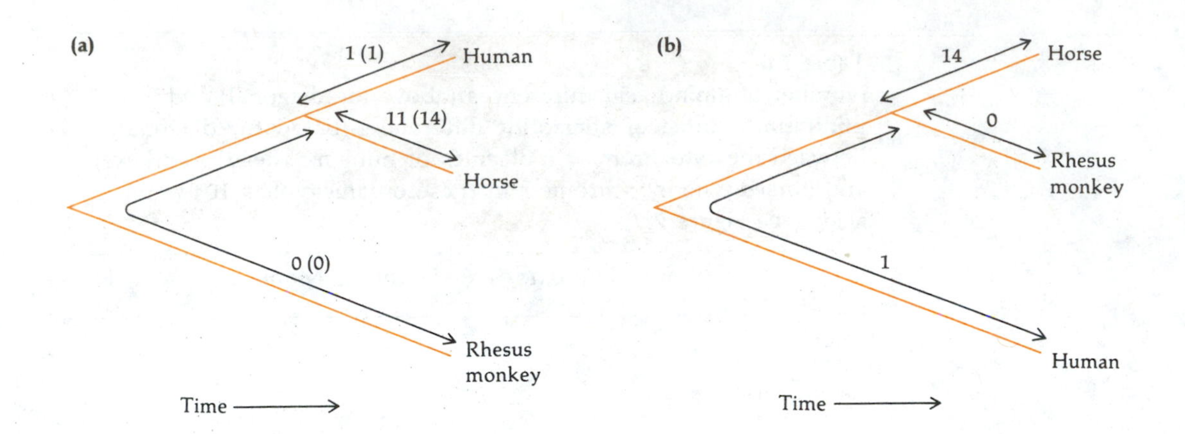

Figure 7.11
Two theoretically possible phylogenies of humans, rhesus monkeys, and horses. The numbers of amino acid (and nucleotide) substitutions required in each branch to account for the cytochrome *c* sequences indicate that neither of these two phylogenies is likely to be correct.

branch, is shown in Figure 7.12. Most of these changes are shown as fractions. It is obvious that a nucleotide change may or may not have taken place, but fractional nucleotide changes cannot occur. However, the values given in Figure 7.12 are those that best satisfy the data in Table 7.7.

The phylogenetic relationships shown in Figure 7.12 correspond fairly well, on the whole, with the phylogeny of the organisms as determined from the fossil record and other sources. There are disagreements, however. For example, chickens appear more closely related to penguins than to ducks and pigeons, and men and monkeys diverge from the other mammals before the marsupial kangaroo separates from the nonprimate placentals. Despite these erroneous relationships, it is remarkable that the study of a single protein should yield such an accurate representation of the phylogeny of 20 organisms as diverse as those in the figure. The amino acid sequences of proteins (and the genetic information contained therein) store considerable evolutionary information.

The reconstruction of phylogenies and the estimation of amounts of genetic change from protein sequence data are based on the assumption that the genes coding for the proteins are *homologous*, i.e., descended from a common ancestor. There are two kinds of homologous relationships among genes: orthologous and paralogous. *Orthologous* genes are descendants of an ancestral gene that was present in the ancestral species from which the species in question have evolved. The evolution of orthologous genes therefore reflects the evolution of the *species* in which they are found. The cytochrome *c* molecules of the 20 organisms shown in Figure 7.12 are orthologous, because they derive from a single ancestral gene present in a species ancestral to all 20 organisms.

Paralogous genes are descendants of a *duplicated* ancestral gene. Paralogous genes, therefore, evolve within the same species (as well as in different species). The genes coding for the α, β, γ, and δ hemoglobin chains in humans are paralogous. The evolution of paralogous genes reflects differences that have accumulated since the genes duplicated.

Table 7.7
Minimum number of nucleotide differences in the genes coding for cytochrome *c* molecules in 20 organisms.

Organism	1	2	3	4	5	6	7	8	9	10	11	12	13	14	15	16	17	18	19	20
1. Human	–	1	13	17*	16	13	12	12	17	16	18	18	19	20	31	33	36	63	56	66
2. Monkey		–	12	16*	15	12	11	13	16	15	17	17	18	21	32	32	35	62	57	65
3. Dog			–	10	8	4	6	7	12	12	14	14	13	30	29	24	28	64	61	66
4. Horse				–	1	5	11	11	16	16	16	17	16	32	27	24	33	64	60	68
5. Donkey					–	4	10	12	15	15	15	16	15	31	26	25	32	64	59	67
6. Pig						–	6	7	13	13	13	14	13	30	25	26	31	64	59	67
7. Rabbit							–	7	10	8	11	11	11	25	26	23	29	62	59	67
8. Kangaroo								–	14	14	15	13	14	30	27	26	31	66	58	68
9. Duck									–	3	3	3	7	24	26	25	29	61	62	66
10. Pigeon										–	4	4	8	24	27	26	30	59	62	66
11. Chicken											–	2	8	28	26	26	31	61	62	66
12. Penguin												–	8	28	27	28	30	62	61	65
13. Turtle													–	30	27	30	33	65	64	67
14. Rattlesnake														–	38	40	41	61	61	69
15. Tuna															–	34	41	72	66	69
16. Screwworm fly																–	16	58	63	65
17. Moth																	–	59	60	61
18. *Neurospora*																		–	57	61
19. *Saccharomyces*																			–	41
20. *Candida*																				–

*The differences between the horse and either human or rhesus monkey given here (17 and 16) are greater than those given in Table 7.6 (15 and 14). The two additional nucleotide substitutions are required when all the organisms included in this table are taken into account.

After W. M. Fitch and E. Margoliash, *Science* 155:279 (1967).

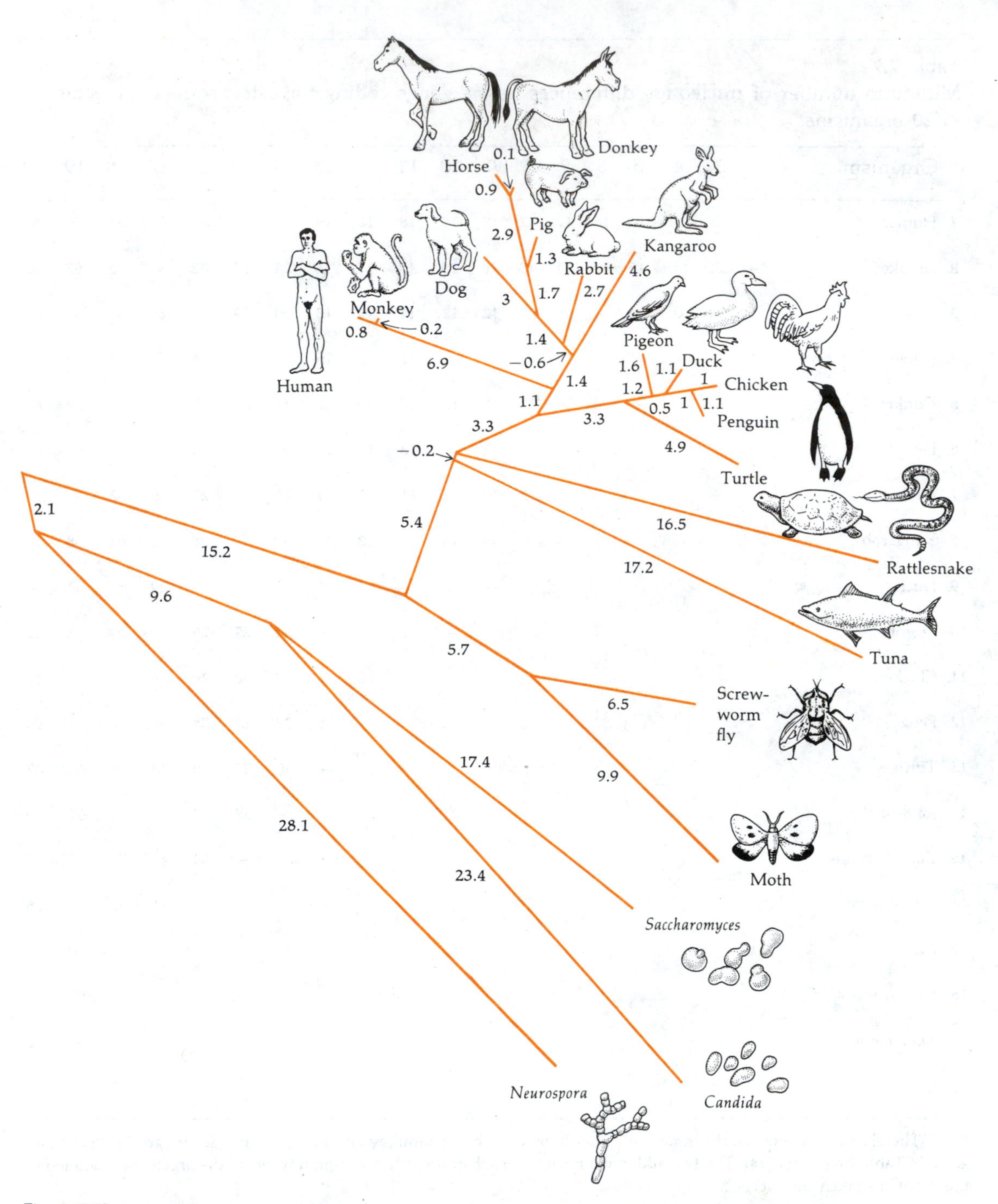

Figure 7.12
Phylogeny of 20 organisms, based on differences in the amino acid sequence of cytochrome *c*. The phylogeny agrees fairly well with evolutionary relationships inferred from the fossil record and other sources. The minimum number of nucleotide substitutions required for each branch is shown. Although fractional numbers of nucleotide substitutions cannot occur, the numbers shown are those that fit best the data in Table 7.7 [After W. M. Fitch and E. Margoliash, *Science* 155:279 (1967).]

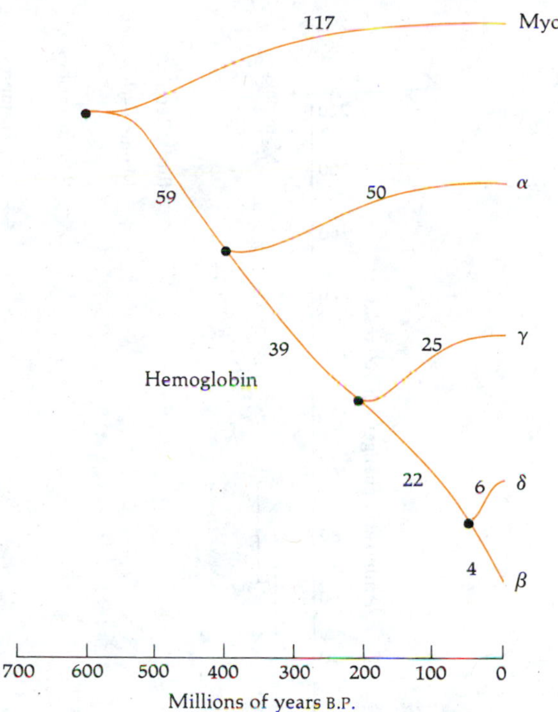

Figure 7.13
Evolutionary history of the globin genes. The dots indicate where the ancestral genes were duplicated, giving rise to a new gene line. The minimum number of nucleotide substitutions required to account for the amino acid differences between the proteins is shown for each branch. The first gene duplication occurred about 600 million years B.P. (before Present), one gene coding for myoglobin and the other being the ancestor of the various hemoglobin genes. About 400 million years ago, the hemoglobin gene became duplicated into one leading to the modern α gene and another that would duplicate again about 200 million years ago into the γ and β genes. The latter duplicated again some 40 million years ago in the ancestral lineage of the higher primates, giving rise to one new gene coding for the δ hemoglobin chain. The time estimates are based on paleontological and morphological studies of the organisms.

Homologies between paralogous genes serve to establish *gene* phylogenies, i.e., the evolutionary history of duplicated genes within a given lineage. Figure 7.13 is a phylogeny of the gene duplications giving rise to the myoglobin and hemoglobin genes found in modern humans. The evolutionary sequence of the duplications is shown, as well as the minimum number of nucleotide changes in each branch.

Cytochrome *c* molecules are slowly evolving proteins. Organisms as different as humans, silkworm moths, and *Neurospora* have a large proportion of amino acids in their cytochrome *c* molecules in common. The evolutionary conservation of this cytochrome makes possible the study of genetic differences among organisms that are only remotely related. However, this same conservation makes cytochrome *c* useless for

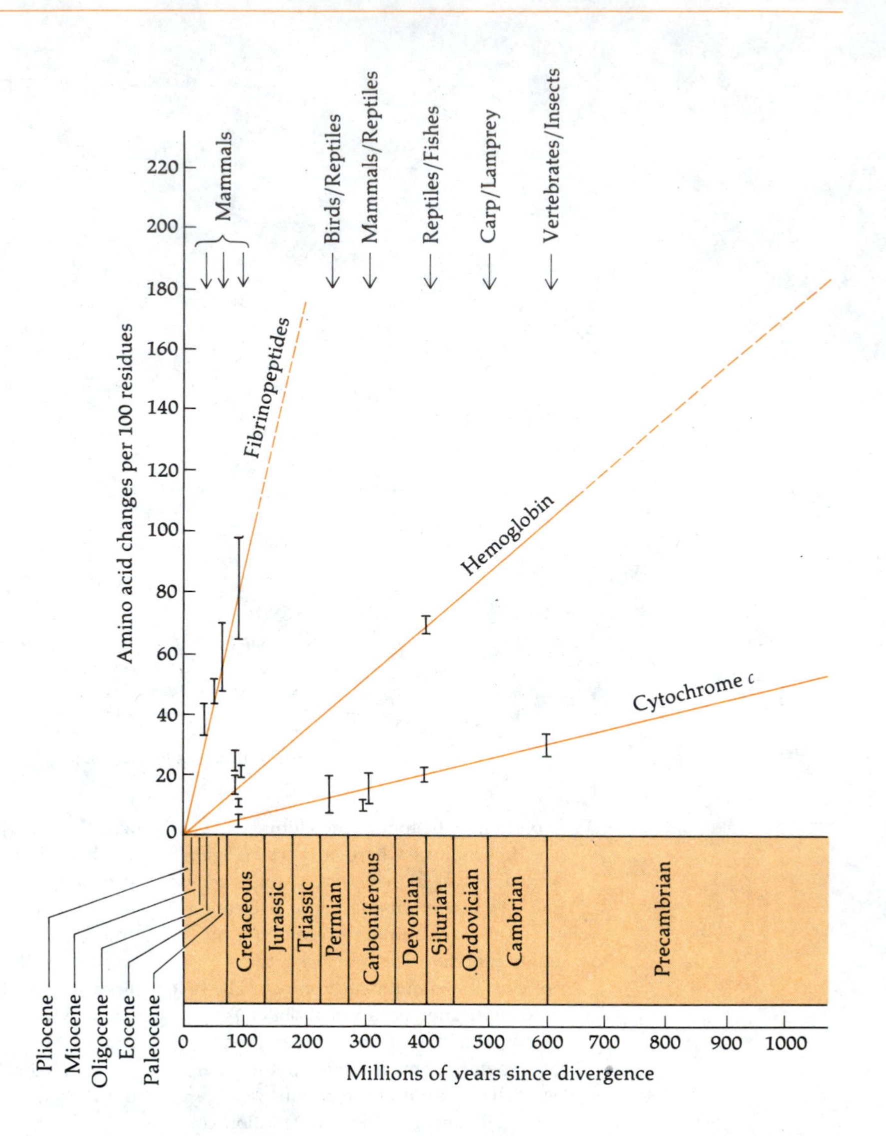

Figure 7.14
Rates of molecular evolution of different proteins. [After R. E. Dickerson, *J. Mol. Evol. 1*:26 (1971).]

determining evolutionary change in closely related organisms, since these may have cytochrome *c* molecules that are completely or nearly identical. For example, the primary structure of cytochrome *c* is identical in humans and chimpanzees, which diverged 10 to 15 million years ago; it differs by only one amino acid between humans and rhesus monkeys, whose most recent common ancestor lived 40 to 50 million years ago.

Fortunately, different proteins evolve at different rates. Phylogenetic relationships among closely related organisms can be inferred by studying the primary sequences of rapidly evolving proteins, such as fibrinopeptides in mammals (Figure 7.14). Carbonic anhydrases are rapidly evolving proteins that are physiologically important in the reversible hydration of CO_2 and in certain secretory processes. A phylogeny of

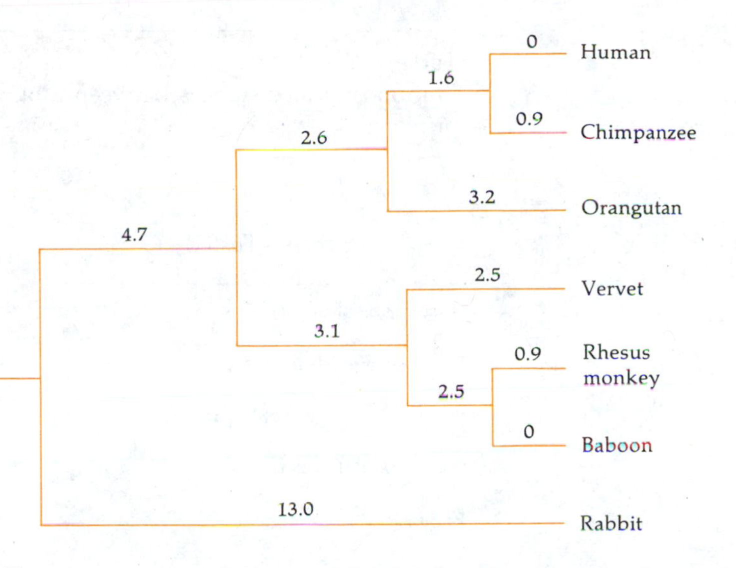

Figure 7.15
Phylogeny of various primates, based on differences in the sequence of 115 amino acids in carbonic anhydrase I. The numbers on the branches are the estimated numbers of nucleotide substitutions that have occurred in evolution. (After R. E. Tashian et al., in *Molecular Anthropology*, ed. by M. Goodman and R. E. Tashian, Plenum Press, New York, 1976, p. 301.)

various primates based on the amino acid sequence of carbonic anhydrase I, as well as the minimum number of nucleotide changes in each branch, is shown in Figure 7.15. Genetic changes in the evolution of closely related species can also be studied by other methods, such as DNA hybridization, immunology, and gel electrophoresis.

Immunology and Electrophoresis

More than 500 sequences or partial sequences of proteins, containing much evolutionary genetic information, are presently known. More sequences are obtained every year, although the procedures for determining the primary sequences of proteins are extremely laborious. Other methods, such as immunological techniques, permit estimating the degree of similarity among proteins with much less work than amino acid sequencing.

The immunological comparison of proteins is performed, in outline, as follows. A protein, say, albumin, is purified from the tissue of an animal, say, a chimpanzee. The protein is injected into a rabbit or some other mammal, which develops an immunological reaction and produces

Table 7.8
Immunological distances between albumins of various Old World primates.

Species Tested	Antiserum to: *Homo*	*Pan*	*Hylobates*
Homo sapiens (human)	0	3.7	11.1
Pan troglodytes (chimpanzee)	5.7	0	14.6
Pan paniscus (pygmy chimpanzee)	5.7	0	14.6
Gorilla gorilla (gorilla)	3.7	6.8	11.7
Pongo pygmaeus (orangutan)	8.6	9.3	11.7
Symphalangus syndactylus (siamang)	11.4	9.7	2.9
Hylobates lar (gibbon)	10.7	9.7	0
Old World monkeys (average of 6 species)	38.6	34.6	36.0

Calculated from data in V. M. Sarich and A. C. Wilson, *Science 158*:1200 (1967).

antibodies against the foreign protein, or *antigen*. The antibodies can be collected by bleeding the rabbit; thereafter, they will react not only against the specific antigen (chimpanzee albumin in the present example), but also against other related proteins (such as albumins from other primates). The greater the similarity between the protein used to immunize the rabbit and the protein being tested, the greater the extent of the immunological reaction. The degree of similarity between the specific antigen and other proteins is expressed as *immunological distance*. Such distances can be converted, if desired, into approximate numbers of amino acid differences.

Table 7.8 gives the immunological distances among humans, apes, and Old World monkeys. Antibodies against albumin obtained from a human (*Homo sapiens*), a chimpanzee (*Pan troglodytes*), and a gibbon (*Hylobates lar*) were prepared independently. The antibodies were then reacted against albumins obtained from humans, six species of apes, and six species of Old World monkeys. The phylogeny obtained from the data in the table is shown in Figure 7.16.

Electrophoresis is another relatively inexpensive method used to estimate protein differences among organisms. With electrophoresis, the number of amino acid differences between two species is not known, but only whether or not two proteins are electrophoretically identical. The simplicity of the method makes feasible the comparison of many proteins.

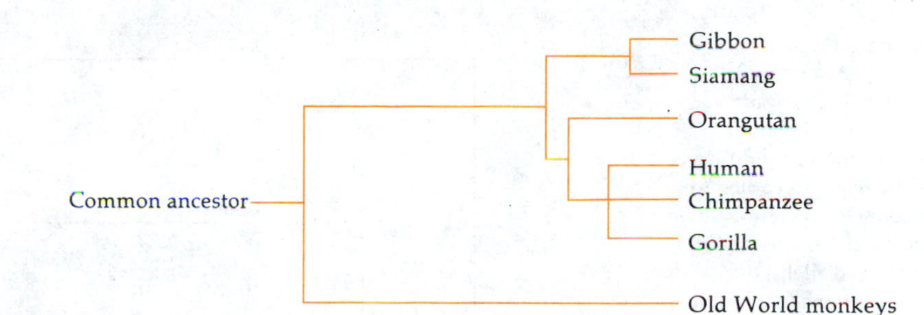

Figure 7.16
Phylogeny of humans, apes, and Old World monkeys, based on immunological differences between their albumins. Humans, chimpanzees, and gorillas appear more closely related to one another than any one of them is to the orangutan—a result confirmed by additional molecular studies. [After V. M. Sarich and A. C. Wilson, *Science 158*:1200 (1967).]

The overall results can be expressed as genetic distance, *D*, using the procedure shown in Box 7.1.

Electrophoresis is ineffective for comparing organisms that are evolutionarily very distant. These are likely to be electrophoretically different at all, or most, loci. Since the number of amino acid differences involved cannot be determined (but only whether or not the two proteins compared have identical electrophoretic migrations), the method fails to determine the degree of differentiation among various species when these differ at all, or nearly all, loci. On the other hand, electrophoretic distances have the advantage of being based on many loci; therefore, unequal rates of evolution in different lineages with respect to one locus may be compensated by other loci. Electrophoresis is, in general, an appropriate method for measuring genetic change among closely related organisms, in which the amino acid sequence of a single protein may fail to show any differences or give misleading results because of the small numbers of substitutions involved.

The phylogeny of the *Drosophila willistoni* group based on a matrix of genetic distances is shown in Figure 7.17. The numbers shown on the branches are expressed in units of genetic distance, *D*, and therefore estimate the average number of electrophoretically detectable allelic substitutions per locus that have occurred in each branch.

Neutrality Theory of Molecular Evolution

The reconstruction of a phylogeny from genetic similarities depends on the assumption that degrees of similarity reflect degrees of phylogenetic propinquity. On the whole, this is a reasonable assumption because

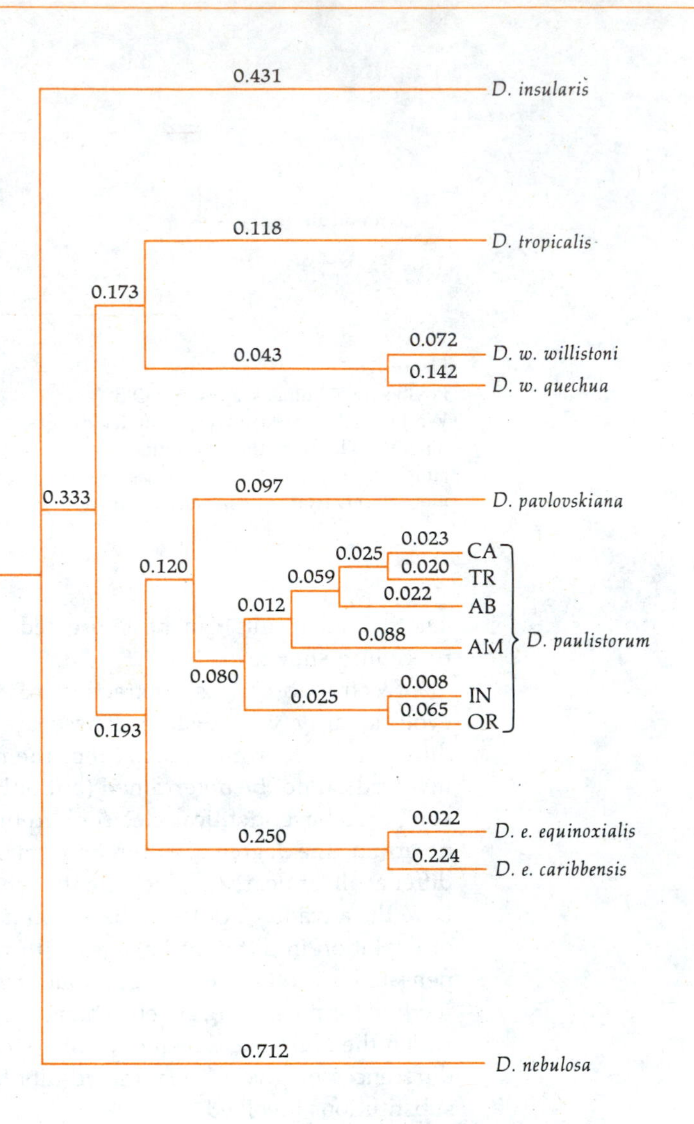

Figure 7.17
Phylogeny of species related to *Drosophila willistoni*, based on electrophoretic differences at 36 gene loci coding for enzymes. The numbers on the branches are the estimated allelic substitutions (detectable by electrophoresis) per locus that have taken place in evolution. There are seven species in the phylogenetic tree. Two species, *D. willistoni* and *D. equinoxialis*, are represented by two subspecies each. *D. paulistorum* is a complex of six semispecies (incipient species), called Central American (CA), Transitional (TR), Andean-Brazilian (AB), Amazonian (AM), Interior (IN), and Orinocan (OR). [After F. J. Ayala et al., *Evolution* 28:576 (1974).]

evolution is a process of gradual change. However, differences in rates of genetic change among lineages may be a source of error. Assume that a given species, A, diverged from the common lineage of two other species, B and C, before these diverged from each other. Assume also that a certain protein has evolved at a much faster rate in the lineage leading to C than in the other two lineages. It might be that the amino acid sequences of the protein would be more similar between A and B than between B and C. The phylogeny inferred from the amino acid sequences of the protein might then be erroneous.

The hypothesis has recently been advanced by Motoo Kimura and others that rates of amino acid substitutions in proteins and of nucleotide substitutions in DNA may be approximately constant because the vast majority of such changes are selectively neutral. New alleles appear in a population by mutation. If alternative alleles have identical fitness, changes in allelic frequencies from generation to generation will occur only by accidental sampling errors from generation to generation, i.e., by genetic drift (Chapter 3). Rates of allelic substitution would be stochastically constant, i.e., they would occur with a constant probability for a given protein. That probability can be shown to be simply the mutation rate for neutral alleles.

The neutrality theory of molecular evolution recognizes that, for any gene, a large proportion of all possible mutants are harmful to their carriers; these mutants are eliminated or kept at very low frequency by natural selection. The evolution of morphological, behavioral, and ecological traits is governed largely by natural selection, because it is determined by the selection of favorable mutants against deleterious ones. It is assumed, however, that a number of favorable mutants, adaptively equivalent to each other, can occur at each locus. These mutants are not subject to selection relative to one another because they do not affect the fitness of their carriers (nor do they modify their morphological, physiological, or behavioral properties). According to the neutrality theory, evolution at the molecular level consists for the most part of the gradual, random replacement of one neutral allele by another that is functionally equivalent to the first. The theory assumes that although favorable mutations occur, they are so rare that they have little effect on the overall evolutionary rate of nucleotide and amino acid substitutions.

Neutral alleles are not defined as having fitnesses that are identical in the mathematical sense. Operationally, neutral alleles are those whose differential contributions to fitness are so small that their frequencies change more owing to drift than to natural selection. Assume that two alleles, A_1 and A_2, have fitnesses 1 and $1 - s$ (where s is a positive number smaller than 1). The two alleles are effectively neutral if, and only if,

$$4N_e s << 1$$

where N_e is the effective size of the population (Chapter 3, page 75).

We now want to find the rate of substitution of neutral alleles, k, per unit time in the course of evolution. The time units can be years or generations. In a random mating population with N diploid individuals,

$$k = 2Nux$$

where u is the neutral mutation rate per gamete per unit time (time measured in the same units as for k) and x is the probability of ultimate fixation of a neutral mutant. The derivation of this equation is straightforward: there are $2Nu$ mutants per unit time, each with a probability x of becoming fixed.

A population of N individuals has $2N$ genes at each autosomal locus. If the alleles are neutral, all genes have the identical probability of becoming fixed, which is simply

$$x = \frac{1}{2N}$$

Substituting this value of x in the previous equation, we obtain

$$k = 2Nu\frac{1}{2N} = u$$

That is, the rate of substitution of neutral alleles is precisely the rate at which the neutral alleles arise by mutation, independently of the size of the population and any other parameters. This is not only a remarkably simple result, but also one with momentous implications if it indeed applies to molecular evolution.

The Molecular Clock of Evolution

If the neutrality theory of molecular evolution were correct for a large number of gene loci, protein and DNA evolution would serve as evolutionary "clocks." The degree of genetic differentiation among species would be a measure of their phylogenetic relatedness; it would thus be justifiable to reconstruct phylogenies on the basis of genetic differences. Moreover, the actual chronological time of the various phylogenetic events could be roughly estimated. Assume that we have a phylogeny such as the one shown in Figure 7.12. If the rate of evolution of cytochrome *c* were constant through time, the number of nucleotide substitutions that have occurred in each branch of the phylogeny would be directly proportional to the time elapsed. Knowing from an outside source (such as the paleontological record) the actual geological time of any one event in the phylogeny would make it possible to determine the times of all the other events by simple proportions. That is, once it is "calibrated" by reference to a single event, the molecular clock can be used to measure the time of occurrence of all the other events in the phylogeny.

The molecular clock postulated by the neutrality theory is, of course, not a metronomic clock, like timepieces in ordinary life, which measure time exactly. The neutrality theory predicts, instead, that molecular evolution is a "stochastic clock," like radioactive decay. The *probability* of change is constant, although some variation occurs. Over fairly long periods of time, a stochastic clock is nevertheless quite accurate. Moreover, each gene or protein would represent a separate clock, providing an independent estimate of phylogenetic events and their time of occurrence.

Table 7.9
Statistical tests of the constancy of evolutionary rates of 7 proteins in 17 species of mammals.

Rates Tested	Chi-Square	Degrees of Freedom	Probability
Overall rates (comparisons among branches over all 7 proteins)	82.4	31	4×10^{-6}
Relative rates (comparisons among proteins within branches)	166.3	123	6×10^{-2}
Total	248.7	154	6×10^{-6}

After W. M. Fitch, in *Molecular Evolution,* ed. by F. J. Ayala, Sinauer, Sunderland, Mass., 1976, p. 160.

Each gene or protein would "tick" at a different rate (the mutation rate to neutral alleles, u, of the gene—see Figure 7.14), but all of them would be timing the same evolutionary events. The joint results of several genes or proteins would provide a fairly precise evolutionary clock.

Is there a molecular clock of evolution? This question can be investigated by examining whether or not the variation in the number of molecular changes that have occurred during equal evolutionary periods is greater than that expected by chance. This would also be a test of the neutrality theory of molecular evolution. Two kinds of test are possible. One kind consists of examining the number of molecular changes between phylogenetic events whose timing is known from the paleontological record and other sources. The other kind does not use actual times, but rather looks at parallel lineages derived from a common ancestor and tests whether the variation in the number of molecular changes along the branches is greater than that expected by chance.

Whether or not the neutrality theory is correct, and how accurate the molecular clock is, are at present controversial matters. The existing evidence suggests that the variation in the rate of molecular evolution is greater than that predicted by the neutrality theory. Nevertheless, molecular evolution appears to occur with sufficient regularity to serve as an evolutionary clock, although not as accurately as if the rate of evolution were stochastically constant, and would have only the variation expected from a Poisson distribution. The results of one test devised by Charles H. Langley and Walter M. Fitch are given in Table 7.9.

The test uses 7 proteins sequenced in 17 mammals, and starts by adding up the proteins one after another and treating them as if they were one single sequence. The minimum number of nucleotide substitutions

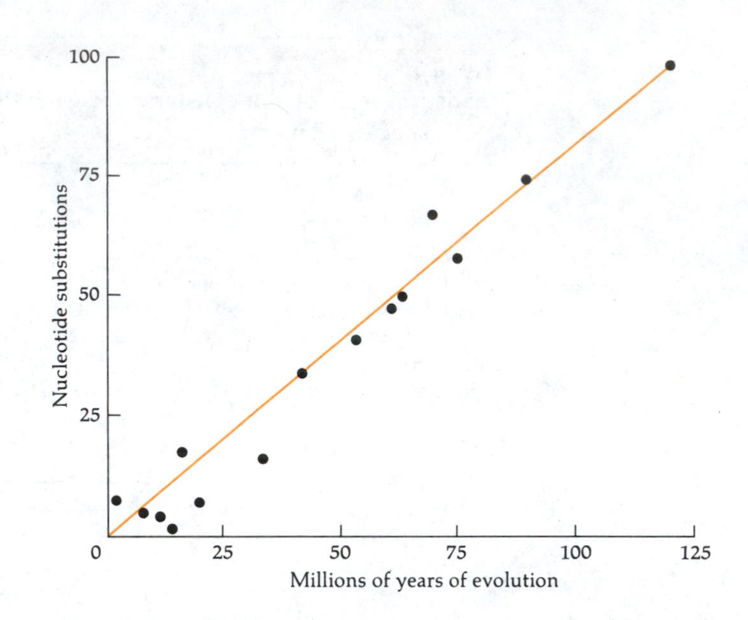

Figure 7.18
Nucleotide substitutions versus paleontological time. The minimum numbers of nucleotide substitutions for seven proteins (cytochrome *c*, fibrinopeptides A and B, hemoglobins α and β, myoglobin, and insulin C-peptide), sequenced in 17 species of mammals, have been calculated for comparisons between pairs of species whose ancestors diverged at the time indicated on the abscissa. The line has been drawn from the origin to the outermost point and corresponds to a rate of 0.41 nucleotide substitution per million years for all seven proteins together. Most points fall near the line, except for some representing comparisons between primates (points below the line at lower left), in which protein evolution seems to have occurred at a lower than average rate. (After W. M. Fitch, in *Molecular Evolution*, ed. by F. J. Ayala, Sinauer, Sunderland, Mass., 1976, p. 160.)

that accounts for the descent of the amino acid sequences from a common ancestor is found; the numbers of substitutions are then assigned to the various branches in the phylogeny. Two independent tests are made. First, the total number of substitutions per unit time is examined for different times; the hypothesis tested is whether the *overall* rate of change is uniform over time. The probability that the variation observed is due to chance is 4×10^{-6}, which is statistically highly significant. The conclusion follows that the proteins have not evolved at a constant rate with a Poisson variance. It is possible, however, that the proteins have all changed their rates *proportionately*—for example, because the rate of molecular evolution is constant per generation rather than per year; variations in generation length might have occurred through time. This possibility is examined by testing whether the rates of evolution of one protein *relative* to those of another are uniform through time. There is a marginally significant deviation from expectation (probability $\approx$ 0.06). The probability that all the variation observed (*total*) is due to chance is extremely small, 6×10^{-6}. The test is particularly valid because it makes no use of paleontological dates. The phylogeny is constructed using the protein data alone; this maximizes the probability of agreement between the data and the hypothesis of stochastically constant rates of molecular evolution. Even so, the data do not fit the hypothesis of constant probability of change with a Poisson variance. However, John H. Gillespie and C. H. Langley have recently shown that the data used for Table 7.9 *are* consistent with the hypothesis that molecular evolution occurs with a constant probability, if it is assumed that the variance of this probability is greater than that expected from a Poisson distribution.

Whether or not molecular evolution is stochastically constant, it appears that the heterogeneity of evolutionary rates is not excessively great. It is, therefore, possible to use genetic data as an approximate evolutionary clock; but in order to avoid large errors, it is necessary to use *average* rates obtained for many proteins and for long periods of time. Figure 7.18 plots the cumulative number of nucleotide substitutions required in 7 proteins against the paleontological dates of divergence in the evolution of 17 mammalian species. The overall correlation is fairly good for all phylogenetic events except those involving some primates, which appear to have evolved at a substantially lower rate than the average. This deviation from the average, observed at the lower left of the graph, illustrates an important point: the more recent the divergence of any two species, the more likely it is that the genetic changes observed will depart from the average evolutionary rate. This is simply because, as time increases, periods of rapid evolution and periods of slow evolution in any one lineage will tend to cancel each other out.

Structural Versus Regulatory Evolution

Our closest living relatives are the great apes—the African chimpanzee and gorilla and the Asian orangutan. Humans are classified in the family Hominidae; chimpanzees, gorillas, and orangutans are classified in the family Pongidae. The lesser Asian apes, gibbons and siamangs, are classified in the family Hylobatidae. The classification of humans and the great apes in different families is justified on biological grounds. As George Gaylord Simpson has written, "*Homo* is both anatomically and adaptively the most radically distinctive of all hominoids, divergent to a degree considered familial by all primatologists." Yet electrophoretic studies indicate that humans and apes are genetically as similar as are closely related species in other groups of organisms (Table 7.10 and Figure 7.19). The average genetic distance between humans and the great apes is $D = 0.354$, or about 35 electrophoretically detectable substitutions per 100 loci. Marie-Claire King and Allan C. Wilson have calculated that

Table 7.10
Genetic differentiation between humans and apes. Genetic identity, *I*, and genetic distance, *D*, are based on 23 gene loci studied electrophoretically.

Species Compared to Humans	*I*	*D*
1. *Pan troglodytes* (chimpanzee)	0.680	0.386
2. *Pan paniscus* (pygmy chimpanzee)	0.732	0.312
3. *Gorilla gorilla* (gorilla)	0.689	0.373
4. *Pongo pygmaeus abelii* (Sumatra orangutan)	0.710	0.347
5. *Pongo pygmaeus pygmaeus* (Borneo orangutan)	0.705	0.350
6. *Hylobates lar* (lar gibbon)	0.489	0.716
7. *Hylobates concolor* (concolor gibbon)	0.429	0.847
8. *Symphalangus syndactylus* (siamang)	0.333	1.099
Averages		
Great apes (1–5)	0.702 ± 0.009	0.354 ± 0.013
All apes (1–8)	0.595 ± 0.055	0.554 ± 0.105

After E. J. Bruce and F. J. Ayala, *Evolution* 33:1040 (1979).

humans and chimpanzees differ in only about 1% of all amino acids in their proteins.

These results lead to a paradox. We perceive ourselves as being considerably different in morphology and ways of life from apes, and this can hardly be attributed exclusively to the fact that humans more readily notice differences involving themselves. However, humans and apes appear to be no more genetically different than are morphologically indistinguishable (sibling) species of *Drosophila* (see Table 7.2). One possible resolution of this paradox is to assume that the estimates of genetic differentiation are biased; there are thousands of structural gene loci, and the few that have been surveyed might not represent a random sample of the whole genome. But the proteins surveyed by electrophoresis, as well as those sequenced or studied with immunological methods, *were* chosen at random (with respect to differentiation between species) and are more or less the same proteins studied in other animal groups. Therefore, it seems that a real discrepancy exists—the rate of organismal evolution appears to

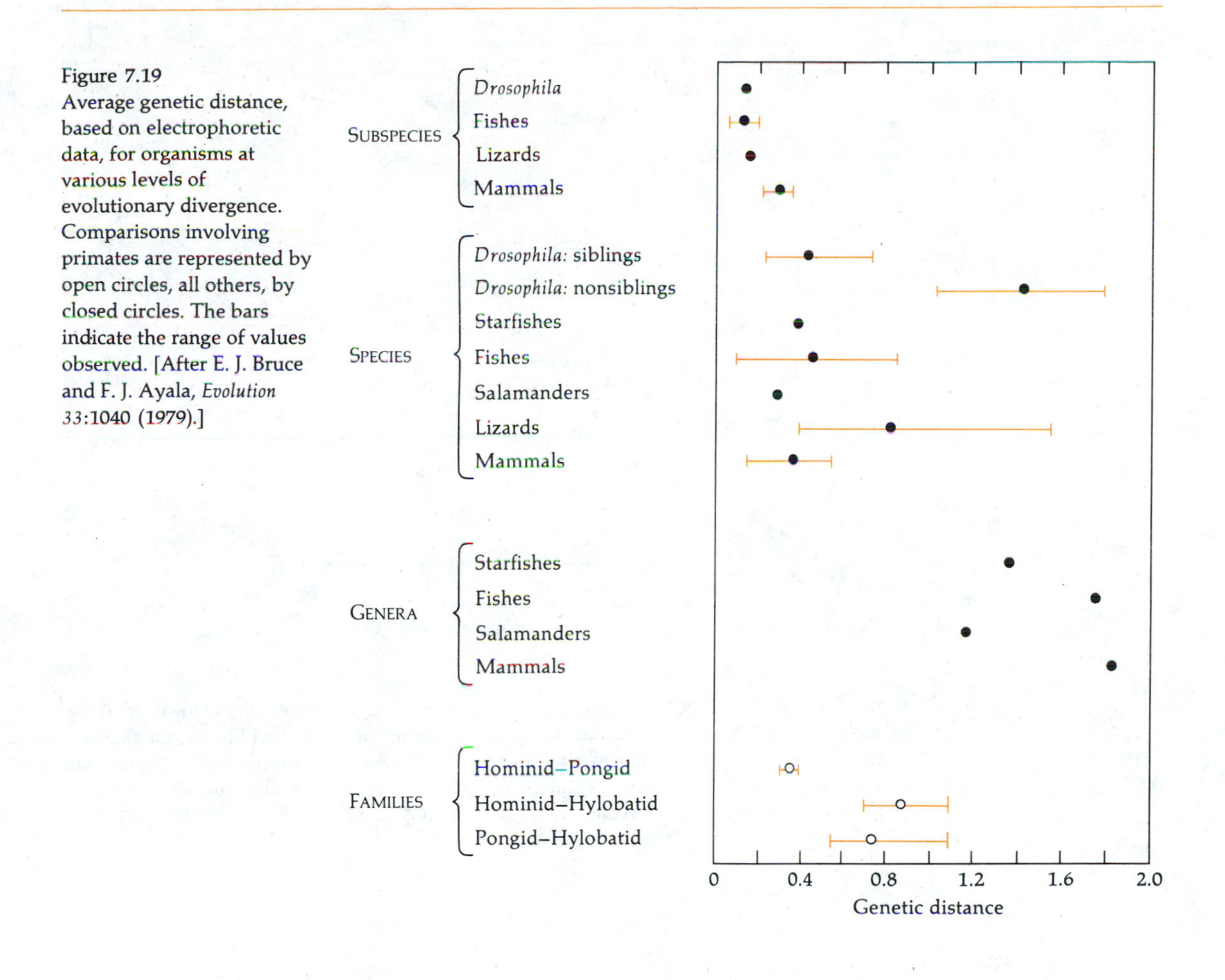

Figure 7.19
Average genetic distance, based on electrophoretic data, for organisms at various levels of evolutionary divergence. Comparisons involving primates are represented by open circles, all others, by closed circles. The bars indicate the range of values observed. [After E. J. Bruce and F. J. Ayala, *Evolution* 33:1040 (1979).]

be greater than the rate of protein evolution in the human lineage (Figure 7.20).

A different resolution of the paradox is possible. This is to postulate that organismal evolution is not determined primarily by changes in structural genes, but by changes in gene regulation. Then, organismal evolution would not necessarily proceed at the same rate as the evolution of structural genes. This hypothesis is supported by other indirect evidence, including the following. (1) Two African toad species, *Xenopus laevis* and *Xenopus borealis*, are morphologically very similar, but their difference in DNA sequence ($\Delta T_s = 12°C$) is greater than that between humans and New World monkeys ($\Delta T_s = 10°C$). (2) Protein evolution proceeds in mammals and in anurans (frogs and toads) at similar rates. Yet there is relatively little morphological differentiation among the more than 3000 known anuran species, while there is considerable morphological divergence among placental mammals (think of an armadillo, a mouse, a whale, and a human). (3) Moreover, frogs—but not mammals—that are

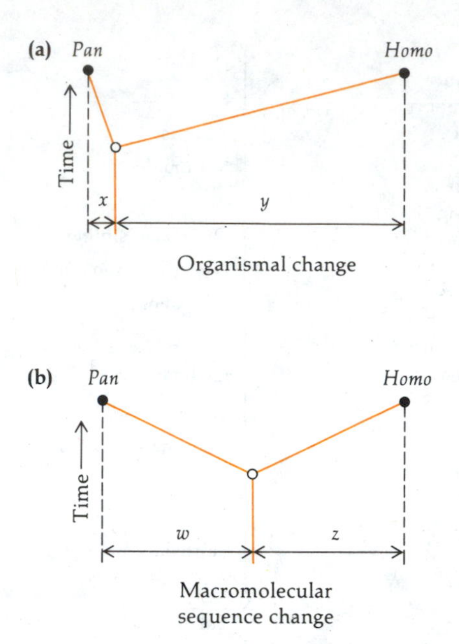

Figure 7.20
Contrast between morphological and molecular evolution since the divergence of the human and chimpanzee lineages. **(a)** More organismal change has taken place in the human lineage (y) than in the chimpanzee lineage (x). **(b)** However, at the protein and nucleic acid levels, the rate of change has been approximately the same in both lineages ($w \approx z$). [After M.-C. King and A. C. Wilson, *Science 188*:107 (1975).]

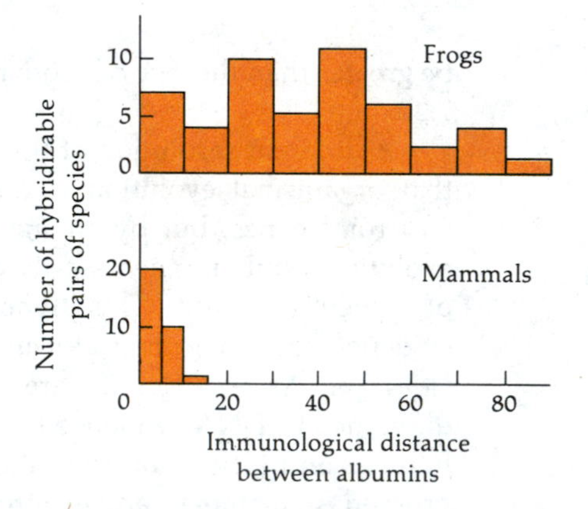

Figure 7.21
Interspecific hybridization as a function of immunological distance between the albumins of pairs of species capable of producing viable hybrids. In this study, 31 such pairs of placental mammals and 50 pairs of frogs were investigated. Immunologically quite different species of frogs are able to hybridize, but not mammals. [After A. C. Wilson et al., *Proc. Natl. Acad. Sci. USA 71*:2843 (1974).]

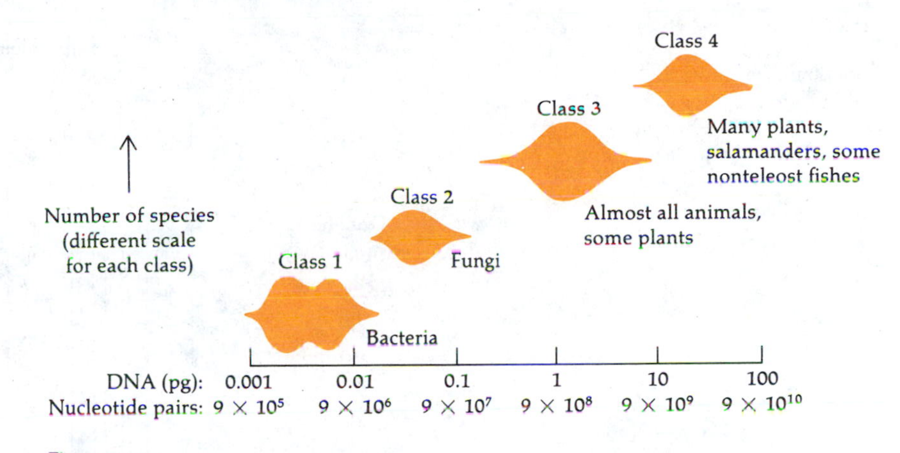

Figure 7.22
Organisms classified according to their amounts of DNA per cell. The amount of DNA is given by weight (1 pg = 10^{-12}g) and by the number of nucleotide pairs. It ranges within one order of magnitude for most organisms within each group. For all organisms, from bacteria to plants and animals, the amount of DNA varies by more than five orders of magnitude. (After R. Hinegardner, in *Molecular Evolution*, ed. by F. J. Ayala, Sinauer, Sunderland, Mass., 1976, p. 179.)

very different at the protein level are able to produce interspecific hybrids (Figure 7.21).

The role of gene regulation in adaptive evolution remains one of the major unsolved issues in evolutionary genetics. The evidence just mentioned suggests that changes in gene regulation may be most important in adaptive evolution, i.e., in the evolution of morphology, behavior, and reproductive isolation. Moreover, recent experimental studies with bacteria, yeast, and *Drosophila* have shown that adaptation to new environmental conditions often occurs by changes in gene regulation, although these may later be followed by changes in structural genes. However, little is known about gene regulation in higher organisms.

Evolution of Genome Size

Evolution consists not only of changes in the nucleotide sequence of the DNA, but also of changes in the *amount* of DNA. The early organisms ancestral to all DNA-containing living beings probably had only a few genes. Today, considerable variation in the amount of DNA per cell exists between organisms of different species. Organisms can be grouped into four broad classes, according to the amount of DNA they carry in each cell (Figures 7.22 and 7.23). The lowest amounts of DNA are found in some viruses, with about 10^4 nucleotide pairs per virus. Bacteria have, on the

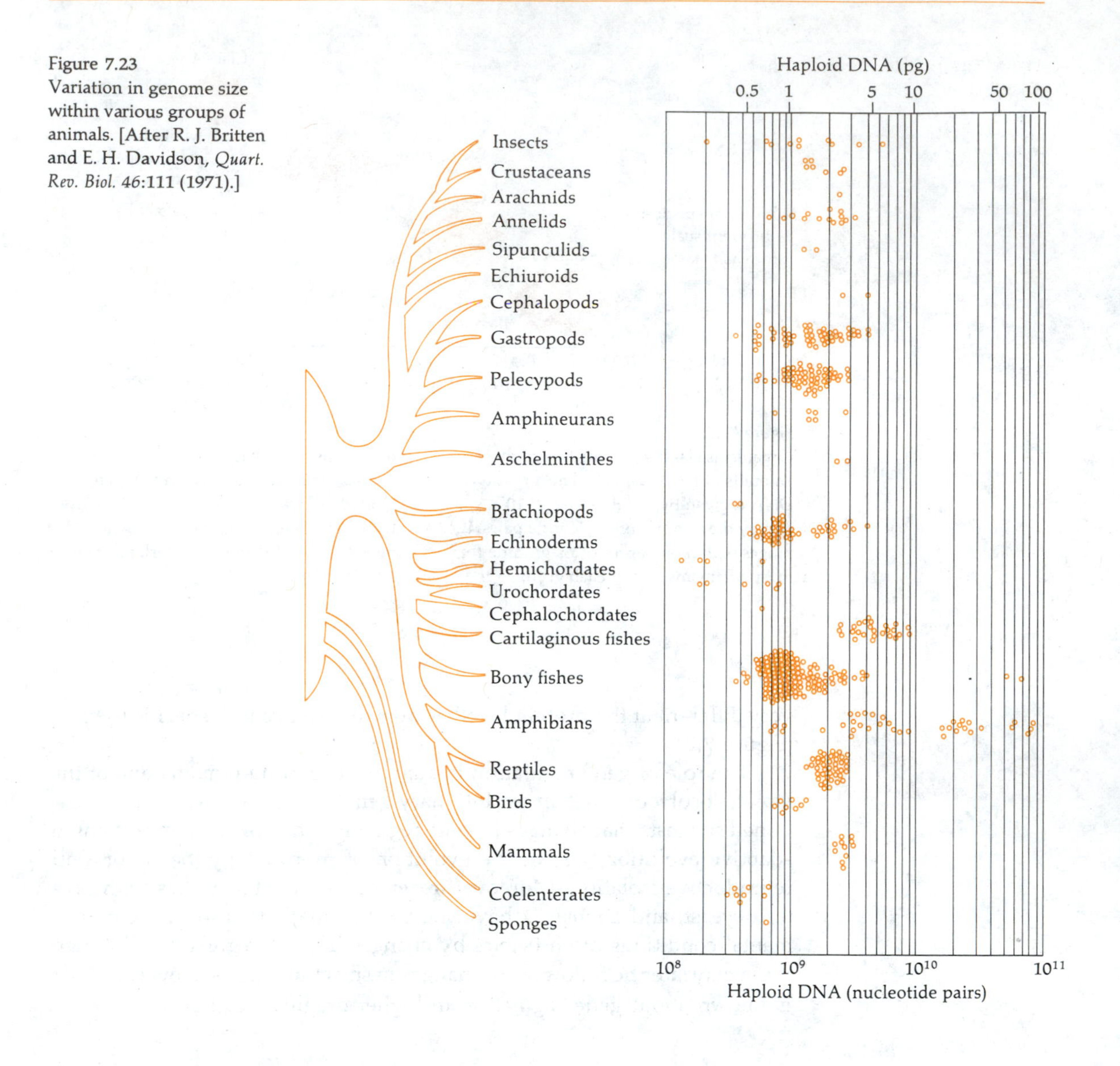

Figure 7.23
Variation in genome size within various groups of animals. [After R. J. Britten and E. H. Davidson, *Quart. Rev. Biol. 46*:111 (1971).]

average, about 4×10^6 nucleotide pairs per cell, and fungi have about ten times as much, or 4×10^7 nucleotide pairs per cell. Most animals and many plants have about 2×10^9 nucleotide pairs per cell, on the average. The most advanced plants, gymnosperms and angiosperms, often have 10^{10} or even more nucleotide pairs per cell. The animals with the largest amounts of DNA are salamanders and some primitive fishes, with more than 10^{10} nucleotide pairs per cell.

A substantial evolutionary increase in the amount of DNA per cell has occurred from bacteria to fungi, to animals, and to plants. More complex

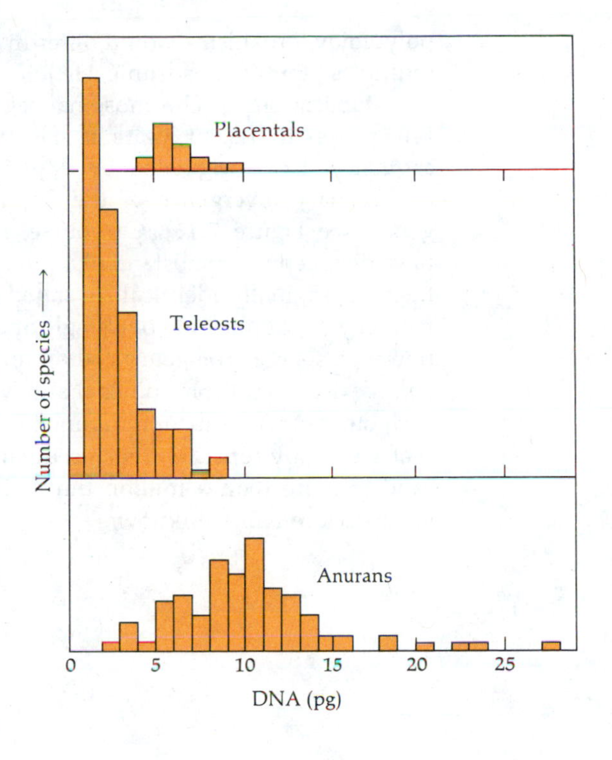

Figure 7.24
Distribution of DNA per cell in certain classes of mammals, fishes, and amphibians. The distributions vary around an intermediate mode. This suggests that evolutionary changes have been numerous and small. [After K. Bachmann, O. B. Goin, and C. J. Goin, in *Evolution of Genetic Systems*, ed. by H. H. Smith, *Brookhaven Symp. Biol.* 23:419 (1972).]

organisms may need more DNA than a bacterium or a mold, but there seems to be no consistent relationship between the amount of DNA in an organism and its complexity of organization. For example, salamanders and flowering plants are not ten times more complex than mammals or birds, although some of the former have ten times more DNA than the latter.

How does the amount of DNA in the nucleus increase during evolution? Polyploidy is a process by which the amount of DNA can increase; when the number of chromosomes per cell is doubled, the amount of DNA is also doubled. Some organisms with very large amounts of DNA, such as some primitive vascular plants (*Psilopsida*), are polyploid. But polyploidy is a rare phenomenon in animals.

Deletions and duplications (Chapter 1) of relatively small segments of DNA appear to be the most general processes by which evolutionary changes in the amount of DNA have taken place. When the genome sizes of many fish, frog, and mammalian species are arranged in a frequency diagram, they are seen to vary around an intermediate mode (Figure 7.24). This indicates that evolutionary changes in the genome size of animals are numerous and individually small, as would be true with duplications and deletions. If changes in the amount of DNA had occurred primarily by polyploidy, organisms would differ in the amount of their DNA by exact multiples (double, quadruple, etc.).

Duplications of chromosomal segments often involve only one or a few genes. In recent years it has been discovered that many DNA sequences have originated by duplication, followed in some cases by evolutionary divergence of the duplicated sequences (e.g., the globin genes—see Figure 7.13). Of course, if the duplicated DNA sequences have diverged very substantially, it may no longer be possible to identify them as originally identical; as stated earlier, all genes must have originated by the duplication of a single or a few original genes. In other cases, however, such as the genes coding for ribosomal RNA or transfer RNA, genes exist in multiple copies that have remained essentially identical to each other in structure and in function. Finally, there are DNA sequences that are highly repetitive, each sequence being present from a few thousand to more than a million times. The role of these highly repetitive sequences remains unknown.

Problems

1. DNA hybridization experiments indicate that the estimated DNA nucleotide differences are 3% between *Drosophila melanogaster* and *D. simulans* and 13% between *D. melanogaster* and *D. funebris* (see Figure 7.7). Draw the most likely configuration of the phylogeny of these three species, indicating the percent nucleotide changes for each segment. Can you determine whether or not the percent change from the last common ancestor of *D. melanogaster* and *D. simulans* to each of these two species has been the same?

2. According to Table 7.5, the estimated DNA nucleotide differences are 9.6% between humans and green monkeys, 15.8% between humans and capuchins, and 16.5% between green monkeys and capuchins. Reconstruct the phylogeny of these three species and indicate the probable percent nucleotide changes for each segment. Why is it possible in this case to determine the percent change from the last common ancestor of humans and green monkeys to each of these two species, whereas you could not make a similar determination in problem 1?

3. The minimum numbers of nucleotide differences in the gene coding for cytochrome *c* between pairs of four species (see Table 7.7) and the configuration of the phylogeny are given at the top of the next page:

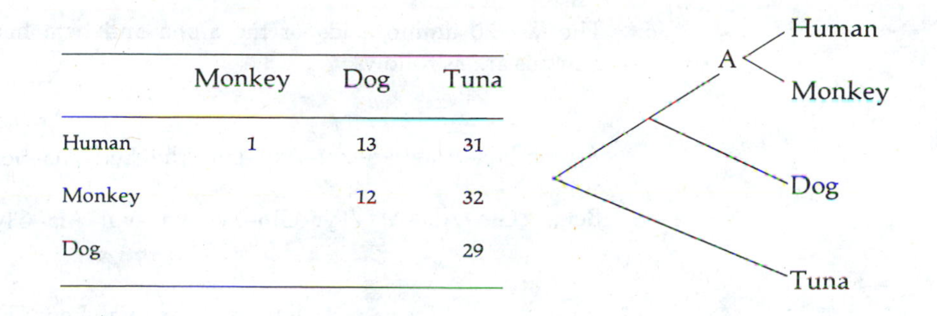

	Monkey	Dog	Tuna
Human	1	13	31
Monkey		12	32
Dog			29

Calculate the minimum numbers of nucleotide differences from species A to human and monkey cytochrome *c* using: (1) only the data for human, monkey, and dog, and (2) only the data for human, monkey, and tuna. Are the two results identical? The inconsistency that you observe is, in fact, a common situation when a problem is "overdetermined" (there are more data than are necessary to find each unknown) and the data are not fully consistent. Evolutionists deal with these inconsistencies by finding out the changes for each segment of a phylogeny that minimize the total amount of discrepancy between the data and the proposed changes. Electronic computers are needed to find such "best solutions" whenever more than a few species are involved.

4. Compare the results obtained in problem 2 with the values shown in Figure 7.8. What is the reason for the discrepancy?

5. Males and females of two closely related species of *Drosophila* were placed together in equal numbers. The following matings of each kind were observed:

♀ *D. bifasciata* × ♂ *D. bifasciata*	229
♀ *D. imaii* × ♂ *D. imaii*	375
♀ *D. bifasciata* × ♂ *D. imaii*	13
♀ *D. imaii* × ♂ *D. bifasciata*	9

Calculate the coefficient of sexual isolation, C_i, between these two species, defined as

$$C_i = p_{11} + p_{22} - q_{12} - q_{21}$$

where p_{11} and p_{22} are the frequencies of the two kinds of homogamic matings (♀ A × ♂ A and ♀ B × ♂ B), and q_{12} and q_{21} are the two kinds of heterogamic matings (♀ A × ♂ B and ♀ B × ♂ A).

6. The last 20 amino acids of the alpha and beta hemoglobin chains in humans are as follows:

Alpha: His–Ala–Ser–Leu–Asp–Lys–Phe–Leu–Ala–Ser–

Beta: Gln–Ala–Ala–Tyr–Gln–Lys–Val–Val–Ala–Gly–

Val–Ser–Thr–Val–Leu–Thr–Ser–Lys–Tyr–Arg

Val–Ala–Asn–Ala–Leu–Ala–His–Lys–Tyr–His

Using the genetic code (Table 1.3), calculate the minimum number of nucleotide changes that have occurred in the evolution of the two segments of DNA coding for the 20 amino acids since the duplication of the alpha and beta genes.

7. The rate of substitution of alleles, $k = 2Nux$ (page 211), may apply to selected alleles as well as to neutral alleles. If a newly arisen allele, A_2, is advantageous relative to the pre-existing allele, A_1, so that the fitness for A_1A_1 is 1, for A_1A_2 is $1 + s$, and for A_2A_2 is $1 + 2s$, the probability of fixation of the new allele is $x = 2N_e s/N$, where N_e is the effective population size and N is the total number of individuals in the population. Assume that the rate of mutation to a new allele is $u = 10^{-5}$ per generation and that the effective population size is 10,000. Calculate the rate of substitution of alleles at three different loci in which (1) $s = 0$, (2) $s = 0.0001$, and (3) $s = 0.01$.

8. Genetic variation at 19 gene loci coding for blood proteins was studied in four primate species: chimpanzee (*Pan troglodytes*), gorilla (*Gorilla gorilla*), gibbon (*Hylobates lar*), and baboon (*Papio cynocephalus*). The results are given below. When only one allele was found, only the number identifying the allele is given; when there was more than one allele, the frequency of each is given in parentheses.

Locus	*Pan*	*Gorilla*	*Hylobates*	*Papio*
1. *Ak*	96	98 (0.20) 100 (0.80)	92	96
2. *Alb*	100	100	100	99
3. *Aph*	100	100	100	100
4. *Cer*	100	98	98	102
5. *Dia*	100	85 (0.67) 95 (0.33)	100 (0.67) 108 (0.33)	95 (0.88) 100 (0.12)
6. *Est-A*	100	101	102	96
7. *Est-B*	100	100	102	95 (0.17) 96 (0.08) 103 (0.75)
8. *G6pd*	100	100	102	102
9. *Got*	100	100	96	96 (0.14) 100 (0.86)
10. *Hb*	100	100	100	100 (0.92) 102 (0.08)
11. *Hpt*	105	107	107	107
12. *Icd*	96	100	100	100 (0.94) 107 (0.06)
13. *Lap*	100	100	100	100
14. *Ldh-A*	96	96	96	96
15. *Ldh-B*	100	100	100	100
16. *Mdh*	100	100	93 (0.62) 100 (0.38)	106
17. *6-Pgd*	97	97 (0.15) 105 (0.85)	94	94
18. *Pgm-1*	96 (0.12) 100 (0.88)	100	100	94
19. *Pgm-2*	100	96	100	102

Calculate the genetic identity, I, and genetic distance, D, and draw the likely configuration of the phylogeny.

Appendix

Probability and Statistics

Probability theory and many other branches of mathematics are used in modern science for developing models of natural phenomena and for predicting results. Mathematical statistics is used for summarizing and analyzing data. In this appendix we present certain concepts and methods of probability theory and statistics that are necessary for an understanding of population genetics. The treatment is quite elementary and will be superfluous for students with college courses in mathematics and statistics, but it will be useful to others. The following topics are introduced: (I) basic concepts of *probability* theory necessary for calculating the numerical expectations from certain kinds of hypotheses; (II) the *chi-square method* of testing hypotheses; (III) the *mean* and the *variance*—two measures frequently used in summarizing data; (IV) the *normal distribution*—a kind of distribution commonly encountered in population genetics.

A.I Probability

Probability can be defined as the number of favorable outcomes of an event divided by the total number of possible outcomes. For example, in a pea plant heterozygous (Rr) for the round-pea allele (R) and the wrinkled-pea allele (r), the probability that a gamete will carry the R allele is $\frac{1}{2}$. In this case, the possible outcomes are R and r, and we designate R the "favorable" one.

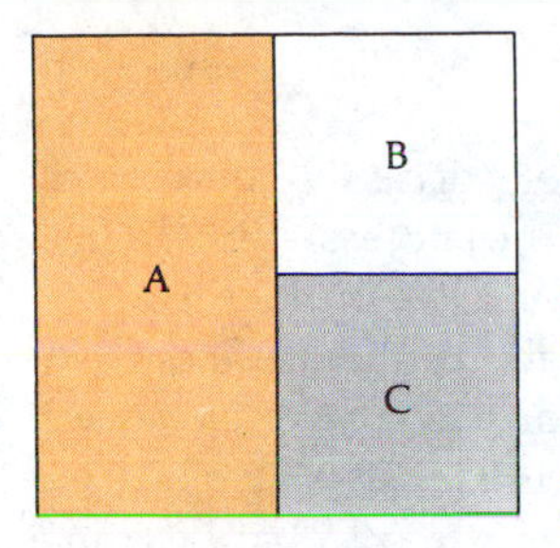

Figure A.1
Additivity of alternative probabilities. If A, B, and C represent three outcomes of a given event, with probabilities measured by their relative areas, the probability that either A or B will occur is the sum of the two corresponding areas.

Sometimes the probability of an outcome is not known *a priori;* it can then be determined empirically by observing the frequency with which it occurs. For example, in a sample of 6346 X chromosomes of *Drosophila,* 8 were found to carry a newly arisen recessive lethal. The probability of mutation to a recessive lethal in one X chromosome is then estimated as $8/6346 = 0.0013$ per generation.

All probabilities must lie between zero and one. A probability of one indicates that the outcome is certain to occur; a probability of zero indicates that the outcome cannot occur.

There are certain laws that apply to the combination of probabilities. The most fundamental ones are called the law of the sum of probabilities and the law of the product of probabilities.

LAW OF THE SUM OF PROBABILITIES The probability that one *or* another of several mutually *exclusive* outcomes of an event will occur is the sum of their individual probabilities. For example, suppose that we represent the probabilities of the only three possible outcomes of an event as areas in a square, as in Figure A.1, where the probability of A is $1/2$, the probability of B is $1/4$, and the probability of C is $1/4$. Then the probability that, say, either A *or* B will occur is $1/2 + 1/4 = 3/4$, as is apparent from the figure. In the progeny of a cross between two *Rr* heterozygous pea plants, the probabilities of the three kinds of progeny are: $1/4$ homozygous *RR*, $1/4$ homozygous *rr*, and $1/2$ heterozygous *Rr.* Then the probability that a progeny will exhibit the dominant phenotype, i.e., that it will be either homozygous *RR* or heterozygous *Rr*, is $1/4 + 1/2 = 3/4$.

LAW OF THE PRODUCT OF PROBABILITIES The probability that several mutually *independent* outcomes will *all* occur is the product of their individual probabilities. Consider, for example, the cross $Rr \times Rr$ mentioned in the previous section. The probability that an individual offspring will receive the *r* allele from one of the parents is $1/2$; the probability that it will receive the *r* allele from the other parent is also $1/2$. Hence the probability that an individual will receive an *r* allele from one parent *and* an *r* allele from the other parent is $1/2 \times 1/2 = 1/4$.

It is important that the outcomes be independent. Consider, for example, the cross shown in Figure 1.5. The F_1 is a cross between a female, heterozygous (w^+/w) for the red-eye allele (w^+) and the white-eye allele (w), with a red-eyed male (w^+/Y). In the F_2, the probability of a white-eyed individual is $1/4$, because this requires that an individual receive the w allele from the mother (an outcome with a probability of $1/2$) and the Y chromosome from the father (an outcome that also has a probability of $1/2$). Now, the probability that an F_2 individual is a male is $1/2$. But it would be erroneous to conclude that the probability that an F_2 individual is both white-eyed *and* male is $(1/4 \times 1/2) = 1/8$. This is because the two outcomes are clearly not independent. For an F_2 individual to have white eyes, it must inherit the Y chromosome from its father; thus, *all* F_2 white-eyed individuals are males.

A caveat. When calculating the probability of successive outcomes, it is important to distinguish the probability of all the successive outcomes *together* from the probability of a given outcome *alone.* Consider, for example, the following question: What is the probability that the first two children born in a family will both be males? For the answer, we need only apply the law of the product of probabilities. Assuming that the probability of having a male child is ½, the answer is $\frac{1}{2} \times \frac{1}{2} = \frac{1}{4}$. Consider, now, the following, slightly different question: What is the probability that the second child in a family in which the first child is a male will also be a male? The answer in this case is ½. Independently of the sex of any previous children, the probability that the next child will be a male is always ½.

A second caveat. Sometimes, outcomes are samples from a "universe" that is limited (this situation is called by statisticians "sampling without replacement"). In such cases one must take into account that the probability of a given outcome depends on the number of previous trials and on the nature of the outcomes of those trials. As an illustration, consider a pack of 52 cards, with 4 aces. We deal the cards successively and show them face up. What is the probability that the first card will be an ace? The answer is 4/52. Assume that the first card was *not* an ace. What is the probability that the second card will be an ace? There are only 51 cards left; therefore this probability is 4/51. Assume, instead, that the first card *was* an ace. Now what is the probability that the second card will be an ace? There are only 3 aces and 51 cards left; the answer is therefore 3/51.

A.II The Chi-Square Method

One useful method for testing whether the results obtained in an experiment are consistent with a hypothesis is the *chi-square* (χ^2) *method.* The χ^2 function is

$$\chi^2 = \sum \frac{(\text{observed} - \text{expected})^2}{\text{expected}}$$

where Σ stands for "summation" and applies to all classes in the experiment.

Consider the experiment in which Mendel crossed a tall plant (*TT*) and a short plant (*tt*). The F_1 cross was $Tt \times Tt$. According to Mendel's hypothesis, the F_2 should consist of tall (*TT* and *Tt*) and short (*tt*) plants in the ratio 3:1. He observed 787 tall plants and 277 short plants. The

Table A.1
Calculation of χ^2 for Mendel's experiment with tall and short pea plants.

Operation	Tall Plants	Short Plants	Total
Observed (O)	787	277	1064
Expected (E)	$1064 \times 3/4 = 798$	$1064 \times 1/4 = 266$	1064
O − E	−11	+11	0
$(O - E)^2$	121	121	
$(O - E)^2/E$	0.15	0.44	$\chi^2 = 0.59$

calculation of chi-square for this experiment is shown in Table A.1. The value of χ^2 is 0.59. Is this value consistent with the hypothesis? That is, can the difference between the observed and expected values be attributed to chance alone? In order to answer this question, we must introduce two concepts: the number of degrees of freedom and the level of significance.

The *number of degrees of freedom* can be simply calculated as the number of classes whose value must be known in order to determine the values of all the classes (once we know the total). In the example given, the number of degrees of freedom is 1, because if we are given the number in one class (e.g., 787 tall plants), we can determine the number in the other class by subtracting the first class from the total: $1064 - 787 = 277$. In general, in experiments of this type, the number of degrees of freedom is the number of classes minus one, or $k - 1$, because the last class can be calculated by subtracting the sum of all the other classes from the total. (We shall see below that in other types of experiments the number of degrees of freedom is not $k - 1$.)

The *level of significance* refers to the risk that we are willing to take of rejecting a hypothesis even if it is true. Differences between observed and expected values may vary as a result of chance, but if the probability of observing a certain degree of discrepancy is too small, we may want to reject the hypothesis, even though it might be true. The level of significance most commonly used is 5%; that is, it is decided that the outcome is not consistent with the hypothesis whenever the probability of observing a certain discrepancy between the observed and expected values owing to chance alone is 5% or less. The chi-square values for various

Table A.2
Values of χ^2 corresponding to various levels of significance and degrees of freedom.

	Level of Significance		
Degrees of Freedom	0.05	0.01	0.001
1	3.84	6.64	10.83
2	5.99	9.21	13.82
3	7.82	11.34	16.27
4	9.49	13.28	18.47
5	11.07	15.09	20.52
6	12.59	16.81	22.46
7	14.07	18.48	24.32
8	15.51	20.09	26.13
9	16.92	21.67	27.88
10	18.31	23.21	29.59

degrees of freedom are given in Table A.2 for levels of significance of 5%, 1%, and 0.1%.

We now return to the question of whether the results of Mendel's experiment are consistent with his hypothesis. The chi-square is 0.59, with one degree of freedom. This is an acceptable amount of discrepancy, because it is smaller than the chi-square value for one degree of freedom at the 5% level of significance, which is 3.84 (Table A.2). Hence, we are willing to accept that the results are consistent with Mendel's hypothesis and that the difference between the observed and expected values is due to chance.

TEST OF INDEPENDENCE It is sometimes desirable to test whether the results of two sets of observations made on the same individuals are independent or not. For example, 256 pairs of twins are classified as monozygotic or dizygotic and also as concordant or discordant with respect to bronchial asthma. A pair of twins is concordant if both suffer from the condition, and it is discordant if one does but the other does not; cases in which neither of the twins exhibits the trait are not included. The results are as follows:

Monozygotic concordant	30
Monozygotic discordant	34
Dizygotic concordant	46
Dizygotic discordant	146

In this case we do not have any hypothesis to tell us what frequency to expect in each class, but we can test whether the two characteristics are independent by means of a 2 × 2 *contingency table.* First we set the table with the *observed* results:

	Concordant	Discordant	Total
Monozygotic	30	34	64
Dizygotic	46	146	192
Total	76	180	256

We can now calculate the *expected* results (if type of twinning and concordance are independent) for each of the four classes by multiplying the corresponding marginal totals and dividing by the grand total. For example, the expected frequency of monozygotic concordant twins is $(64 \times 76)/256 = 19.00$. The 2 × 2 table of expected results is:

	Concordant	Discordant	Total
Monozygotic	19	45	64
Dizygotic	57	135	192
Total	76	180	256

The χ^2, calculated as in previous examples, is 12.08. Although there are four classes, the number of degrees of freedom in this case is 1, not 3. This can be seen by noting that it is enough to know one of the four values in the 2 × 2 contingency table in order to calculate the other three, by subtraction from the marginal totals (e.g., if the number of monozygotic concordant twins is 30, the number of monozygotic discordant twins must be 34, in order for the total number of monozygotic twins to be 64, and so on). Contingency tables may have any number of rows (r) and columns (c), not counting the "Totals," of course. In general, the number of degrees of freedom is $(r - 1)(c - 1)$.

Because the chi-square, 12.08, is larger than the chi-square for one degree of freedom at the 5% level of significance, we conclude that type of twinning and concordance or discordance with respect to bronchial asthma are not independent. (Lack of independence might be due to the existence of a genetic component for susceptibility to bronchial asthma.)

TESTING THE HYPOTHESIS OF HARDY-WEINBERG EQUILIBRIUM
Two alleles exist at the *Pgm* locus in human populations. The number of individuals of each genotype in a sample of 200 Caucasians is

Pgm^1/Pgm^1	108
Pgm^1/Pgm^2	86
Pgm^2/Pgm^2	6
Total	200

We want to find out whether the observed numbers occur with the frequencies predicted by the Hardy-Weinberg law. First we calculate the frequency, p, of the Pgm^1 allele:

$$p = \frac{(108 \times 2) + 86}{400} = \frac{302}{400} = 0.755$$

The expected frequencies and the expected numbers of the three genotypes are

Genotype	Frequency	Number
Pgm^1/Pgm^1	$p^2 = 0.570$	114
Pgm^1/Pgm^2	$2pq = 0.370$	74
Pgm^2/Pgm^2	$q^2 = 0.060$	12

Calculating the chi-square as above, we obtain $\chi^2 = 5.26$. What is the number of degrees of freedom? It is 1—not 2, as it might seem by analogy with the cases of Mendelian segregation considered above. The reason is that we have used the data to calculate the allele frequency, $p = 0.755$. Given this value and the total size of the sample, it is sufficient to know the number of individuals in one of the three genotypic classes in order to calculate the numbers in the other two classes.

This leads to another rule (equivalent to the one given earlier) for calculating the number of degrees of freedom: The number of degrees of freedom is the number of classes minus the number of independent values obtained from the data that are used for calculating the expected numbers. In the cases of Mendelian segregation considered earlier, the total number of individuals was the only value obtained from the data; with it and the Mendelian laws, we were able to calculate the expected numbers for each phenotypic class. In the Hardy-Weinberg problem, however, we calculate *two* values from the data: the total number of individuals and p. Note that $\chi^2 = 5.26$ is statistically significant at the 5% level of significance with one degree of freedom, but not with two degrees of freedom. If we had erroneously assumed that there were two degrees of freedom, we would not have rejected the hypothesis that the three genotypes are in Hardy-Weinberg equilibrium.

A caveat. The chi-square method is an approximate method that is good only if the total sample and the *expected* numbers in each class are large, but not if they are small. The practical rules are as follows: (1) If there is only one degree of freedom, the expected numbers in each class should be at least 5. (2) If the number of degrees of freedom is greater than 1, the expected numbers in each class should be at least 1. There are however, procedures that can be followed whenever these rules are broken.

If the number of degrees of freedom is 1 and one of the classes is smaller than 5, one must apply *Yates's correction.* This consists in making the differences (Observed − Expected) one-half unit closer to zero and then calculating the chi-square as before. This has been done in Table A.3 for the results of a cross between a wild-type heterozygous rabbit (c^+/c^a) and an albino rabbit (c^a/c^a). Without Yates's correction, $\chi^2 = 4$, which is statistically significant at the 0.05 level of significance. *With* Yates's correction, $\chi^2 = 3.06$, which is *not* significant. We conclude that the results are consistent with the expectations.

If the number of degrees of freedom is greater than 1, but there are classes with expected values less than 1, we can combine the smaller classes with each other, so that the expected values in all new classes are 1 or larger. However, we must take into account that the number of classes after combining them is the number to be used for calculating the number of degrees of freedom. Table A.4 gives the results of a study in which the chromosome arrangements were ascertained in a sample of 50 *Drosophila pseudoobscura* larvae. First we calculate the frequency of each arrangement in the population:

$$p(AR) = \frac{(16 \times 2) + 22 + 6}{100} = 0.60$$

Table A.3
Calculation of χ^2, with and without Yates's correction, for the results of a cross between a heterozygous (c^+/c^a) and an albino (c^a/c^a) rabbit.

Operation	Wild Type	Albino	Total
Observed (O)	12	4	16
Expected (E)	8	8	16
O − E	4	−4	0
$(O - E)^2$	16	16	
$(O - E)^2/E$	2	2	$\chi^2 = 4$
With Yates's correction:			
O − E	3.5	−3.5	0
$(O - E)^2$	12.25	12.25	
$(O - E)^2/E$	1.53	1.53	$\chi^2 = 3.06$

$$q(CH) = \frac{(4 \times 2) + 22 + 0}{100} = 0.30$$

$$r(TL) = 1 - 0.60 - 0.30 = 0.10$$

The expected frequencies of the genotypes can be calculated using the square expansion, $(p + q + r)^2$; the expected number of individuals is obtained by multiplying the total number in the sample (50) by the expected frequencies. This has been done in Table A.4. Three independent values are calculated from the data in order to obtain the expected values: p, q, and the total number of individuals (r is not independently obtained from the data, but is simply calculated as the difference $1 - p - q$). Since there are six classes in the data, the number of degrees of freedom is therefore 6 − 3 independent values = 3. The chi-square is 8.67, which is statistically significant at the 0.05 level with three degrees of freedom. In the lower part of Table A.4, the two classes with the lower expected numbers have been combined. Now we have five classes and, therefore, 5 − 3 = 2 degrees of freedom. The $\chi^2 = 1.81$ is, however, not statistically significant at the 0.05 level.

Table A.4
Calculation of χ^2, with and without combination of small classes, for a Hardy-Weinberg test.

Operation	*AR/AR*	*AR/CH*	*CH/CH*	*AR/TL*	*CH/TL*	*TL/TL*	Total
Observed (O)	16	22	4	6	0	2	50
Expected (E)	18	18	4.5	6	3	0.5	50
O − E	−2	+4	−0.5	0	−3	+1.5	0
$(O - E)^2$	4.00	16.00	0.25	0	9	2.25	
$(O - E)^2/E$	0.22	0.89	0.06	0	3	4.50	$\chi^2 = 8.67$
Combining small classes:							
Observed	16	22	4	6	2		50
Expected	18	18	4.5	6	3.5		50
O − E	−2	+4	−0.5	0	−1.5		0
$(O - E)^2$	4	16	0.25	0	2.25		
$(O - E)^2/E$	0.22	0.89	0.06	0	0.64		$\chi^2 = 1.81$

A.III Mean and Variance

Assume that we have a sample of individuals measured with respect to some trait, such as height. Two parameters that summarize the information are the arithmetic mean and the variance. The mean is said to be a measure of "central tendency," the variance, a measure of "dispersion."

The *arithmetic mean,* or simply the *mean,* of a distribution is the average of the value in all individuals of the sample:

$$\overline{X} = \frac{\Sigma X}{N}$$

where $\overline{X}$ is the mean, ΣX represents the sum of all the values observed, and N is the number of individuals.

Assume that we measure the height, to the nearest centimeter, of 10 students and obtain the following values: 170, 174, 177, 178, 178, 179, 179, 180, 181, and 184 cm. The mean height of the individuals in this sample is

$$\overline{X} = \frac{170 + 174 + 177 + 178 + 178 + 179 + 179 + 180 + 181 + 184}{10}$$

$$= 178.0 \text{ cm}$$

The *variance* is the sum of the squares of the differences between each value and the mean, divided by the number of individuals minus one, or

$$s^2 = \frac{\Sigma(X - \overline{X})^2}{N - 1}$$

where s^2 is the variance and the other symbols are as in the previous formula: X represents each one of the values, $\overline{X}$ is the mean, and N is the number of individuals. The variance in height among the 10 students in our sample is

$$s^2 = \frac{(-8)^2 + (-4)^2 + (-1)^2 + 0^2 + 0^2 + 1^2 + 1^2 + 2^2 + 3^2 + 6^2}{9}$$

$$= \frac{132}{9} = 14.67 \text{ cm}^2$$

It is often convenient to calculate the variance using the following formula, which is mathematically equivalent to the previous one:

$$s^2 = \frac{\Sigma X^2 - N\overline{X}^2}{N - 1}$$

For the example in question,

$$s^2 = \frac{170^2 + 174^2 + 177^2 + 178^2 + 178^2 + 179^2 + 179^2 + 180^2 + 181^2 + 184^2 - 10(178^2)}{9}$$

$$= \frac{316{,}972 - 316{,}840}{9} = 14.67 \text{ cm}^2$$

The variance is measured in squared units because it is based on the sum of *squared* differences. A related measure of dispersion which, however, is given in the same units as the mean is the *standard deviation,* which is simply the square root of the variance. Hence

$$s = \sqrt{\frac{\Sigma(X - \overline{X})^2}{N - 1}}$$

Number of individuals:	1	0	0	1	5	7	7	22	25	26	27	17	11	17	4	4	1
Height in inches:	58	59	60	61	62	63	64	65	66	67	68	69	70	71	72	73	74

Figure A.2
Height distribution in 175 men recruited for the Army around the turn of the century. [From A. F. Blakeslee, *J. Hered.* 5:511 (1914).]

where s is the standard deviation. For the height example discussed above, the standard deviation is $s = 3.83$ cm.

A.IV The Normal Distribution

For many quantitative traits, such as height, weight, and number of eggs laid, the frequency distributions among individuals are bell-shaped. The number of individuals is greater at the intermediate values and gradually decreases toward the extremes. An example is shown in Figure A.2. The *normal distribution* is a mathematical curve that corresponds to such bell-shaped distributions.

The normal distribution has some interesting properties based on its mean and its standard deviation. The most useful application concerns the number of observations that are expected to fall within certain values (Figure A.3). In a normal distribution, 50% of all observations fall between values separated by 0.67 standard deviation on each side of the mean ($\overline{X} \pm 0.67s$; dark-shaded area), 67% of the observations fall within $\overline{X} \pm s$, and 95% of the observations fall within $\overline{X} \pm 1.96s$ (light- and dark-shaded areas).

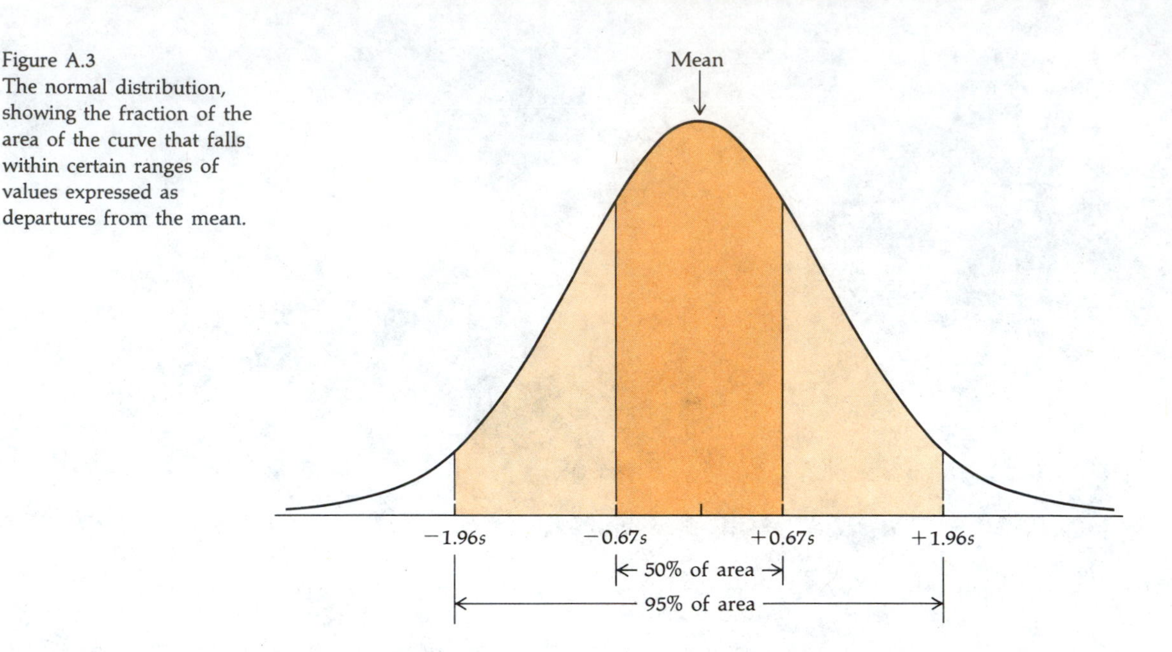

Figure A.3
The normal distribution, showing the fraction of the area of the curve that falls within certain ranges of values expressed as departures from the mean.

As an example, consider the height of the army recruits shown in Figure A.2. The mean and the standard deviation for that sample of 175 men are $\overline{X} = 67.3$ inches and $s = 2.7$ inches. The heights corresponding to $\overline{X} \pm s$ are 64.6 and 70.0 inches. The number of individuals taller than 64.6 inches but shorter than 70.0 inches is 117, which turns out to be exactly 67% of the 175 individuals in the sample. The heights corresponding to $\overline{X} \pm 1.96s$ are 62.0 and 72.6 inches. The number of individuals within this range is 163, or 93% of the total; 95% of the individuals should theoretically fall within this range. Although a sample of 175 individuals is not very large, the agreement between the observed and expected numbers is quite good.

Glossary

Adaptation A structural or functional characteristic of an organism that allows it to cope better with its environment; the evolutionary process by which organisms become adapted to their environment.

Adaptive value A measure of the reproductive efficiency of an organism (or genotype) compared with other organisms (or genotypes); also called *selective value*.

Additive genes Genes that interact but that show no dominance, if they are alleles, or no epistasis, if they are not alleles.

Additive variance Genetic variance due to additive genes.

Allele One of two or more alternative forms of a gene, each possessing a unique nucleotide sequence; different alleles of a given gene are usually recognized, however, by the phenotypes rather than by comparison of their nucleotide sequences.

Allopatric Of populations or species that inhabit separate geographic regions (cf. *sympatric*).

Allozymes Alternative enzyme forms encoded by different alleles at the same locus.

Amino acids The building blocks of proteins; several hundred are known, but only 20 are normally found in proteins.

Antibody A protein, synthesized by the immune system of a higher organism, that binds specifically to the foreign molecule (*antigen*) that induced its synthesis.

Artificial selection The process of choosing the parents of the following generation on the basis of one or more genetic traits (cf. *natural selection*).

Assortative mating Nonrandom selection of mates with respect to one or more traits; it is positive (negative) when individuals with the same form of a trait mate more (less) often than would be predicted by chance (cf. *random mating*).

Autosome A chromosome other than a sex chromosome.

Bottleneck A period when a population becomes reduced to only a few individuals.

Chromosomal mutation A change in the structure or number of the chromosomes; also called *chromosomal aberration* or *chromosomal abnormality*.

Chromosomal polymorphism The presence in a population of more than one gene sequence for a given chromosome.

Chromosome A threadlike structure, found in the nuclei of cells, that contains the genes arranged in linear sequence; a whole DNA molecule comprising the genome of a prokaryotic cell; a DNA molecule complexed with histones and other proteins in eukaryotic cells.

Chromosome complement The group of chromosomes in a normal gametic or zygotic nucleus; it may consist of one (monoploid nucleus), two (diploid nucleus), or more (polyploid nucleus) chromosome sets.

Chromosome set The normal gametic complement of chromosomes of a diploid individual (cf. *chromosome complement*).

Cline A gradient in the frequencies of genotypes or phenotypes along a stretch of territory.

Coadaptation The harmonious interaction of genes; the selection process by which harmoniously interacting genes become established in a population.

Codominant Of alleles whose gene products are both manifest phenotypically.

Codon A group of three adjacent nucleotides in an mRNA molecule that code either for a specific amino acid or for polypeptide chain termination during protein synthesis.

Coefficient of selection The intensity of selection, as measured by the proportional reduction in the gametic contribution of one genotype compared with another.

Consanguinity The sharing of at least one recent common ancestor.

Continuous variation Variation, with respect to a certain trait, among phenotypes that cannot be classified into a few clearly distinct classes, but rather that differ very little one from the next.

Crossing over The exchange of chromatid segments between homologous chromatids during meiosis; if different alleles are present on the chromatids, crossing over can be detected by the formation of genetically recombinant chromatids.

Darwinian fitness The relative fitness of one genotype compared with another, as determined by its relative contribution to the following generations.

Deficiency A chromosomal mutation characterized by the loss of a chromosome segment; also called *deletion* (cf. *duplication*).

Deletion See *deficiency*.

Deoxyribonucleic acid (DNA) A polynucleotide in which the sugar residue is deoxyribose and which is the primary genetic material of all cells.

DNA See *deoxyribonucleic acid*.

Dominant An allele, or the corresponding trait, that is manifest in all heterozygotes (cf. *recessive*).

Drift See *random genetic drift*.

Duplication A chromosomal mutation characterized by the presence of two copies of a chromosome segment in the haploid genome (cf. *deficiency*).

Effective population size The number of reproducing individuals in a population.

Electromorphs Allozymes that can be distinguished by electrophoresis.

Electrophoresis A technique for separating molecules based on their differential mobility in an electric field.

Fitness The reproductive contribution of an organism or genotype to the following generations (cf. *Darwinian fitness*).

Founder effect Genetic drift due to the founding of a population by a small number of individuals.

Frequency-dependent selection Natural selection whose effects depend on the frequency of the genotypes or the phenotypes.

Gamete A mature reproductive cell capable of fusing with a similar cell of opposite sex to give a zygote; also called *sex cell*.

Gene In the genome of an organism, a sequence of nucleotides to which a specific function can be assigned; e.g., a nucleotide sequence coding for a polypeptide, a nucleotide sequence specifying a tRNA, or a nucleotide sequence required for the proper transcription of another gene.

Gene flow The exchange of genes (in one or both directions) at a low rate between two populations, due to the dispersal of gametes or of individuals from one population to another; also called *migration*.

Gene pool The sum total of the genes in a breeding population.

Genetic drift See *random genetic drift*.

Genetic variance The fraction of the phenotypic variance that is due to differences in the genetic constitution of individuals in a population.

Genotype The sum total of the genetic information contained in an organism; the genetic constitution of an organism with respect to one or a few gene loci under consideration (cf. *phenotype*).

Hardy-Weinberg law A principle by which genotypic frequencies can be predicted on the basis of gene frequencies, under the assumption of random mating.

Hemizygous Of genes present only once in the genotype.

Heritability In the broad sense, the fraction of the total phenotypic variance that remains after exclusion of the variance due to environmental effects. In the narrow sense, the ratio of the additive genetic variance to the total phenotypic variance.

Heterogametic Of the sex whose gametes differ with respect to the sex chromosomes.

Heterogamic Of matings between individuals from different populations or species.

Heterosis Superiority of the heterozygote over the homozygotes with respect to one or more characters; also called *hybrid vigor*.

Heterozygosity The condition of being heterozygous; the proportion of heterozygous individuals at a locus, or of heterozygous loci in an individual.

Heterozygote A cell or organism having two different alleles at a given locus on homologous chromosomes.

Homogametic Of the sex whose gametes do not differ with respect to the sex chromosomes.

Homogamic Of matings between individuals from the same population or species.

Homologous Of chromosomes or chromosome segments that are identical with respect to their constituent genetic loci and their visible structure; in evolution, of genes and structures that are similar in different organisms owing to their having inherited them from a common ancestor.

Homozygosity The condition of being homozygous; the proportion of homozygous individuals at a locus, or of homozygous loci in an individual.

Homozygote A cell or organism having at a given locus the same allele on homologous chromosomes.

Identical by descent Of two genes that are identical in nucleotide sequence because they are both derived from a common ancestor.

Identical in structure Of two genes that are identical in nucleotide sequence, regardless of whether or not they are both derived from a common ancestor.

Inbreeding Mating between relatives.

Inbreeding coefficient The probability that the two genes (alleles) at a locus are identical by descent.

Inbreeding depression Reduction in fitness or vigor due to the inbreeding of normally outbreeding organisms.

Incipient species Populations that are too distinct to be considered as subspecies of the same species, but not sufficiently differentiated to be regarded as different species; also called *semispecies*.

Inversion A chromosomal mutation characterized by the reversal of a chromosome segment.

Inversion polymorphism The presence of two or more chromosome sequences, differing by inversions, in the homologous chromosomes of a population.

Isolating mechanism See *reproductive isolating mechanism*.

Lethal A gene or chromosomal mutation that causes death (in all carriers if it is dominant, but only in homozygotes if it is recessive) before reproductive age.

Linkage A measure of the degree to which alleles of two genes assort independently at meiosis or in genetic crosses.

Linkage disequilibrium Nonrandom association of alleles at different loci in a population.

Locus The place at which a particular mutation or

a gene resides in a genetic map; often used interchangeably with *mutation* or *gene*.

Macroevolution Evolution above the species level, leading to the formation of genera, families, and other higher taxa; also called *transpecific evolution*.

Mating system The pattern of mating in sexually reproducing organisms; two types of mating systems are *random mating* and *assortative mating*.

Meiosis Two successive divisions of a nucleus following one single replication of the chromosomes, so that the resulting four nuclei are haploid.

Mendelian population An interbreeding group of organisms sharing in a common gene pool.

Messenger RNA (mRNA) An RNA molecule whose nucleotide sequence is translated into an amino acid sequence on ribosomes during polypeptide synthesis.

Metric character A trait that varies more or less continuously among individuals, which are therefore placed into classes according to measured values of the trait; also called *quantitative character*.

Migration See *gene flow*.

Mitosis The division of a nucleus following replication of the chromosomes, so that the resulting daughter nuclei have the same number of chromosomes as the parent nucleus.

Modifier gene A gene that interacts with other genes by modifying their phenotypic expression (cf. *regulatory gene, structural gene*).

mRNA See *messenger RNA*.

Multiple alleles The occurrence in a population of more than two alleles at a locus.

Multiple-factor inheritance The determination of a phenotypic trait by genes at more than one locus.

Natural selection The differential reproduction of alternative genotypes due to variable fitness (cf. *artificial selection*).

Nonrandom mating A mating system in which the frequencies of the various kinds of matings with respect to some trait or traits are different from those expected according to chance.

Nucleus A membrane-enclosed organelle of eukaryotes that contains the chromosomes.

Orthologous genes Homologous genes that have become differentiated in different species derived from a common ancestral species (cf. *paralogous genes*).

Outbreeding A mating system in which matings between close relatives do not usually occur.

Overdominance The condition when the heterozygote exhibits a more extreme manifestation of the trait (usually, fitness) than does either of the homozygotes.

Paralogous genes Homologous genes that have arisen through a gene duplication and that have evolved in parallel within the same organism (cf. *orthologous genes*).

Parthenogenesis The production of an embryo from a female gamete without participation of a male gamete.

Phenotype The observable characteristics of an individual, resulting from the interaction between the genotype and the environment in which development occurs.

Phenotypic variance The variance among individuals with respect to some phenotypic trait or traits (cf. *genetic variance*).

Polygenic Of traits determined by many genes, each having only a slight effect on the expression of the trait.

Polymorphism The presence of several forms (of a trait or of a gene) in a population; the proportion of polymorphic gene loci in a population.

Polypeptide A chain of amino acids covalently bound by peptide linkages.

Polyploid A cell, tissue, or organism having three or more complete chromosome sets.

Protein A polymer composed of one or more polypeptide subunits and possessing a characteristic three-dimensional shape imposed by the sequence of its component amino acid residues.

Quantitative character See *metric character*.

Quantum speciation The rapid rise of a new species, usually in small isolates, with the founder effect and random genetic drift playing important roles; also called *saltational speciation*.

Race A population or group of populations distinguishable from other such populations of the same species by the frequencies of genes, chromosomal arrangements, or hereditary phenotypic characteristics. A race that has received a taxonomic name is a subspecies.

Random genetic drift Variation in gene frequency from one generation to another due to chance fluctuations.

Random mating Random selection of mates with respect to one or more traits; also called *panmixia* (cf. *assortative mating*).

Random sample A sample obtained in such a way that each individual in the population, or each gene in the genome, has the same chance of being selected.

Recessive An allele, or the corresponding trait, that is manifest only in homozygotes (cf. *dominant*).

Recombination The creation of a new association of DNA molecules (chromosomes) or parts of DNA molecules (chromosomes).

Regulatory gene In the broad sense, any gene that regulates or modifies the activity of other genes. In the narrow sense, a gene that codes for an allosteric protein that (alone or in combination with a corepressor) regulates the genetic transcription of the structural genes in an operon by binding to the operator (cf. *modifier gene, structural gene*).

Reproductive isolating mechanism (RIM) Any biological property of an organism that interferes with its interbreeding with organisms of other species.

Reproductive isolation The inability to interbreed due to biological differences.

Ribonucleic acid (RNA) A polynucleotide in which the sugar residue is ribose and which has uracil rather than the thymine found in DNA.

RIM See *reproductive isolating mechanism*.

RNA See *ribonucleic acid*.

Saltational speciation See *quantum speciation*.

Secondary sex ratio See *sex ratio*.

Selection See *natural selection* and *artificial selection*.

Selection coefficient See *coefficient of selection*.

Selection differential In artificial selection, the difference in mean phenotypic value between the individuals selected as parents of the following generation and the whole population.

Selection gain In artificial selection, the difference in mean phenotypic value between the progeny of the selected parents and the parental generation.

Selective value See *adaptive value*.

Self-fertilization The union of male and female gametes produced by the same individual.

Selfing Breeding by self-fertilization.

Semispecies See *incipient species*.

Sex cell See *gamete*.

Sex chromosomes Chromosomes that are different in the two sexes and that are involved in sex determination (cf. *autosome*).

Sex linkage Linkage of genes that are located in the sex chromosomes.

Sex ratio The number of males divided by the number of females (sometimes expressed in percent) at fertilization (primary sex ratio), at birth (secondary sex ratio), or at sexual maturity (tertiary sex ratio).

Sickle-cell anemia A human disease characterized by defective hemoglobin molecules and due to homozygosity for an allele coding for the beta chain of hemoglobin.

Somatic cells All body cells except the gametes and the cells from which these develop.

Speciation The process of species formation.

Species Groups of interbreeding natural populations that are reproductively isolated from other such groups.

Structural gene A gene that codes for a polypeptide (cf. *modifier gene, regulatory gene*).

Subspecies A population or group of populations distinguishable from other such populations of the same species by the frequencies of genes, chromosomal arrangements, or hereditary phenotypic characteristics. Subspecies sometimes exhibit incipient reproductive isolation, although not sufficiently to make them different species.

Supergene A DNA segment that contains a number of closely linked genes affecting a single trait or an array of interrelated traits.

Sympatric Of populations or species that inhabit, at least in part, the same geographic region (cf. *allopatric*).

Translocation A chromosomal mutation characterized by a change in position of a chromosome segment.

Transspecific evolution See *macroevolution*.

Variance A measure of variation, calculated as the sum of the squares of the differences between the value of each individual and the mean of the population.

Zygote The diploid cell formed by the union of egg and sperm nuclei within the cell.

Bibliography

Chapter 1

Ayala, F. J., and J. K. Kiger, Jr., *Modern Genetics*, Benjamin/Cummings, Menlo Park, Calif., 1980.

Clausen, J., D. D. Keck, and W. M. Hiesey, Experimental studies on the nature of species. I. Effects of varied environments on western North American plants, Carnegie Institution of Washington Publ. No. 520, Washington, D.C., 1940, pp. 1–452.

Dunn, L. C., ed., *Genetics in the 20th Century*, Macmillan, New York, 1951.

Kornberg, A., *DNA Replication*, W. H. Freeman, San Francisco, 1980.

Luria, S. E., S. J. Gould, and S. Singer, *A View of Life*, Benjamin/Cummings, Menlo Park, Calif., 1981.

Morgan, T. H., Sex-limited inheritance in *Drosophila*, *Science 32*:120–122 (1910).

Sutton, W. S., The chromosomes in heredity, *Biol. Bull.* 4:231–251 (1903).

Watson, J. D., *Molecular Biology of the Gene*, 3rd ed., W. A. Benjamin, Menlo Park, Calif., 1976.

Watson, J. D., and F. H. C. Crick, Molecular structure of nucleic acids. A structure for deoxyribose nucleic acid, *Nature 171*:737–738 (1953).

White, M. J. D., *Animal Cytology and Evolution*, 3rd ed., Cambridge University Press, Cambridge, 1973.

Chapter 2

Ayala, F. J., J. R. Powell, and Th. Dobzhansky, Polymorphism in continental and island populations of *Drosophila willistoni*, *Proc. Natl. Acad. Sci. USA 68*:2480–2483 (1971).

Ayala, F. J., J. R. Powell, and M. L. Tracey, Enzyme variability in the *Drosophila willistoni* group: V. Genetic variation in natural populations of *Drosophila equinoxialis*, *Genet. Res., Camb., 20*:19–42 (1972).

Brown, A. H. D., Enzyme polymorphism in plant populations, *Theor. Pop. Biol. 15*:1–42 (1979).

Cockerham, C. C., Analyses of gene frequencies, *Genetics 74*:679–700 (1973).

Cox, E. C., and T. C. Gibson, Selection for high mutation rates in chemostats, *Genetics 77*:169–184 (1974).

Coyne, J. A., A. A. Felton, and R. C. Lewontin, Extent of genetic variation at a highly polymorphic esterase locus in *Drosophila pseudoobscura, Proc. Natl. Acad. Sci. USA 75*:5090–5093 (1978).

Dobzhansky, Th., and B. Spassky, Genetics of natural populations. XXXIV. Adaptive norm, genetic load and genetic elite in *D. pseudoobscura, Genetics 48*:1467–1485 (1963).

Finnerty, V., and G. Johnson, Post-translational modification as a potential explanation of high levels of enzyme polymorphism: Xanthine dehydrogenase and aldehyde oxidase in *Drosophila melanogaster, Genetics 91*:695–722 (1979).

Fisher, R. A., *The Genetical Theory of Natural Selection*, Clarendon Press, Oxford, 1930.

Gillespie, J. H., A general model to account for enzyme variation in natural populations: V. The SAS–CFF model. *Theor. Pop. Biol. 14*:1–45 (1978).

Gottlieb, L. D., Electrophoretic evidence and plant systematics, *Ann. Missouri Bot. Gard. 64*:161–180 (1977).

Harris, H., and D. A. Hopkinson, Average heterozygosity in man, *J. Human Genet. 36*:9–20 (1972).

Johnson, G., Genetic polymorphism among enzyme loci, in *Physiological Genetics*, Academic Press, New York, 1979, pp. 239–273.

Koehn, R. K., and W. F. Eanes, Molecular structure and protein variation within and among populations, in *Evolutionary Biology*, Vol. 11, ed. by M. K. Hecht, W. C. Steere, and B. Wallace, Plenum Publishing Corp., 1978, pp. 39–100.

Lerner, I. M., and W. I. Libby, *Heredity, Evolution, and Society*, 2nd ed., W. H. Freeman, San Francisco, 1976.

Lewontin, R. C., *The Genetic Basis of Evolutionary Change*, Columbia University Press, New York, 1974.

Milkman, R., Electrophoretic variation in *Escherichia coli* from natural sources, *Science 182*:1024–1026 (1973).

Mukai, T., and O. Yamaguchi, The genetic structure of natural populations of *Drosophila melanogaster*. XI. Genetic variability of local populations, *Genetics 76*:339–366 (1974).

Nevo, E., Genetic variation in natural populations: Patterns and theory, *Theor. Pop. Biol. 13*:121–177 (1978).

Selander, R. K., Genic variation in natural populations, in *Molecular Evolution*, ed. by F. J. Ayala, Sinauer, Sunderland, Mass., 1976, pp. 21–45.

Wright, S., *Evolution and the Genetics of Populations*, Vol. 4, *Variability Within and Among Natural Populations*, University of Chicago Press, Chicago, 1978.

Chapter 3

Adams, J., and R. H. Ward, Admixture studies and the detection of selection, *Science 180*:1137–1143 (1973).

Bodmer, W., and L. L. Cavalli-Sforza, *Genetics, Evolution, and Man*, W. H. Freeman, San Francisco, 1976.

Crow, J. F., and M. Kimura, *An Introduction to Population Genetics Theory*, Harper & Row, New York, 1970.

Dobzhansky, Th., *Genetics of the Evolutionary Process*, Columbia University Press, New York, 1970.

Dobzhansky, Th., and O. Pavlovsky, An experimental study of interaction between genetic drift and natural selection, *Evolution 7*:198–210 (1953).

Glass, H. B., and C. C. Li, The dynamics of racial intermixture: An analysis based on the American Negro, *Amer. J. Human Genet. 5*:1–20 (1953).

Hardy, G. H., Mendelian proportions in a mixed population, *Science 28*:49–50 (1908).

Hartl, D. L., *Principles of Population Genetics*, Sinauer, Sunderland, Mass., 1980.

Lande, R., The maintenance of genetic variability by mutation in a polygenic character with linked loci, *Genet. Res., Camb. 26*:221–235 (1976).

Levin, D. A., and H. W. Kerster, Gene flow in seed plants, *Evol. Biol. 7*:139–220 (1974).

Li, C. C., *Population Genetics*, University of Chicago Press, Chicago, 1955.

Mayr, E., *Populations, Species, and Evolution*, Harvard University Press, Cambridge, Mass., 1970.

Mettler, L. E., and T. G. Gregg, *Population Genetics and Evolution*, Prentice-Hall, Englewood Cliffs, NJ, 1969.

Mourant, A. E., *The Distribution of the Human Blood Groups*, Blackwell, Oxford, 1954.

Nei, M., *Molecular Population Genetics and Evolution*, American Elsevier, New York, 1975.

Richardson, R. H., Effects of dispersal, habitat selection and competition on a speciation pattern of *Drosophila* endemic to Hawaii, in *Genetic Mechanisms of Speciation in*

Insects, ed. by M. J. D. White, Australia and New Zealand Book Co., Sydney, 1974, pp. 140–164.

Schaffer, H. E., D. G. Yardley, and W. W. Anderson, Drift or selection: A statistical test of gene frequency variation over generations, *Genetics 87*:371–379 (1977).

Slatkin, M., Gene flow and genetic drift in a species subject to frequent local extinctions. *Theor. Pop. Biol. 12*:253–262 (1977).

Spiess, E., *Genes in Populations*, J. Wiley, New York, 1977.

Stern, C., Weilhelm Weinberg (1862–1937) biography, *Genetics 47*:1–5 (1962).

Turelli, M., Random environments and stochastic calculus, *Theor. Pop. Biol. 12*:140–178 (1976).

Chapter 4

Allison, A. C., Polymorphism and natural selection in human populations, *Cold Spring Harbor Symp. Quant. Biol. 29*:139–149 (1964).

Anderson, W. W., Polymorphism resulting from the mating advantage of rare male genotypes, *Proc. Natl. Acad. USA 64*:190–197 (1969).

Antonovics, J., and D. A. Levin, The ecological and genetic consequences of density-dependent regulation in plants, in *Ann. Rev. Ecol. Syst. 11*:411–452 (1980).

Ayala, F. J., and C. A. Campbell, Frequency-dependent selection, *Ann. Rev. Ecol. Syst. 5*:115–138 (1974).

Bajema, C. J., ed., *Natural Selection in Human Populations*, J. Wiley, New York, 1971.

Battaglia, B., Balanced polymorphism in *Tisbe reticulata*, a marine copepod, *Evolution 12*:358–364 (1958).

Bundgaard, J., and F. B. Christiansen, Dynamics of polymorphisms: I. Selection components in an experimental population of *Drosophila melanogaster, Genetics 71*:439–460 (1972).

Charlesworth, B., Selection in populations with overlapping generations. I. The use of Malthusian parameters in population genetics, *Theor. Pop. Biol. 1*:352–370 (1970).

Clarke, B., Density-dependent selection, *Am. Nat. 106*:1–13 (1972).

Ewens, W. J., Testing the generalized neutrality hypothesis, *Theor. Pop. Biol. 15*:205–216 (1979).

Felsenstein, J., The theoretical population genetics of variable selection and migration, *Ann. Rev. Genet. 10*:253–280 (1976).

Gromko, M. H., What is frequency-dependent selection?, *Evolution 31*:435–442 (1977).

Haldane, J. B. S., and S. D. Jayakar, Polymorphism due to selection of varying direction, *J. Genet. 58*:237–242 (1963).

Hamrick, J. L., Y. B. Linhart, and J. B. Mitton, Relationships between life history characteristics and electrophoretically detectable genetic variation in plants, in *Ann. Rev. Ecol. Syst. 10*:173–200 (1979).

Hedrick, P. W., M. E. Ginevan, and E. P. Ewing, Genetic polymorphism in heterogeneous environments, *Ann. Rev. Ecol. Syst. 7*:1–32 (1976).

Kettlewell, H. B. D., The phenomenon of industrial melanism in the Lepidoptera, *Ann. Rev. Entom. 6*:245–262 (1961).

Kidwell, J. F., M. T. Clegg, F. M. Stewart, and T. Prout, Regions of stable equilibria for models of differential selection in the two sexes under random mating, *Genetics 85*:171–183 (1977).

Kojima, K., and K. M. Yarbrough, Frequency-dependent selection at the esterase-6 locus in *Drosophila melanogaster, Proc. Natl. Acad. Sci. USA 57*:645–649 (1967).

O'Donald, P., Theoretical aspects of sexual selection: Variation in threshold of female mating response, *Theor. Pop. Biol. 15*:191–204 (1979).

Petit, C., and L. Ehrman, Sexual selection in *Drosophila, Evol. Biol. 3*:177–223 (1969).

Prout, T., The relation between fitness components and population prediction in *Drosophila, Genetics 68*:127–167 (1971).

Schull, W. J., ed., *Genetic Selection in Man*, University of Michigan Press, Ann Arbor, 1963.

Spiess, E. B., Low frequency advantage in mating of *Drosophila pseudoobscura* karyotypes, *Amer. Nat. 102*: 363–379 (1968).

Sved, J. A., and F. J. Ayala, A population cage test for heterosis in *Drosophila pseudoobscura, Genetics 66*:97–113 (1970).

Wallace, B., *Topics in Population Genetics*, W. W. Norton, New York, 1968.

Chapter 5

Allard, R. W., The mating system and microevolution, *Genetics 79*:115–126 (1975).

Antonovics, J., A. D. Bradshaw, and R. G. Turner, Heavy metal tolerance in plants, *Adv. Ecol. Res. 7*:1–85 (1971).

Ayala, F. J., Relative fitness of populations of *Drosophila serrata* and *Drosophila birchii*, *Genetics 51*:527–544 (1965).

Ayala, F. J., and W. W. Anderson, Evidence of natural selection in molecular evolution, *Nature New Biology 241*:274–276 (1973).

Baker, W. K., Linkage disequilibrium over space and time in natural populations of *Drosophila montana*, *Proc. Natl. Acad. Sci. USA 72*:4095–4099 (1975).

Bodmer, W. F., and L. L. Cavalli-Sforza, *Genetics, Evolution, and Man*, W. H. Freeman, San Francisco, 1976.

Cavalli-Sforza, L. L., Population structure and human evolution, *Proc. Roy. Soc., Ser. B 164*:362–379 (1966).

Charlesworth, D., and B. Charlesworth, Theoretical genetics of Batesian mimicry, *J. Theor. Biol., 55*:305–337 (1976).

Clarke, C. A., and P. M. Sheppard, The evolution of mimicry in the butterfly *Papilio dardanus*, *Heredity 14*:163–173 (1960).

Clegg, M. T., R. W. Allard, and A. L. Kahler, Is the gene the unit of selection? Evidence from two experimental plant populations, *Proc. Natl. Acad. Sci. USA 69*: 2474–2478 (1972).

Endler, J. A., *Geographic Variation, Speciation, and Clines*, Princeton University Press, Princeton, 1977.

Fisher, R. A., *The Theory of Inbreeding*, 2nd ed., Oliver and Boyd, London, 1949.

Ford, E. B., *Ecological Genetics*, 3rd ed., Chapman & Hall, London, 1971.

Garn, S. M., *Human Races*, C. C. Thomas, Springfield, Ill., 1961.

Hirszfeld, L., and H. Hirszfeld, Essai d'application des méthodes sérologiques au problème des races, *Anthropologie 29*:505–537 (1919).

Jones, D. F., The attainment of homozygosity in inbred strains of maize, *Genetics 9*:405–418 (1924).

Kretchmer, N., Lactose and lactase, *Scientific American*, October, 1972, pp. 70–78.

Moroni, A., *Historical Demography, Human Ecology, and Consanguinity*, International Union for the Scientific Study of Population, Liège, 1969.

Nabours, R. K., I. Larson, and N. Hartwit, Inheritance of color patterns in the grouse locust, *Acridium arenosum* Burmeister (Tettigidae), *Genetics 18*:159–171 (1933).

Powell, J. R., H. Levene, and Th. Dobzhansky, Chromosomal polymorphism in *Drosophila pseudoobscura* used for diagnosis of geographic origin, *Evolution 26*:553–559 (1973).

Prakash, S., and R. C. Lewontin, A molecular approach to the study of genic heterozygosity in natural populations. III. Direct evidence of coadaptation in gene arrangements of *Drosophila*, *Proc. Natl. Acad. Sci. USA 59*:398–405 (1968).

Roberts, D. F., Body weight, race, and climate, *Amer. J. Phys. Anthro. 11*:533–558 (1953).

Stern, C., *Principles of Human Genetics*, 3rd ed., W. H. Freeman, San Francisco, 1973.

Turner, J. R. G., Butterfly mimicry: The genetical evolution of an adaptation, *Evol. Biol. 10*:163–206 (1977).

Weir, B. S., R. W. Allard, and A. L. Kahler, Analysis of complex allozyme polymorphisms in a barley population, *Genetics 72*:505–523 (1972).

Chapter 6

Bodmer, W. F., and L. L. Cavalli-Sforza, Intelligence and race, *Scientific American*, October, 1970, pp. 19–29.

Cavalli-Sforza, L. L., and M. W. Feldman, The evolution of continuous variation: III. Joint transmission of genotype, phenotype and environment, *Genetics 90*:391–425 (1978).

East, E. M., Studies on size inheritance in *Nicotiana*, *Genetics 1*:164–176 (1916).

Falconer, D. S., *Introduction to Quantitative Genetics*, Ronald Press, New York, 1961.

Feldman, M. W., and R. C. Lewontin, The heritability hangup, *Science 190*:1163–1168 (1975).

Hill, W. G., Estimation of heritability by regression using collateral relatives; linear heritability estimation, *Genet. Res., Camb. 32*:265–274 (1978).

Johannsen, W., Über Erblichkeit in Populationen und in reinen Linien, G. Fischer, Jena, 1903.

Karlin, S., Models of multifactorial inheritance, *Theor. Pop. Biol. 15*:308–438 (1979).

Kempthorne, O., Logical, epistemological and statistical aspects of nature-nurture data interpretation, *Biometrics 34*:1–23 (1978).

Lande, R., Statistical tests for natural selection on quantitative characters, *Evolution 31*:442–444 (1977).

Lewontin, R. C., Race and intelligence, *Bulletin of the Atomic Scientists*, March, 1970, pp. 2–8.

Lush, J. L., *Animal Breeding Plans*, Iowa State University Press, Ames, 1945.

Mather, K., Polygenic inheritance and natural selection, *Biol. Rev. 18*:32–64 (1943).

Nilsson-Ehle, H., Kreuzungsuntersuchungen an Hafer und Weizen, *Lunds Univ. Aarskr. N. F. Atd.*, Ser. 2, Vol. 5, No. 2, 1909, pp. 1–122.

Pollak, E., O. Kempthorne, and T. B. Bailey, Jr., eds., *International Conference on Quantitative Genetics*, Iowa State University Press, Ames, 1977.

Chapter 7

Ayala, F. J., Genetic differentiation during the speciation process, *Evol. Biol. 8*:1–78 (1975).

Ayala, F. J., ed., *Molecular Evolution*, Sinauer, Sunderland, Mass., 1976.

Ayala, F. J., M. L. Tracey, D. Hedgecock, and R. C. Richmond, Genetic differentiation during the speciation process in *Drosophila*, *Evolution 28*:576–592 (1974).

Bachmann, K., O. B. Goin, and C. J. Goin, Nuclear DNA amounts in vertebrates, *Brookhaven Symp. Biol. 23*: 419–450 (1972).

Bruce, E. J., and F. J. Ayala, Phylogenetic relationships between man and the apes: Electrophoretic evidence, *Evolution 33*:1040–1056 (1979).

Bush, G. L., Modes of animal speciation, *Ann. Rev. Ecol. Syst. 6*:339–364 (1975).

Carson, H. L., Speciation and the founder principle, *Stadler Symp. 3*:51–70 (1971).

Davidson, E. H., and R. J. Britten, Regulation of gene expression: Possible role of repetitive sequences, *Science 204*:1052–1059 (1979).

Dickerson, R. E., The structure of cytochrome *c* and the rates of molecular evolution, *J. Mol. Evol. 1*:26–45 (1971).

Dobzhansky, Th., F. J. Ayala, G. L. Stebbins, and J. W. Valentine, *Evolution*, W. H. Freeman, San Francisco, 1977.

Fitch, W. M., Molecular evolutionary clocks, in *Molecular Evolution*, ed. by F. J. Ayala, Sinauer, Sunderland, Mass., 1976, pp. 160–178.

Fitch, W. M., and E. Margoliash, Construction of phylogenetic trees, *Science 155*:279–284 (1967).

Gillespie, J. H., and C. H. Langley, Are evolutionary rates really variable? *J. Mol. Evol. 13*:27–34 (1979).

Gottlieb, L. D., Genetic confirmation of the origin of *Clarkia lingulata*, *Evolution 28*:244–250 (1974).

Grant, V., *Plant Speciation*, Columbia University Press, New York, 1971.

Hinegardner, R., Evolution of genome size, in *Molecular Evolution*, ed. by F. J. Ayala, Sinauer, Sunderland, Mass., 1976, pp. 179–199.

Hood, L., J. H. Campbell, and S. C. R. Elgin, The organization, expression, and evolution of antibody genes and other multigene families, *Ann. Rev. Genet. 9*:305–354 (1975).

Hoyer, B. H., and R. B. Roberts, Studies on nucleic acid interactions using DNA-agar, in *Molecular Genetics*, Part II, ed. by J. H. Taylor, Academic Press, New York, 1967, pp. 425–479.

Kimura, M., Evolutionary rate at the molecular level, *Nature 217*:624–626 (1968).

King, J. L., and T. H. Jukes, Non-Darwinian evolution, *Science 164*:788–798 (1969).

King, M. C., and A. C. Wilson, Evolution at two levels: Molecular similarities and biological differences between humans and chimpanzees, *Science 188*:107–116 (1975).

Kohne, D. E., J. A. Chiscon, and B. H. Hoyer, Evolution of primate DNA sequences, *J. Human Evol. 1*:627–644 (1972).

Laird, C. D., and B. J. McCarthy, Magnitude of interspecific nucleotide sequence variability in *Drosophila*, *Genetics 60*:303–322 (1968).

Lewis, H., Speciation in flowering plants, *Science 152*:167–172 (1966).

Maynard Smith, J., "Haldane's dilemma" and the rate of evolution, *Nature 219*:1114–1116 (1968).

Nei, M., Genetic distance between populations, *Amer. Nat. 106*:283–291 (1972).

Nevo, E., Y. J. Kim, C. R. Shaw, and C. S. Thaeler, Genetic variation, selection, and speciation in *Thomomys talpoides* pocket gophers, *Evolution 28*:1–23 (1974).

Ohno, S., *Evolution by Gene Duplication*, Springer-Verlag, New York, 1970.

Sarich, V. M., and A. C. Wilson, Immunological time scale for hominid evolution, *Science 158*:1200–1203 (1967).

Stebbins, G. L., *Variation and Evolution in Plants*, Columbia University Press, New York, 1950.

Tashian, R. E., M. Goodman, R. E. Ferrell, and R. J. Tanis, Evolution of carbonic anhydrase in primates and other mammals, in *Molecular Anthropology*, ed. by M. Goodman and R. E. Tashian, Plenum Press, New York, 1976, pp. 301–319.

Templeton, A. R., The theory of speciation *via* the founder principle, *Genetics 94*:1011–1038 (1980).

White, M. J. D., *Modes of Speciation*, W. H. Freeman, San Francisco, 1977.

Wilson, A. C., C. R. Maxson, and V. M. Sarich, Two types of molecular evolution: Evidence from studies of interspecific hybridization, *Proc. Natl. Acad. Sci. USA 71*:2843–2847 (1974).

Answers to Problems

Chapter 2: Genetic Structure of Populations

2.1 The genotypic frequencies are 634/1110 = 0.571; 391/1110 = 0.352; and 85/1110 = 0.077. The frequency of allele *1* is 0.571 + (0.352/2) = 0.747; of allele *2*, 0.077 + (0.352/2) = 0.253.

2.2 The total number of alleles is 219 × 2 = 438.

$$\text{Frequency of allele } 1\text{:} \quad \frac{(9 \times 2) + 135}{438} = 0.349$$

$$\text{Frequency of allele } 3\text{:} \quad \frac{(75 \times 2) + 135}{438} = 0.651$$

2.3 The expected frequency of heterozygotes is $1 - (f_1^2 + f_2^2)$, where f_1 and f_2 are the frequencies of the two alleles. Hence, the expected frequencies of heterozygotes are:

Problem 1: $1 - (0.747^2 + 0.253^2) = 0.378$
Problem 2: $1 - (0.349^2 + 0.651^2) = 0.454$

The chi-square test for Problem 1 is:

	Heterozygotes	Homozygotes	Total
Observed (O)	391	719	1110
Expected (E)	419.6	690.4	1110
$(O - E)^2/E$	1.95	1.18	$\chi^2 = 3.13$

For Problem 2, $\chi^2 = 23.3$.

The chi-square at the 0.05 level with 1 degree of freedom is 3.84 (see Table A.2). Hence, the observed and expected number of heterozygotes are not significantly different for Problem 1, but they are for Problem 2.

2.4

	Genotypic frequencies			Allelic frequencies		Hetero-zygotes		
	M	MN	N	*M*	*N*	obs.	exp.	χ^2
Eskimos	0.83	0.16	0.01	0.91	0.09	89	93	0.21
Pueblo Indians	0.59	0.33	0.08	0.76	0.24	46	51	0.77
Russians	0.40	0.44	0.16	0.62	0.38	215	230	1.85
Swedes	0.36	0.47	0.17	0.60	0.40	564	576	0.48
Chinese	0.33	0.49	0.18	0.58	0.42	500	501	0.004
Japanese	0.32	0.47	0.21	0.56	0.44	519	542	1.92
Belgians	0.29	0.50	0.21	0.54	0.46	1559	1540	0.47
English	0.29	0.47	0.24	0.52	0.48	200	211	1.15
Egyptians	0.28	0.49	0.23	0.52	0.48	245	251	0.29
Ainu	0.18	0.50	0.32	0.43	0.57	253	247	0.29
Fijians	0.11	0.445	0.445	0.33	0.67	89	88	0.02
Papuans	0.07	0.24	0.69	0.19	0.81	48	62	4.58

The observed and expected number of heterozygotes statistically agree in every case, except for the Papuans.

2.5

	Allelic frequencies			
	96	*100*	*104*	*108*
January	0.000	0.976	0.024	0.000
February	0.011	0.989	0.000	0.000
March	0.003	0.997	0.000	0.000
April	0.001	0.978	0.020	0.001
May	0.002	0.980	0.018	0.000
June	0.000	0.962	0.038	0.000
July	0.000	0.974	0.024	0.002
August	0.002	0.990	0.008	0.000
September	0.000	0.990	0.010	0.000
October	0.000	0.997	0.003	0.000
November	0.000	0.971	0.029	0.000
Total	0.0011	0.9805	0.0180	0.0004

The chi-square test for the total of all monthly samples is not significant:

	Homozygotes	Heterozygotes	Total
Observed	2250	87	2337
Expected	2247.8	89.2	2337
$(O - E)^2/E$	0.002	0.055	$\chi^2 = 0.057$

2.6

	Frequencies				Exp. freq. of heterozygotes	Exp. no. of heterozygotes
	ST	*AR*	*CH*	*TL*		
Keen	0.345	0.239	0.394	0.023	0.670	177
Piñon Flat	0.385	0.245	0.334	0.036	0.679	141
Andreas Canyon	0.572	0.260	0.136	0.031	0.584	165

2.7 Chimpanzees: At the *Pgm-1* locus, the frequency of allele *96* is 0.13 and that of allele *100* is 0.87. The expected heterozygosity is 0.226. For all 22 loci, the observed heterozygosity is the value observed at the *Pgm-1* locus (6/23 = 0.261) divided by the total number of loci, or 0.261/22 = 0.012; the expected heterozygosity is 0.226/22 = 0.010; the proportion of polymorphic loci is 1/22 = 0.045.

For the gorillas, the observed heterozygosity is 0.055, the expected heterozygosity is 0.047, and the proportion of polymorphic loci is 0.136.

Chapter 3: Processes of Evolutionary Change

3.1 *Allelic frequencies:* $M = 0.560$; $N = 0.440$
Expected genotypic frequencies:

$MM = p^2 = (0.560)^2 = 0.314$; $0.314 \times 1100 = 345$ individuals
$MN = 2pq = 0.493$; 542 individuals
$NN = q^2 = 0.194$; 213 individuals
$\chi^2 = [(356 - 345)^2/345] + [(519 - 542)^2/542] + [(225 - 213)^2/213]$
$= 2.00$, with 2 degrees of freedom: not significant.

3.2

Type of mating			Frequency of progeny type		
♂	♀	Frequency	M	MN	N
M	M	0.099	0.099	—	—
M	MN	0.155	0.077	0.077	—
M	N	0.061	—	0.061	—
MN	M	0.155	0.077	0.077	—
MN	MN	0.243	0.061	0.122	0.061
MN	N	0.096	—	0.048	0.048
N	M	0.061	—	0.061	—
N	MN	0.096	—	0.048	0.048
N	N	0.038	—	—	0.038
Totals		1.004	0.314	0.494	0.195

The genotypic frequencies are the same as in the previous problem.

3.3 Since red-green color blindness is sex linked, the frequency of red-green color blindness in men represents the frequency, q, of the red-green color blind allele; $q = 0.08$ and $p = 1 - 0.08 = 0.92$. The expected frequencies of the three possible genotypes among women are: $(0.92)^2 = 0.846$; $2(0.92)(0.08) = 0.147$; and $(0.08)^2 = 0.006$.

3.4 $q = 0.0001$; hence, $p = 0.9999$. These are also the frequencies of the two male genotypes. For the three female genotypes: $q^2 = (10^{-4})^2 = 10^{-8}$; $2pq = 0.0002$; $p^2 = 0.9998$.

3.5 If the population is at Hardy-Weinberg equilibrium, the frequency of the disease among newborns is q^2. Hence, $q = \sqrt{10^{-5}} = 0.003$, and $2pq = 2 \times 0.003 \times 0.997 = 0.006$.

3.6 $q^2 = 4/10{,}000$, so $q = 0.02$ and $p = 0.98$. Consequently, $p^2 = 0.9604$ and $2pq = 0.0392$.

3.7 In Hiroshima and Nagasaki: $2pq = 0.0009$; but $p = 1 - q$, therefore $2q(1 - q) = 2q - 2q^2 = 0.0009$, or $2q^2 - 2q + 0.0009 = 0$. Using the quadratic equation:

$$x = \frac{-b \pm \sqrt{b^2 - 4ac}}{2a} = \frac{2 \pm \sqrt{4 - 0.0072}}{4} = \frac{2 \pm 1.9982}{4}$$

we obtain, as the two solutions, $q = 0.00045$ and $p = 0.99955$.

One can obtain the same result as follows. Because $2pq$ is very small, we know that q is very small and therefore p is nearly one. Therefore, $2q \simeq 2pq = 0.0009$; hence, $q \simeq 0.00045$.

In the rest of Japan: $2pq = 0.014$; $2q \simeq 0.014$; hence, $q \simeq 0.007$, $p \simeq 0.993$.

3.8 The F_1 will consist exclusively of bw/bw^+ individuals; the F_2 of 25% bw/bw, 50% bw/bw^+, and 25% bw^+/bw^+ individuals. These are the equilibrium frequencies, as can be confirmed by noticing that $p = q = 0.5$, and therefore $p^2 = 0.25$, $2pq = 0.50$, $q^2 = 0.25$.

3.9 The total allelic frequencies are:

$$A \quad \frac{400/1000 + (640 \times 2 + 320)/2000}{2} = 0.6$$

$$a \quad \frac{600/1000 + (320 + 40 \times 2)/2000}{2} = 0.4$$

Thus, the equilibrium genotypic frequencies are: *Males:* $A = 0.6$, $a = 0.4$. *Females:* $AA = 0.36$, $Aa = 0.48$, $aa = 0.16$.

3.10 The frequency of the a allele in homozygotes is q^2. The frequency of a in the population is q. Therefore, the proportion of a alleles found in homozygotes is $q^2/q = q$.

3.11 The proportion is q in each case. That is, Tay-Sachs 0.003; cystic fibrosis 0.0004.

3.12 The frequencies of the four blood groups in the populations are:

Population	AB	B	A	O	Total
English	0.0304	0.0856	0.4172	0.4668	1.0000
Chinese	0.1010	0.2710	0.3200	0.3080	1.0000
Pygmies	0.0998	0.2907	0.3033	0.3062	1.0000
Eskimos	0.0145	0.0351	0.5372	0.4132	1.0000

The calculations for the English population are as follows: $r = \sqrt{0.4668} = 0.68323$; $q + r = \sqrt{0.0856 + 0.4668} = 0.74324$, and hence $q = 0.74324 - 0.68323 = 0.06001$; $p + r = \sqrt{0.4172 + 0.4668} = 0.94021$, and hence $p = 0.94021 - 0.68323 = 0.25698$. Therefore, $D = 1 - p - q - r = -0.00022$. The corrected allelic frequencies are

$$p^* = p(1 + D/2) = 0.25698(0.99989) = 0.25695$$
$$q^* = q(1 + D/2) = 0.06001(0.99989) = 0.06000$$
$$r^* = (r + D/2)(1 + D/2) = (0.68312)(0.99989) = 0.68304$$

The corrected allelic frequencies, as well as the D values, for the four populations are:

Population	I^A	I^B	i	D
English	0.25695	0.06000	0.68304	−0.00022
Chinese	0.23768	0.20611	0.55622	0.00159
Pygmies	0.22735	0.21924	0.55340	0.00006
Eskimos	0.33179	0.02674	0.64147	−0.00083

3.13 The frequency of his^+ and his^- will be, respectively:

$$p = \frac{v}{u + v} = \frac{4 \times 10^{-8}}{2 \times 10^{-6} + 4 \times 10^{-8}} = 0.02; \; q = 0.98$$

3.14 $p = \dfrac{6 \times 10^{-7}}{2 \times 10^{-5} + 6 \times 10^{-7}} = 0.03$; $q = 0.97$. The allele equilibrium frequencies are calculated in the same way for haploids and diploids.

3.15 The frequency, p_t, after t generations is $p_t = p_0(1 - u)^t$. If we assume that the initial frequency is $p_0 = 1.00$, we obtain

$$p_{10} = (1 - 10^{-6})^{10} = 0.99999$$
$$p_{1000} = (1 - 10^{-6})^{1000} = 0.999$$
$$p_{100,000} = (1 - 10^{-6})^{100,000} = 0.905$$

3.16 The relevant equation is $(1 - m)^t = \dfrac{p_t - P}{p_0 - P}$, with $t = 10$, $p_0 = 0.000$, $p_t = 0.045$, $P = 0.422$.

Therefore, $(1 - m)^{10} = \dfrac{0.045 - 0.422}{0.000 - 0.422} = 0.8934$; $m = 0.112$, or 1.12% per generation.

3.17 If only drift is involved, the probability of fixation of a given allele at a given time is its frequency at that time. Therefore, we infer that the initial frequency of A_1 is $p = 0.220$, and that of A_2 is $q = 0.780$.

3.18 $s = \sqrt{\dfrac{pq}{2n}} = \sqrt{\dfrac{0.220 \times 0.780}{2 \times 20}} = 0.0655$

The expected range of allele frequencies of allele A_1 in 95% of the populations is $p \pm 2s$, that is, between 0.089 and 0.351.

3.19 In the first ranch, $N_e = \dfrac{4 \times 100 \times 400}{100 + 400} = 320$. In the neighbor's ranch, $N_e = \dfrac{4 \times 1 \times 500}{501} = 4$

Chapter 4: Natural Selection

4.1 The general formula is

$$\Delta q = \frac{pq[p(w_2 - w_1) + q(w_3 - w_2)]}{\overline{w}}$$

where $\overline{w} = p^2w_1 + 2pqw_2 + q^2w_3$. (Remember that $p + q = 1$, and $p^2 + 2pq + q^2 = 1$).

Selection against recessive homozygotes (Table 4.4): $w_1 = 1$, $w_2 = 1$, $w_3 = 1 - s$. Therefore, $\overline{w} = p^2(1) + 2pq(1) + q^2(1 - s) = 1 - sq^2$.

$$\Delta q = \frac{pq[p(1 - 1) + q(1 - s - 1)]}{1 - sq^2} = \frac{-spq^2}{1 - sq^2}$$

Selection against a dominant allele (Table 4.8): $w_1 = 1 - s$, $w_2 = 1 - s$, $w_3 = 1$. Therefore, $\overline{w} = p^2(1 - s) + 2pq(1 - s) + q^2(1) = 1 - sp^2 - 2pqs = 1 - s(p^2 + 2pq) = 1 - s(1 - q^2)$ $= 1 - s + sq^2$.

$$\Delta q = \frac{pq[p(1 - s - (1 - s)) + q(1 - (1 - s))]}{1 - s + sq^2} = \frac{spq^2}{1 - s + sq^2}$$

$$\Delta p = -\Delta q = \frac{-spq^2}{1 - s + sq^2}$$

Selection with no dominance (Table 4.9); $w_1 = 1$, $w_2 = 1 - s/2$, $w_3 = 1 - s$. Therefore, $\overline{w} = p^2(1) + 2pq(1 - s/2) + q^2(1 - s) = (p^2 + 2pq + q^2) - spq - sq^2 = 1 - sq(p + q) = 1 - sq$.

$$\Delta q = \frac{pq[p(1-(1-s/2)) + q(1-s-(1-s/2))]}{1-sq} = \frac{pq[p(-s/2)+q(-s/2)]}{1-sq}$$
$$= \frac{pq[-s/2(p+q)]}{1-sq} = \frac{-spq/2}{1-sq}$$

Selection with overdominance (Table 4.10): $w_1 = 1 - s$, $w_2 = 1$, $w_3 = 1 - t$. Therefore, $\overline{w} = p^2(1-s) + 2pq(1) + q^2(1-t) = 1 - sp^2 - tq^2$.

$$\Delta q = \frac{pq[p(1-(1-s)) + q(1-t-1)]}{1-sp^2-tq^2} = \frac{pq(sp-tq)}{1-sp^2-tq^2}$$

Selection against heterozygotes (Table 4.12): $w_1 = 1$, $w_2 = 1 - s$, $w_3 = 1$. Therefore, $\overline{w} = p^2(1) + 2pq(1-s) + q^2(1) = 1 - 2spq$.

$$\Delta q = \frac{pq[p(1-s-1) + q(1-(1-s))]}{1-2spq} = \frac{pq[p(-s)+q(s)]}{1-2spq} = \frac{spq(q-p)}{1-2spq}$$

4.2

	DD	*Dd*	*dd*	Total	Frequency of *d*
1. Initial freq.	0.16	0.48	0.36	1.00	0.60
2. Fitness	1	1	0.47		
3.	0.16	0.48	0.17	0.81	
4.	0.20	0.59	0.21	1.00	0.51
5.					$\Delta q = -0.09$

(1) $\Delta q = \dfrac{-spq^2}{1-sq^2} = \dfrac{-0.53 \times 0.10 \times 0.81}{1 - 0.53 \times 0.81} = -0.075$

(2) $\Delta q = \dfrac{-0.53 \times 0.90 \times 0.01}{1 - 0.53 \times 0.01} = -0.0048$

The rate of change is largest when $q = 0.60$, smaller when $q = 0.90$, and smallest when $q = 0.10$ (see Figure 4.2).

4.3 We can use the formula for Δq in the general model of selection (Table 4.13), with fitnesses $w_1 = 1$, $w_2 = 1 - hs$, $w_3 = 1 - s$; and $\overline{w} = p^2 + 2pq(1-hs) + q^2(1-s) = 1 - 2hspq - sq^2$.

$$\Delta q = pq\frac{p(1-hs-1) + q(1-s-1+hs)}{1-2hspq-sq^2} = -spq\frac{hp+(1-h)q}{1-2hspq-sq^2}$$

4.4 The allele is a dominant lethal; therefore, at equilibrium, $p \simeq u = 10^{-5}$.

4.5 The allele is a recessive lethal; therefore, at equilibrium $q \simeq \sqrt{u} = \sqrt{10^{-5}} = 0.0032$.

The equilibrium frequency is much higher for a recessive allele, because the recessive allele is protected from selection in the heterozygotes.

4.6 $q^2 = 0.001$; therefore, $q = \sqrt{0.001} = 0.032$, $p = 0.968$, and $2pq = 0.061$. At equilibrium, $q = \sqrt{u}$ and $q^2 = u$. Therefore, if the mutation rate is doubled, the frequency of sterile individuals at equilibrium will also double: $q^2 = 0.002$, $q = \sqrt{0.002} = 0.045$, $2pq = 2 \times 0.955 \times 0.045 = 0.086$.

4.7 Refer to Table 4.10. The fitnesses are *AA*: $(1-s)$; *Aa*: 1; *aa*: 0. At equilibrium, $q = s/(s+t) = s/(1+s) = 0.333$; therefore, $s = 0.50$. The fitnesses are 0.50, 1, and 0.

4.8 When heterozygotes are interbred, the zygotes are produced in the ratios 1:2:1.

Low Density	V^VV^V	V^VV^M	V^MV^M	Total
Observed number	904	2023	912	3839
Observed frequency	0.235	0.527	0.238	1.000
Expected frequency	0.25	0.50	0.25	
Survival (observed/expected)	0.940	1.054	0.952	
Relative fitness (survival/1.054)	0.892	1	0.903	

High Density: For the heterozygote, the observed frequency 1069/1751 = 0.611, and the survival efficiency 0.611/0.5 = 1.222. For the V^VV^V homozygote, observed frequency 353/1751 = 0.202, survival 0.202/0.25 = 0.808, fitness 0.808/1.222 = 0.661. For the V^MV^M homozygote, observed frequency 329/1751 = 0.188, survival 0.188/0.25 = 0.752, fitness 0.752/1.222 = 0.615.

The fitnesses of the two homozygotes are lower at the high than at the low density.

4.9 In Experiment 1, the fitnesses are 1, 1, and 0.50. For Experiment 2, they are 0.50, 1, and 1, respectively, for *FF*, *FS*, and *SS*. This is a case of frequency-dependent selection. Because the fitnesses of the homozygotes are inversely related to their frequency, we would expect a stable polymorphic equilibrium. The fitnesses appear to be symmetric and, therefore, the best guess is that the stable equilibrium will occur when the allelic frequencies are 0.50.

4.10 This is a case of selection against a dominant allele. Using the formula for Δp given in Table 4,8, we obtain:

(1) $\Delta p = (-spq^2)/(1 - s + sq^2) = (-0.53 \times 0.40 \times 0.36)/(1 - 0.53 + 0.53 \times 0.36) = -0.12$

(2) $\Delta p = -0.048$

(3) $\Delta p = -0.010$

4.11 The Hb^S allele is incompletely dominant with respect to sickling; with respect to fitness, it is recessive in malaria-free regions, but overdominant where malaria exists.

Chapter 5: Inbreeding, Coadaptation, and Geographic Differentiation

5.1 Since C is the daughter of A, there is only one common "ancestor" in this case and hence there is only one path: D-A-C-D, with 3 steps. Therefore, $n = 3 - 1 = 2$, $F = (1/2)^n = 1/4$.

5.2 There are two paths: (1) G-C-A-D-F-G, $n = 5 - 1 = 4$; and (2) G-C-B-D-F-G, $n = 5 - 1 = 4$. Therefore, $F = (1/2)^4 + (1/2)^4 = 1/8$.

5.3 In this case, there are four common ancestors and, therefore, four paths, each with $n = 5$. $F = 4(1/2)^5 = 1/8$.

5.4 The frequency of heterozygotes is $2pq - 2pqF$: (a) 0.32; (b) 0.192; (c) 0.064.

5.5 The allelic frequencies are: $p = \dfrac{28 + (24/2)}{28 + 24 + 48} = 0.40$; $q = 0.60$. Therefore, $2pq = 0.48$. The observed frequency of heterozygotes is: $0.24 = 2pq - 2pqF = 0.48 - 0.48F$. Therefore, $0.48F = 0.24$ and $F = 0.50$.

5.6 At equilibrium, $q^2 = u = 4 \times 10^{-4}$; $q = \sqrt{u} = 0.02$. The inbreeding coefficient for the progeny of first-cousin matings is $F = 1/16$. Therefore, the incidence is $q^2 + pqF = 4 \times 10^{-4} + (0.98)(0.02)(1/16) = 0.0016$.

If $u = 8 \times 10^{-4}$, $q^2 = 0.0008$; $q = 0.0283$. Therefore, the incidence is $0.0008 + (0.9717)(0.0283)(1/16) = 0.0025$.

5.7 $2pq = 0.4928$; $2pq - 2pqF = 0.4435$. Therefore, $2pqF = 0.4928 - 0.4435 = 0.0493$; $F = 0.0493/0.4928 = 0.100$.

5.8 The population of the large lake is ten times as large as the population of the small lake; therefore, the gametic frequencies immediately after mixing are: *AB* 10/11; *ab* 1/11; *Ab* 0; *aB* 0. The initial linkage disequilibrium is: $d = (10/11)(1/11) = 10/121 = 0.0826$. The value of d will decrease each generation of random mating according to: $d_1 = (1 - c)d_0$, where $c = 0.5$ if the two loci assort independently. After five generations of random mating: $d_5 = (1 - c)^5 d_0 = (0.5)^5(0.0826) = 0.0026$.

5.9 If $c = 0.10$, $d_5 = (1 - 0.10)^5(0.0826) = 0.0488$. The number of generations, t, required to reduce d to 0.0026, can be obtained as follows: $0.0026 = (0.90^t)(0.0826)$; $0.90^t = 0.0026/0.0826 = 0.0315$; $t(\log 0.90) = \log 0.0315$; $t = (-1.5020)/(-0.0458) = 32.8$, or about 33 generations.

5.10 Represent the active and null alleles by A and a at locus 1, B and b at locus 2. The gametic frequencies are: $AB = 31/474 = 0.0654$; $ab = 0.1540$; $Ab = 0.2046$; and $aB = 0.5759$. Therefore, $d = (0.0654)(0.1540) - (0.2046)(0.5759) = -0.1078$.

5.11 $d_4 = (0.453)(0.019) - (0.076)(0.452) = -0.0257$
$d_{14} = (0.407)(0.004) - (0.098)(0.491) = -0.0465$
$d_{26} = (0.354)(0.003) - (0.256)(0.387) = -0.0980$
The increase in linkage disequilibrium is probably due to natural selection favoring certain gametic combinations. The increase in linkage disequilibrium is facilitated by the fact that self-fertilization reduces recombination (because it increases the frequency of homozygotes relative to the Hardy-Weinberg expectations).

Chapter 6: Quantitative Characters

6.1 (a) The F_1 cross is $S_2s_2\ S_3s_3\ s_4s_4 \times S_2s_2\ S_3s_3\ s_4s_4$. In the F_2 progeny, only those having at least one S_2 and one S_3 allele will be colored. The expected F_2 progenies are

9/16 have both S_2 and S_3: colored
3/16 have S_2 but not S_3 }
3/16 have S_3 but not S_2 } 7/16 colorless
1/16 have neither S_2 nor S_3 }

(b) The F_1 cross is $S_2s_2\ S_3s_3\ S_4S_4 \times S_2s_2\ S_3s_3\ S_4S_4$. In the F_2 progeny, all have S_4; hence only those that are homozygous for both s_2 and s_3 will be colorless, i.e., 1/16 of the total.

6.2 Artificial selection may cease to be effective because the stock becomes fixed (homozygous) for all favorable alleles present in the stock. Different stocks may have become fixed for favorable alleles at different loci. When these are intercrossed, the favorable alleles of both stocks become available for selection.

6.3 Assuming that the two parental strains are homozygous for alternate alleles affecting size, the only variability present in either strain is environmental. Similarly, the F_1's are all alike genetically and will show only environmental variation. But in the F_2 generation, segregation occurs and genetic variability appears, in addition to the environmental variability.

6.4 When the difference between the parental strains is due to *three pairs of genes*, 1/64 of the F_2 progeny are expected to have the same genotype as one of the parents (the 10 g plant, in this case).

6.5 The F_2 plants may be heterozygous at 3, 2, 1, or 0 gene loci. The F_3 families will be genetically more or less variable depending on the number of loci at which their F_2 parent was heterozygous. None of the F_3 families is expected to be more variable than the F_2 itself, because no F_2 individual can be heterozygous at more than three loci (which is the number of loci at which the F_1 was heterozygous).

6.6 (1) The backcross is *Aa Bb* (F_1 intermediate) × *AA BB* (long-eared). The progeny is 1/4 *AA BB* (long-eared).
(2) If the difference is due to three pairs of genes, $(1/2)^3 = 1/8$ of the backcross progeny will be long-eared.
(3) $(1/2)^4 = 1/16$ long-eared progeny.

6.7 36 is approximately 1/16 of 560. Assume that white seeds are homozygous for the recessive alleles at two loci (i.e., *aabb*) and that the black-seed parent was homozygous for the dominant allele at both loci (*AABB*). Then 1/16 of the F_2 progeny will be homozygous *aabb*.

106 is approximately 3/16 of 560. Under the assumptions just made, 3/16 of the F_2 progeny will be homozygous for one recessive allele, but not the other (say, *A*_*bb*).

The results can be explained if genotypes having the *B* allele are always black, those having the *A* allele but homozygous for the *b* allele are gray, and those homozygous for *a* and *b* are white.

6.8

	Yield	Firmness	Size
$\overline{X}$	380	4.4	11.3
σ	91	0.6	3.0
$D = (\overline{X} + 2\sigma) - \overline{X}$	182	0.12	6.0
H	0.48	0.46	0.20
$G = H \times D$	87.4	0.05	1.2

6.9

Heavy mice	*Light mice*
$\overline{X} = 21.5$	$\overline{X} = 21.5$
$D = 27.5 - 21.5 = 6$	$D = 15.5 - 21.5 = -6$
$G = 22.7 - 21.5 = 1.2$	$G = 18.1 - 21.5 = -3.4$
$H = G/D = 1.2/6 = 0.20$	$H = G/D = (-3.4)/(-6) = 0.56$

6.10 $V_i = (4^2 + 7^2 + 5^2 + \cdots + 3^2 + 7^2)/10 = 32.4$

$V_f = (12^2 + 4^2 + 9^2 + \cdots + 9^2 + 9^2)/10 = 89.1$

$H = (V_f - V_i)/V_f = 0.636$

Chapter 7: Speciation and Macroevolution

7.1 Because *D. melanogaster* is much more similar to *D. simulans* than to *D. funebris*, the most likely configuration of their phylogeny is

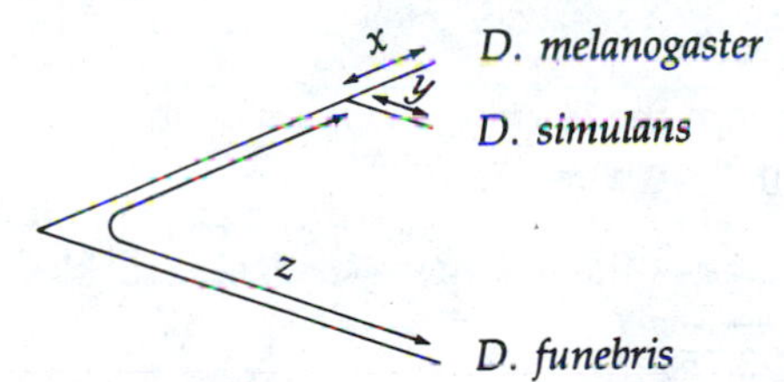

In order to solve for x, y, and z, we need three genetic distances, but we have only two. Therefore, the best we can do is to assume $x = y$ and then we obtain $x = 1.5\%$; $y = 1.5\%$; $z = 13 - 1.5 = 11.5\%$ nucleotide changes. We cannot determine whether or not the percent change from the last common ancestor of *D. melanogaster* and *D. simulans* to each of these species is the same, because we do not have data for the comparison between *D. simulans* and *D. funebris*.

7.2 Let H stand for humans, GM for green monkeys, and C for capuchins. The most likely configuration of their phylogeny is

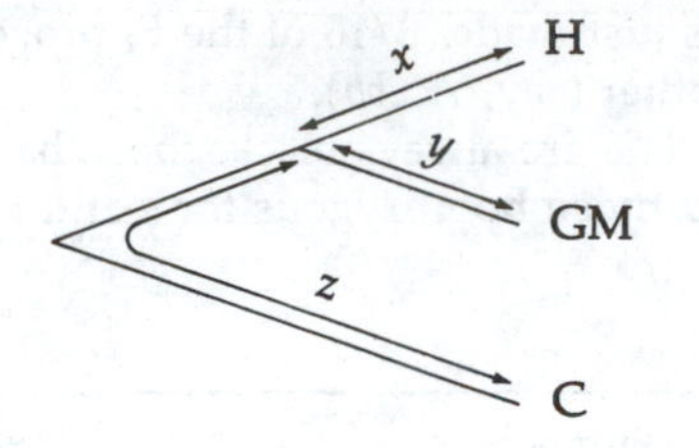

$x + y = 9.6$, $x + z = 15.8$, $y + z = 16.5$. Solving, we obtain $x = 4.45$, $y = 5.15$, $z = 11.35$. We are able to determine the values of x and y in this case because we have the three distances, H-C, H-GM, and C-GM.

7.3 Let H = human, M = monkey, D = dog, T = tuna.

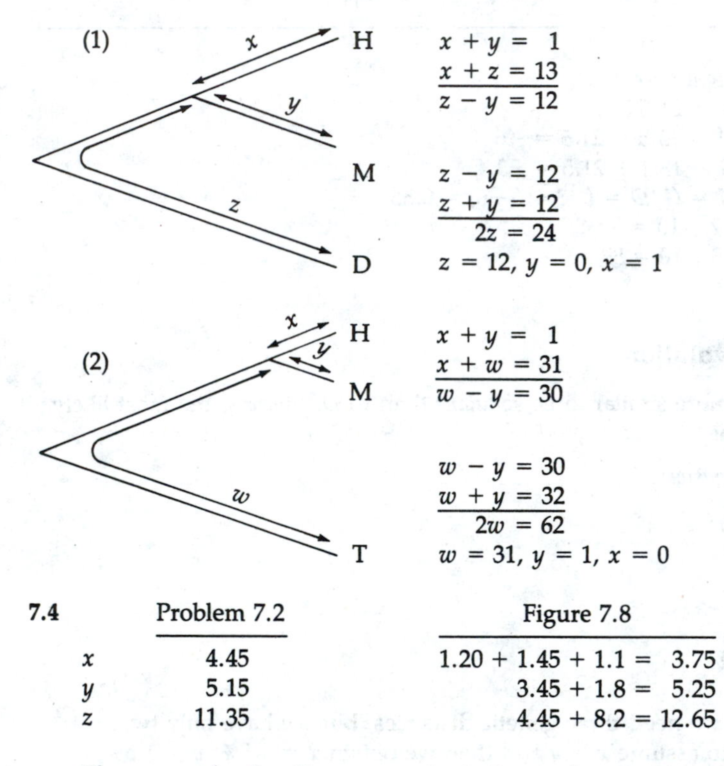

(1)

$$\begin{aligned} x + y &= 1 \\ x + z &= 13 \\ \hline z - y &= 12 \end{aligned}$$

$$\begin{aligned} z - y &= 12 \\ z + y &= 12 \\ \hline 2z &= 24 \end{aligned}$$

$z = 12$, $y = 0$, $x = 1$

(2)

$$\begin{aligned} x + y &= 1 \\ x + w &= 31 \\ \hline w - y &= 30 \end{aligned}$$

$$\begin{aligned} w - y &= 30 \\ w + y &= 32 \\ \hline 2w &= 62 \end{aligned}$$

$w = 31$, $y = 1$, $x = 0$

7.4

	Problem 7.2	Figure 7.8
x	4.45	1.20 + 1.45 + 1.1 = 3.75
y	5.15	3.45 + 1.8 = 5.25
z	11.35	4.45 + 8.2 = 12.65

The reason for the discrepancy is similar to the one given in the chapter for Problem 7.3. Figure 7.8 has been constructed using the data for all seven species and minimizing the discrepancies between all the data available and the values given in the figure.

7.5 $C_i = p_{11} + p_{22} - q_{12} - q_{21} = \dfrac{229 + 375 - 13 - 9}{229 + 375 + 13 + 9} = \dfrac{582}{626} = 0.93$

7.6 The amino acids that are different, with their codons and the minimum number of nucleotide changes, are:

Alpha	Beta	Minimum number of nucleotide changes
His:CAU,CAC	Gln:CAA,CAG	1
Ser:UCU,UCC,UCA,UCG,AGU,AGC	Ala:GCU,GCC,GCA,GCG	1
Leu:UUA,UUG,CUU,CUC,CUA,CUG	Tyr:UAU,UAC	2
Asp:GAU,GAC	Gln:CAA,CAG	2
Phe:UUU,UUC	Val:GUU,GUC,GUA,GUG	1
Leu:UUA,UUG,CUU,CUC,CUA,CUG	Val:GUU,GUC,GUA,GUG	1
Ser:UCU,UCC,UCA,UCG,AGU,AGC	Gly:GGU,GGC,GGA,GGG	1
Ser:UCU,UCC,UCA,UCG,AGU,AGC	Ala:GCU,GCC,GCA,GCG	1
Thr:ACU,ACC,ACA,ACG	Asn:AAU,AAC	1
Val:GUU,GUC,GUA,GUG	Ala:GCU,GCC,GCA,GCG	1
Thr:ACU,ACC,ACA,ACG	Ala:GCU,GCC,GCA,GCG	1
Ser:UCU,UCC,UCA,UCG,AGU,AGC	His:CAU,CAC	2
Arg:CGU,CGC,CGA,CGG,AGA,AGG	His:CAU,CAC	1
	Total	16

7.7 (1) When $s = 0$, $k = u$; therefore, $k = 10^{-5}$.
(2) $k = 2Nux = 2Nu(2N_e s/N) = 4N_e us = 4 \times 10^{-5}$.
(3) $k = 4 \times 10^{-3}$.
Notice that when the population size and the mutation rate are constant, the rate of substitution of an allele is proportional to the selection coefficient of that allele.

7.8 The method for calculating I and D is given in Box 7.1. The detailed calculations for the comparison between chimpanzee (A) and gorilla (B) are as follows:

Locus	$\Sigma a_i b_i$	Σa_i^2	Σb_i^2
1	0	1	$.2^2 + .8^2 = .68$
2	1	1	1
3	1	1	1
4	0	1	1
5	0	1	$.67^2 + .33^2 = .5578$
6	0	1	1
7	1	1	1
8	1	1	1
9	1	1	1
10	1	1	1
11	0	1	1
12	0	1	1
13	1	1	1
14	1	1	1
15	1	1	1
16	1	1	1

17	.15	1	$.15^2 + .85^2 = .7450$
18	.88	$.12^2 + .88^2 = .7888$	1
19	0	1	1
Total	11.03	18.7888	17.9828
Average	$I_{ab} = 0.5805$	$I_a = 0.9889$	$I_b = 0.9465$

Therefore, $I = I_{ab}/\sqrt{I_a I_b} = 0.6000$; $D = -\ln I = 0.5108$. Proceeding similarly for the other pairwise comparisons, we obtain the following matrix, with I given above the diagonal and D below the diagonal.

	Pan	Gorilla	Hylobates	Papio
Pan	—	0.6000	0.4844	0.3764
Gorilla	0.5108	—	0.5756	0.4466
Hylobates	0.7248	0.5523	—	0.5050
Papio	0.9771	0.8061	0.6833	—

Looking at the genetic distance values, we notice that the smallest is $D = 0.5108$ between Pan and Gorilla. The average distance between these two species and Hylobates is $D = (0.7248 + 0.5523)/2 = 0.6385$, while between them and Papio is $D = (0.9771 + 0.8061)/2 = 0.8916$. Therefore, the likely configuration of the phylogeny is:

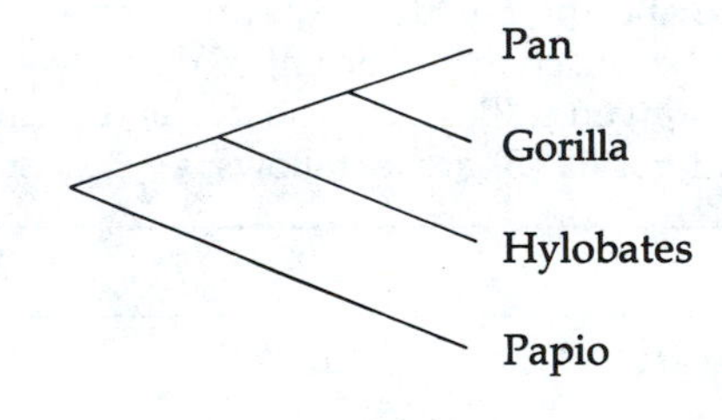

Index